Routes of Learning

Routes of Learning

Highways, Pathways, and Byways
in the History of Mathematics

IVOR GRATTAN-GUINNESS

The Johns Hopkins University Press
Baltimore

Printed in the United States of America on acid-free paper
2 4 6 8 9 7 5 3 1

The Johns Hopkins University Press
2715 North Charles Street
Baltimore, Maryland 21218-4363
www.press.jhu.edu

Library of Congress Cataloging-in-Publication Data
Grattan-Guinness, I.
Routes of learning : highways, pathways, and byways in the history of mathematics /
Ivor Grattan-Guinness.
p. cm.
Includes bibliographical references and index.
ISBN-13: 978-0-8018-9247-9 (hardcover : alk. paper)
ISBN-10: 0-8018-9247-3 (hardcover : alk. paper)
ISBN-13: 978-0-8018-9248-6 (pbk. : alk. paper)
ISBN-10: 0-8018-9248-1 (pbk. : alk. paper)
1. Mathematics—History. I. Title.
QA21.G696 2009
510.9—dc22 2008046474

A catalog record for this book is available from the British Library.

Special discounts are available for bulk purchases of this book. For more information, please contact Special Sales at 410-516-6936 or specialsales@press.jhu.edu.

The Johns Hopkins University Press uses environmentally friendly book materials, including recycled text paper that is composed of at least 30 percent postconsumer waste, whenever possible. All of our book papers are acid-free, and our jackets and covers are printed on paper with recycled content.

To my wife, Enid, in recognition of so much care and assistance over so many years

CONTENTS

Intentions

Historians have their special topics, periods, and regions of concern; my own lie in the development of the calculus and mathematical analysis from around 1730 to 1930, symbolic logics and set theory from 1820 to 1940, and mechanics and mathematical physics from 1750 to 1850. But I have also treated the history of mathematics in broader ways: writing a general history and editing two others, and considering its own history and philosophical concerns; examining in general and in particular cases both the history of mathematics education and the utility of the history for current mathematics education; and wandering into some little-studied topics, especially numerology. These three forays determine the structure of this book, in which I reprint some of the resulting articles.

By way of introduction I launch the book with an account in chapter 1 of my entrée into this unusual field. Then comes the first part, "Highways in the History of Mathematics," a name that refers primarily to historical work *itself,* not to the mathematics that is its chief concern. In chapter 2 I explain in detail the distinction between "history" and "heritage" as two legitimate *but fundamentally different* ways of handling the mathematics of the past. Then comes in chapter 3 an essay on the history of the history of mathematics in the twentieth century—and a very odd history it is, with a large and long dip in the middle. Despite a substantial renaissance, especially in the last 25 years, various topics and aspects are still somewhat neglected; several are highlighted in chapter 4. The next chapter considers the usefulness of general histories of mathematics, with reference to my own trio mentioned above.

The rest of this part treats relationships between the history of mathematics and neighboring disciplines. One might expect good companionship to obtain, but quite often indifference, patronization, or even hostility hold sway. Parts of the explanation are proposed in the title of chapter 6, "Too Mathematical for Historians, Too Historical for Mathematicians"; much of the material here was inspired by experiences arising from my presidency of the British Society for the History of Mathematics from 1986 to 1988. The next chapter looks at some aspects of the practice of the history of science itself, based on my experience of editing the journal *Annals of Science* from 1974 to 1981. Finally, chapter 8 introduces some distinctions in theory change in order to use the word "revolution" more carefully than is often the case in the history of the sciences, mathematics included.

For reasons recorded in chapter 1, educational questions have been important in my interests from the start. The second part of the book, "Pathways in Mathematics Education," treats some parts of the history of mathematics that have also been very influential and popular in the teaching of the subject. Chapter 9 addresses the use of history in education in a general way, together with some examples. Then comes a chapter interpreting Zeno's paradox of Achilles and the tortoise in a nonstandard way in regarding it as a valid argument. We stay Greek in chapter 11 with an appraisal of the mathematics that is actually practiced in Euclid's *Elements* as opposed to its uses as algebra in teaching and even in interpretations by historians in the past: the issue of history versus heritage is marked here, and indeed this case helped me much to formulate the discussion in chapter 2. We turn to arithmetic in chapter 12, with issues of history and teaching addressed at various levels of sophistication. Finally comes the calculus in chapter 13, where the four principal ways of formulation are summarized, and the transfer of notations from one tradition to another is lamented; in particular, *"dy/dx"* has suffered much abuse.

A multisided subject like mathematics has accrued many obscure and curious corners. Much of the final part, "Byways in Mathematics and Its Culture," burrows into numerology, which is dismissed nowadays as superstition. This judgment may be rational in itself, but heritage is confused with history when it is used to disregard the history of the topic; on the contrary, it seems clear that numerology was given high status in antiquity and later epochs, and history should allow for this perhaps regrettable situation. Chapter 14 considers the role of numerology in the formulation and practice of Christianity, while the next chapter notes that Christianity played a modest role in mathematics after around 1750, in contrast to its earlier high status and its contemporary importance in other sciences, such as biology and geology. Then chapter 16 briefly reviews the place of numerology in the music of Mozart and Beethoven; and a kind of corollary is explored in the next chapter, where Mozart and Lagrange are shown to have been contemporary critics of aspects of Descartes's philosophy in the 1780s, in very different contexts.

The conductor Sir Thomas Beecham sometimes ended his concerts with a "lollipop" or two, attractive short orchestral pieces to send his audience away in a good mood. I emulate this practice in the final short part, where I present in chapter 18 four nice little theorems concerning or involving the planar triangle; none of them is well known, and in some cases neither are their histories. So we are on the boundary between knowledge and ignorance, where we have been quite regularly in the meantime.

The ensemble of theories and their relationships form a network that I try to characterize in the title of this book. There are both mathematical routes from one theory to other theories and historical routes from some stage in the development of a theory to later (occasionally earlier) stages. In addition, when foundational issues are handled, the homonym comes into play because the roots of the pertaining theories are considered. As for learning, when mathematical research is at hand, it occurs at the level of scholarship, while with mathematical education, it appears at the level of instruction.

Conditions on the Boundaries

Some routes lie at the edge. One of the well-known ways of bluffing your way in mathematics is to introduce the question "What happens on the boundary?" into a technical conversation; there is a good chance that, whatever the subject of discussion, the question will be a good one. This amusing property inspires the title of this section, which relate to two points of view that form important themes running through the book. First, the history of mathematics lies on the boundaries of mathematics, mathematics education, the philosophy of mathematics, and the history of science and does not often penetrate their interiors. Second, there are several topics lying on its own boundary that deserve to gain more attention; the one emphasized here is numerology. Sadly, in these contexts the question "What happens on the boundary?" does not always stimulate interest!

The notion of boundary also lies behind the photograph that adorns the back cover of this book. Showing a splatter in a mud pool at Waiotapu, New Zealand, it is an image of a dynamical system, a partial differential equation with boundary conditions.

Housework

Having summarized the content of these articles, I must indicate the ways in which they have been preserved and modified in this book. On the one hand, I have maintained the individual character of each article, even retaining the acknowledgments. On the other hand, I have taken the liberty of rephrasing the occasional clumsiness, correcting some recognized mistakes, and modifying a few views where I now demur from the initial formulation. Occasionally, I have added a little pertinent material that could have appeared in the original text had I known or thought of it at the time, or that now describes a current situation that differs significantly from the one stated in the original article. When I want to make clear that an addition or change of text has been made, I enclose the new material in curled brackets {}. Usually it takes the form of, or includes, a pertinent historical writing on a specific point that has appeared since my article was originally published. I have not tried to update the articles in a comprehensive way; on the contrary, they have their own place in recent history, which should be conserved. But when I have presented a particular point or case study in more than one article, I have reduced or even eliminated the measure of repetition.

I have also imposed a uniform style on the book. Each chapter carries a number and is divided into numbered sections. To help bind the chapters together I have inserted cross-references between them; the form adopted is "§6.4," which cites section 4 of chapter 6. A few chapters are tightly organized, with internal cross-references of their own; they are rendered there as "¶3.6," citing a subsection that would be cited in other chapters as, say, "§2.3.6" to refer to subsection 6 of section 3 in chapter 2. Another important move toward uniformity is that the references in every chapter are given in a bibliography at its end; in some cases, the original version provided the references in footnotes.

Each chapter has an unnumbered footnote on the first text page. It begins with the details of the original publications of the article, together with indications of changes made here when they are sufficiently noteworthy to merit an outline. Copyright information is also provided there when the holder has requested it. In a few cases I have added remarks on the motivation or circumstances of the original writing.

Acknowledgments

As mentioned earlier, I have preserved in the chapters the original expressions of thanks. I welcomed the acceptance of the articles by the various editors and publishers at the time of first publication, and I record this welcome again for the permissions granted for the reprintings in this book. I also thank Trevor Lipscombe at the Johns Hopkins University Press for proposing the book to me in the first place, and his colleagues for their expert handling of the manuscript. Finally and above all, I wish to record my indebtedness for all the secretarial and other forms of help that I have received over more than 40 years from my wife, Enid; as a small recompense, I dedicate this book to her.

Routes of Learning

CHAPTER ONE

Searching for Reasons

My Way In and Onward

A short account is given of my entrée into the history of mathematics as a field for research. A central motive was dislike of my student experience; a central guide was Karl Popper's philosophy.

1. Perplexities and Their Resolution

Historians often discuss or at least mention influence, but they often think only of its positive forms; yet negative influence can be just as significant. My own entry into the history of mathematics was motivated by a reaction against the experience of learning, or rather being taught, mathematics as an undergraduate at Oxford from 1959 to 1962. As usual in education, the caloric theory was assumed: everything taught by the expert was learned by the beginner. The result on me was to appreciate the existence of remarkable bodies of knowledge; but proper understanding of them was scarred by the "perfect" expert delivery. Two questions came to mind:

1. Heuristic: what was the, or a, motivation for these theories?
2. Historical: How had they come about; surely not by falling out of the sky like rain onto blackboards or pages?

I put these questions to some of the lecturers: silence attended the historical one, but to the first I was told that further theories could be learned afterward. The need to use serious heuristic material when launching sophisticated theories on inexperienced students such as me never crossed the expert minds—a situation encouraged by the doubtlessly incoherent forms in which I posed my questions to them.

Upon graduation I spent some time in industry discovering that two of the main secrets about the British defense industry were the amount of money and effort

The original version of this essay was first published as "Cannot Put It Down: Spiralling into the History of Mathematics," *The British Society for the History of Mathematics Newsletter*, no. 40 (1999), 29–35. Compare also the interview with Chris Weeks in *History and Pedagogy of Mathematics Newsletter*, no. 63 (November 2006), 1–6.

spent on it, and the ineffectiveness of the defense provided against Soviet aircraft.[1] Then I took a teaching appointment in order to have time to think about those two questions and maybe research them. But how, and with whom? What exactly were the questions anyway? A directed walk around the University of London early in 1964 led me to the Department of History of Science at University College, where the head, Douglas McKie (1896–1967), required me to apply to be a postgraduate before I could determine the nature of the study applied for. A decisive stroke of good fortune attended the interview: McKie was not available, so it was conducted by a historian of medicine, J. S. Wilkie (1906–1982). His powers of diagnosis were extraordinary: no specialist in mathematics at all, he agreed that I could study under their visiting staff member J. F. Scott (1892–1971) but advised that it would be a "disaster," and that a far better option was coming available at the London School of Economics (LSE) in the form of a new master of science (economics) course in the philosophy of science, run by the department headed by Karl Popper (1902–1994). Upon inquiry, I found that it included an obligatory part on "Popperian brainwash," as it was amusingly called; and among the options was one on mathematical logic, close to the set theory that I had already partially learned and much liked.

An intellectually stimulating two years followed, part time from teaching; especially striking to this deeply confused mathematics graduate and new teacher was hearing Popper stress that science was guesswork, and that philosophical problems often took their stimulus from outside philosophy. Socially speaking, however, was another matter. An incoherent seminar paper by me on time was compounded with my strong dislike of the atmosphere, where the LSE epithet about Popper that *The Open Society and Its Enemies* (1945) was written by one of its enemies seemed regrettably apt; negative influence arose again, about how a staff member should conduct himself in front of students.[2] As a logical corollary, my relations with I. Lakatos (1922–1974), who was there for my second year, were best when nonexistent (Grattan-Guinness 1979).

1. I joined a group that was forming a concept called "lethality," a quantifiable relation between a missile system and various types of target. During the Cuba crisis we happened to be analyzing possible sources of malfunction of one of our systems over the sea; this produced insights about possible outcomes of that crisis caused by similar malfunction over Cuban waters, which I have not seen discussed. Some colleagues did not come in on the latter days of the crisis, preferring to be with their families at the end.

Concerning the ineffectiveness of mathematical education, some of my work there involved using the Rayleigh distribution, which embarrassed me since mathematical statistics had been completely omitted from the Oxford curriculum. The results of our work were written up as extended reports, a kind of academic activity left untaught at Oxford.

2. In his last years and at his invitation, I renewed contact with Popper, now much older and widowered; the change in personality since the previous quarter century was quite remarkable (Grattan-Guinness 1998).

Now armed with a philosophical context in which to set my heuristic and historical questions, the obvious move was to attempt a doctorate on some aspect of the history of mathematics in which both historical and modern educational issues were active. {Here the Oxford experience came into play. The starkest example of perplexity occurred in the first term, when the course on pure calculus happened to be followed immediately by the one on fluid mechanics; epsilontics was succeeded by differentials, with no cross references in either direction. For example *"dy/dx"* changed at the hour from a whole symbol to the ratio *dy* ÷ *dx* of differentials (§13.7.1). What was going on? Such mysteries led me to choose the history of the calculus and mathematical analysis as my first port of call.} During the master of science degree, I had already tried the history of the links between set theory and measure theory as a test case, preparing for an agreed question area on one of the examination papers (but I was to find that it had been omitted). However, I had been already amazed to learn both that many techniques and definitions in mathematical analysis, and also set theory itself, had been stimulated by problems in Fourier series, and equally that no teacher or book on the subject ever mentioned these links (because, I soon realized, the teachers and authors themselves were ignorant of them). Now scales were falling from the eyes: mathematics turned out to be a human endeavor, with people working on interesting problems and connections and even making mistakes sometimes! What an extraordinary transformation, or rather creation, of my understanding; it took only a few months to take place, and consequences have been emerging ever since.

2. First Researches

The mathematicians mainly responsible for the topics that I was studying were Joseph Fourier (1768–1830) and A. L. Cauchy (1789–1857); so in 1966 I decided to start researching on them. Learning of the existence of many Fourier manuscripts in Paris led me to seek advice from D. T. Whiteside. Then in full flood with his Newton edition, nevertheless he took time to provide me with much detailed advice on handling manuscripts, and on the history of mathematics in general; he also suggested that I contact J. R. Ravetz at Leeds University, whose writings on Fourier had escaped my amateurish literature searches. In the end Ravetz agreed to cooperate on a book on Fourier, focusing not only on the series but also on heat diffusion, which had been Fourier's initial motivation, as well as treating his fascinating life (Grattan-Guinness 1972a).

Work on Cauchy ran in parallel; I conceived a doctorate on his mathematical analysis and its prehistory in the eighteenth century, as it had struck me quickly that his use of the theory of limits was confused in the historical literature with the refinements to it made later by Karl Weierstrass (1815–1897) and followers. When I registered in the (new) Department of Mathematics at the London School of Economics headed by A. C. Offord,[3] Ravetz also kindly agreed to be the external supervisor. The thesis was published with a few changes as *The Development of the Foundations of Mathematical Analysis from Euler to Riemann* (1970a).

In addition, Ravetz placed me in touch with Sir Edward Collingwood (1900–1970), an eminent analyst also deeply interested in the history of mathematics, especially in Georg Cantor (1845–1918). He was to be my external examiner, and he reinforced my growing attraction to the history of set theory, which had arisen as an extension of mathematical analysis as Weierstrass had formed it. Collingwood's investigations had led him to think that there were important materials in the former home of Weierstrass's follower Gustav Mittag-Leffler (1846–1927), and he personally sponsored a research visit for me there in 1968. During the few hours in the week when the Institut Mittag-Leffler was then open I found not only the Cantoriana, especially correspondence with Philip Jourdain (1879–1919) (Grattan-Guinness 1971a), but also as a major bonus a far larger collection of letters between Jourdain and Bertrand Russell (1872–1970). A second visit in 1970 with the Institut under new management was much more satisfactory, and in many places there I now found more on Cantor, masses of sources on Weierstrass, quite a lot for Sonya Kovalevskaya (1850–1891), and some on Henri Poincaré (1854–1912) (Grattan-Guinness 1971c).

Collingwood and Offord also backed me at the Royal Society for a grant to research Cantor's career in Halle University (Grattan-Guinness 1971b) in the German Democratic Republic; this led my wife and I to some fascinating weeks in 1969 in a country which then was not easy or even advisable to visit. The trip also led to a lasting friendship with the country's leading historian of science, Hans Wussing. By then I had met J. W. Dauben, a graduate student at Harvard University working on a doctorate on Cantor; and my Russell findings suggested to me that I focus on them (Grattan-Guinness 1977) and leave the main Cantor effort to Dauben. The Russell Archives had just been formed at McMaster University in Canada (Russell was still alive); I told them of my new findings, and so started a close professional link, which is still active.

Through Collingwood I came to know Cecily Tanner (1900–1992) (Grattan-

3. Offord's eminence in measure theory tempted me to switch to its history, but in the end I stuck to Cauchy. This was a lucky decision, since at the time T. W. Hawkins was researching that history in his own doctorate at the University of Wisconsin, which appeared in print as *Lebesgue's Theory of Integration* (1970).

Guinness 1993), the eldest daughter of W. H. Young (1863–1942) and G. C. Young (1868–1944), British pioneers in both set theory and measure theory from the 1900s and personal friends of Cantor. The contact had continued in the descendants, for through Tanner I met Cantor's grandchildren, and so gained access to his Nachlass (Grattan-Guinness 1970b). Tanner also conserved her parents' Nachlässe, which I helped her to organize; so I came to their story, the first major case in mathematics of collaboration between husband and wife (Grattan-Guinness 1972b, 1972c). In addition, she possessed their library, and access to it greatly simplified my research on set theory.[4]

3. Contacts Abroad

The chain of friends extended to Germany, for Tanner arranged with J. E. Hofmann (1900–1973) that I be invited to the next meeting in his series on the history of mathematics, held in November 1969 at the Mathematisches Forschungsinstitut in Oberwolfach, Germany. For the first time I met with a group of active figures in the field and was able to receive a body of informed advice about the merits or otherwise of my research work, on further topics to investigate, and on consequences of history for education. Hofmann's assistant at the meeting was C. J. Scriba, who like Wussing has been a major mentor ever since, especially at the later meetings that I have been able to attend and on two occasions codirect with him.

Since those formative years my researches have largely convoluted out of Fourier and Cauchy: extensively into the extraordinary mathematical community in France over the period 1780–1840 of which they formed part, especially its calculus and mathematical analysis and its mechanics and mathematical physics (Grattan-Guinness 1990); historically back into the calculus and mechanics of the eighteenth century; historically forward into set theory and mathematical logic, especially Cantor and Russell (Grattan-Guinness 2000); sideways into aspects of British mathematics (partly stimulated by doctoral students); and philosophically into the bearing of history upon education (Grattan-Guinness 1973) and to efforts to explain historical method, which mathematicians seem often congenitally incapable of grasping at all. In the 1990s invitations to more general editing and writing have been fulfilled (ch. 5): an encyclopedia on the history and philosophy of mathematics (Grattan-Guinness 1994) prepared with a rather large number of collaborators; and a general history (Grattan-Guinness 1997), which I dedicated to the memory of

4. The *Nachlässe* of both Cecily and her parents are housed in Liverpool University Archives.

Popper because his philosophy of criticism had much guided its conception and execution.

But the dilemma of the British situation has always been stark: helpful and willing individuals around, and fantastic libraries especially in London;[5] but no substantial body of support, especially not in the neighboring communities of mathematicians, mathematical educators, or historians of science. I stopped work on the doctorate three times, mainly for doubt of the merit or purpose, which bored everyone else; it was mainly the rich libraries nearby that stimulated the restarts. Collingwood died in office as the president of the London Mathematical Society, where he had been hoping to initiate historical meetings—we even sketched out one on real- and complex-variable analysis—but he made no progress at all.

Thus when John Dubbey launched the British Society for the History of Mathematics in 1971, I thought the idea was hopelessly optimistic. Its continued existence has proved me wrong and him right, as I am delighted to acknowledge: the annual research day meetings for doctoral students seem particularly valuable. Nevertheless, through no fault of its own the society tends also to be the British ghetto for research in the field, although history courses are now taught in quite a few universities (Grattan-Guinness 1992). My research life has always centered largely on two kinds of journey: to the center of London, to use the libraries; and to a station or airport, to travel abroad and tell foreigners about the findings and give lecture courses to their students.

BIBLIOGRAPHY

Grattan-Guinness, I., 1970a. *The development of the foundations of mathematical analysis from Euler to Riemann*, Cambridge, Mass.: MIT Press.

Grattan-Guinness, I., 1970b. "An unpublished paper by Georg Cantor: *Principien einer Theorie der Ordnungstypen. Erste Mittheilung*," *Acta mathematica, 124*, 65–107.

Grattan-Guinness, I., 1971a. "The correspondence between Georg Cantor and Philip Jourdain," *Jahresbericht der Deutschen Mathematiker-Vereinigung, 73*, part 1, 111–130.

Grattan-Guinness, I., 1971b. "Towards a biography of Georg Cantor," *Annals of science, 27*, 345–391 and plates xxv–xxviii.

Grattan-Guinness, I., 1971c. "Materials for the history of mathematics in the Institut Mittag-Leffler," *Isis, 62*, 363–374.

Grattan-Guinness, I., 1972a. In collaboration with J. R. Ravetz; *Joseph Fourier 1768–1830. A*

5. In my work on the French community (Grattan-Guinness 1990), I found virtually everything that they published in London quite easily (though with a complicated distribution of locations), often in more than one library. I am quite sure that I located this literature far more efficiently than anyone could in Paris libraries.

survey of his life and work, based on a critical edition of his monograph on the propagation of heat, presented to the Institut de France in 1807, Cambridge, Mass.: MIT Press.

Grattan-Guinness, I., 1972b. [Edited with R. C. H. Tanner.] W. H. and G. C. Young, *The theory of sets of points*, 2nd ed., New York: Chelsea. [New introduction and appendix.]

Grattan-Guinness, I., 1972c. "A mathematical union: William Henry and Grace Chisholm Young," *Annals of science, 29*, 105–186.

Grattan-Guinness, I., 1973. "Not from nowhere: History and philosophy behind mathematical education," *International journal of mathematical education in science and technology, 4*, 421–453.

Grattan-Guinness, I., 1977. *Dear Russell—dear Jourdain: A commentary on Russell's logic, based on his correspondence with Philip Jourdain*, London: Duckworth; New York: Columbia University Press.

Grattan-Guinness, I., 1979. Letter to the editor, *Mathematical intelligencer, 1*, 247–248.

Grattan-Guinness, I., 1990. *Convolutions in French mathematics, 1800–1840: From the calculus and mechanics to mathematical analysis and mathematical physics*, 3 vols., Basel, Switzerland: Birkhäuser; Berlin: Deutscher Verlag der Wissenschaften.

Grattan-Guinness, I., 1992. "A tale of a tub: On the Society's first 21 years," in *The British Society for the History of Mathematics, 1971–1992, Newsletter*, no. 21, 1–9. [See also chapter 6 below; and *Annals of science, 50* (1993), 483–490.]

Grattan-Guinness, I., 1993. "Cecily Tanner," *The British Society for the History of Mathematics Newsletter*, no. 23 (1993), 10–15. [Also in *Newsletter of the Association of Women in Mathematics*, 23, no. 6 (1993) 21–24.]

Grattan-Guinness, I., 1994. (Ed.), *Companion encyclopaedia of the history and philosophy of the mathematical sciences*, 2 vols., London: Routledge.

Grattan-Guinness, I., 1997. *The Fontana history of the mathematical sciences: The rainbow of mathematics*, London: Fontana. [Also printed on proper paper as *The Norton history of the mathematical sciences*, New York and London: Norton, 1998.]

Grattan-Guinness, I., 1998. "Karl Popper for and against Bertrand Russell," *Russell, new ser., 18*, 25–42.

Grattan-Guinness, I., 2000. *The search for mathematical roots, 1870–1940: Logics, set theories and the foundations of mathematics from Cantor through Russell to Gödel*, Princeton, N.J.: Princeton University Press.

PART I

HIGHWAYS IN THE HISTORY OF MATHEMATICS

CHAPTER TWO

The Mathematics of the Past

Distinguishing Its History from Our Heritage

Mathematics shows much more durability in its attention to concepts and theories than do other sciences: for example, Galen may not be of much use to modern medicine, but one can still read and use Euclid. One might expect that this situation would make mathematicians sympathetic to history, but quite the opposite is the case. Their normal attention to history is with heritage; that is, how did we get here? Old results are modernized to show their current place; but the historical context is ignored and thereby often distorted. By contrast, the historian is concerned with what happened in the past, whatever be the modern situation. Each approach is perfectly legitimate, but they are often confused. In this chapter, I discuss the difference between them, with examples: these include Euclid, set theory, limits, and applied mathematics in general.

For the centenary of Jean Cavaillès (1903–1944), historian and philosopher of mathematics, and self-chosen victim of Nazi millenarianism

> However eager to tell us how scientists of the seventeenth century used their inheritance from the sixteenth, the scholars seem to regard as irrelevant anything a scientist today might think about any aspects of science, including his own debt to the past or reaction against it.
>
> —*C. A. Truesdell III (1968, foreword)*

> You think that the world is what it looks like in fine weather at noonday; I think that it seems like in the early morning when one first wakes from deep sleep.
>
> —*A. N. Whitehead to B. Russell (Russell 1956, 41)*

> As all historians know, the past is a great darkness, and filled with echoes. Voices may reach us from it; but [. . .] try as we may, we cannot always decipher them in the clearer light of our own day.
>
> —*Margaret Atwood (1985, end)*

1. The Pasts and the Futures

1.1. The Basic Distinction

The growth of interest and work in the history of mathematics during the last three decades or so has led to reactions among mathematicians. Some of them have been welcoming and indeed have contributed their own historical research; but many others have been cautious, even contemptuous, about the work produced by practicing historians, especially on account of the historians' apparently limited knowledge of mathematics.[1] By the latter they usually mean some modern version of the mathematics in question, and the failure of historians to take due note of it.

There is a deep and general distinction involved here, locatable in any branch of mathematics, any period, any culture, and possibly involving teaching or popularization of mathematics as well as its research. It seems to be sensed by people working in history, whether they come to the subject with mainly a historical or a mathematical motivation. However, it has not been much discussed in the literature; even the survey of historiography jumps across it (May 1976).

I use the words "history" and "heritage" to name two interpretations of a mathematical theory; the corresponding actors are "historians" and "inheritors" (or "heirs") respectively. The word "notion" serves as the umbrella term to cover a theory (or definition, proof method, technique, algorithm, notation[s], a whole branch of mathematics, etc.), and the letter N to symbolize it. A sequence of notions in recognized order in the development of a mathematical theory is notated N_0, N_1, N_2, . . .

By "history" I refer to the details of the development of N: its prehistory and concurrent developments; the chronology of progress, as far as it can be determined; and maybe also the impact in the immediately following years and decades. History addresses the question "What happened in the past?" and gives descriptions; maybe it also attempts explanations of some kinds, in order to answer the corresponding "why?" question (¶3.10). History should also address the dual questions "What did not happen in the past?" and "Why not?"; there false starts, missed opportunities (Dyson 1972), sleepers, and repeats are noted and maybe explained. The (near) absence of later notions from N is registered, as well as their eventual arrival; *dif-*

The original version of this essay was first published in *Historia mathematica, 31* (2004), 161–185, after many years of cogitation. Reprinted by permission of Elsevier Rightslink.

1. Another point of division between the two disciplines is techniques and practices specific to historical work, such as the finding, examining and deploying manuscript sources and of large-scale bibliographies. The latter are rehearsed, for the precomputer age, in May (1973, 3–41). They are not directly relevant to this paper.

ferences between N and seemingly similar more modern notions are likely to be emphasized.

By "heritage" I refer to the impact of N on later work, both at the time and afterward, especially the forms that it may take, or be embodied in, in later contexts.[2] Some modern form of N is usually the main focus, with attention paid to the course of its development. Here the mathematical relationships will be noted, but historical ones in the above sense will hold much less interest. Heritage addresses the question "How did we get here?" and often the answer reads like "the royal road to me." The modern notions are inserted into N when appropriate, and thereby N is unveiled (a nice word proposed to me by Henk Bos): similarities between N and its more modern notions are likely to be emphasized; the present is photocopied onto the past.

Both kinds of activity are quite legitimate and indeed important in their own right; in particular, mathematical research often seems to be conducted in a heritagelike way (¶3.1), whether the predecessors produced their work long ago or very recently. *The confusion of the two kinds of activity is not legitimate,* either taking heritage to be history (frequently the mathematicians' view—and historians' sometimes!) or taking history to be heritage (the occasional burst of excess enthusiasm by a historian): indeed, such conflations may well mess up both categories, especially the historical record. In the case of sequences of notions, a pernicious case arises when N_1 is a logical consequence or a generalization of N_0, and the claim is made that a knower of N_0 knew N_1 also (May 1975a): an example is given in ¶3.5.

A philosophical difference is that inheritors tend to focus on knowledge alone (theorems as such, and so on), while historians also seek motivations, causes, and understanding in a more general sense. The distinction sometimes made by historians of science between "internal" and "external" history forms part of this difference. Each category is explicitly metatheoretical, though history may demand the greater finesse in the handling of different levels of theory.

A third category of writing is when a theory is laid out completely time-free with all developments omitted, historical or otherwise; for example, as a strictly axiomatized theory. This kind of writing is also quite legitimate; it tells us that "we are here." A similar fourth category is large-scale bibliographies, including classifications and indexing by topic. These categories are neither history nor heritage, although they may well involve both.[3] Apart from noting that they too will be influenced by his-

2. In my first lectures on this topic I used the word "genealogy" to name this concept. I now prefer "heritage," partly on semantic grounds and partly for its attractive similarity with "history" as another three-syllable word in English beginning with *h*.

3. A current project to classify the primary literature as reviewed in the *Jahrbuch ueber die Fortschritte der Mathematik* (1867–1942) imposes a modern division into topics and subtopics. My

tory though probably without the knowledge of the practitioners (¶5.4), I shall not consider them further here.

1.2. Some Literature

Two prominent types of writing in which heritage is the main guide are review articles and lengthy reports. Names, dates, and references are given frequently, and chronology (of publication) may well be checked quite scrupulously; but motivations, cultural background, processes of genesis, and historical complications are usually left out. A golden period in report writing was at the turn of the nineteenth and twentieth centuries, especially in German, with two main locations: the reports, often lengthy, in the early volumes of the *Jahresberichte* of the Deutsche Mathematiker-Vereinigung (1892–) and the articles comprising the *Encyklopädie der mathematischen Wissenschaften* (1898–1935) with its unfinished extension into the French *Encyclopédie des sciences mathématiques* (1904–1916?) (Gispert 1999). Some of these texts are quite historical.[4]

Among modern examples of heritage-oriented historical writings, Jean Dieudonné's lengthy account of algebraic and differential topology in the twentieth century is (impressively) typical (Dieudonné 1989), and several of the essays in the Bourbaki history have the same character (Bourbaki 1974). André Weil's widely read advice (1980) on how to do history is largely driven by needs of heritage and even dismissive of history, especially concerning the relative importance of judgments of the mathematics of the past (¶2). An interesting slip is his use of "history of mathematics" and "mathematical history" as synonyms, whereas the expressions denote quite different subjects (Grattan-Guinness 1997, 759–761).

2. An Example

The distinction between history and heritage has been cast above in as general a manner as possible; any piece of mathematics from any culture will be susceptible to it. Here is an example, mathematically simple but historically very important (a contrast that itself manifests the distinction).

---

efforts to handle the early articles on mechanics were quite unsatisfying: heritage dominated a task intrinsically historical, at least for the early decades of that period.

4. See Dauben (1999) on the journals for the history of mathematics at that time.

Book 2, proposition 4 of Euclid's *Elements* comprises this theorem about "completing the square":

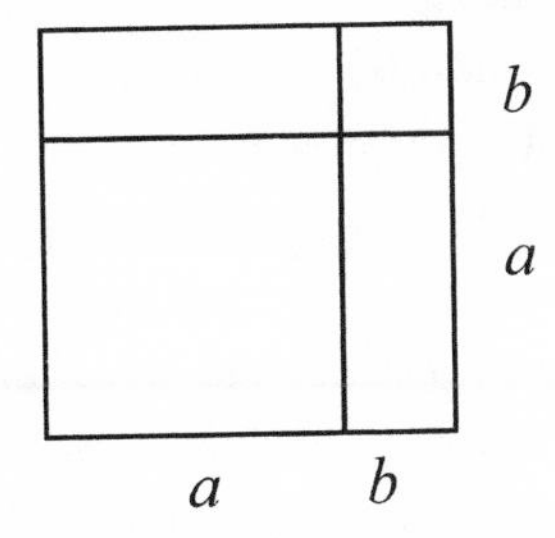

Figure 2.1

From the late nineteenth century onward an influential historical interpretation developed, in which Euclid was taken to be a "geometric algebraist," handling geometrical notions and configurations but actually practicing common algebra. (Compare the remarks in ¶1.2 on history and heritage at that time.) Under this interpretation the diagram is rendered as

$$(a + b)^2 = a^2 + 2ab + b^2. \tag{2.1}$$

However, historical disquiet should rise.

First, (2.1) is a piece of algebra, which Euclid did not use, even covertly: his diagram does not carry the letters a and b.[5] His theorem concerned geometry, about the large square being composed of four parts, with rectangles to the right and above the smaller square and a little square off in the northeast corner; indeed, he specifically defined as "the gnomon" the L-shape formed by the three small regions (*Elements*, book 2, definition 2), known also for its use in sundials and the measurement of time. All these geometrical relationships, essential to the theorem, are lost in the single sign "+" in (2.1).

Further, a and b are associated with numbers, and thereby with lengths and their multiplication. But Euclid worked with lines, regions, solids, and angles, not any arithmeticized analogues such as lengths, areas, volumes, or degrees; he never multiplied geometrical magnitudes of any kind (in important contrast to his arithmetic

5. A characterization of algebra is needed. "The determination of unknowns" is a necessary but not a sufficient condition, for under it most mathematics is algebra. I would also require the *explicit* representation of knowns and unknowns by special words and/or symbols, and articulation of operations upon them (such as addition or concatenation), relationships between them (such as inequalities and expansions), and their basic laws. On the specification of ancient "algebra" see Høyrup (2002, ch. 7).

books 7–9, where he multiplied integers in the usual way). Hence a^2 is already a historical distortion (Grattan-Guinness 1996, or ch. 9).[6]

For reasons such as this, the algebraic reading of Euclid has been discredited by specialist historians in recent decades. By contrast, it is still advocated by mathematicians, such as Weil (1980), who even claimed that group theory is *necessary* to understand book 5 (introducing ratios, and forming propositions and other theorems involving geometrical magnitudes) and book 7 (introducing basic properties of positive integers). An interesting practitioner of the reading of Euclid as a geometric algebra was T. L. Heath, whose translation and edition of Euclid, first published in the 1900s, is still the major source in English (Euclid, *Elements*). I am assured by Greek specialists that his translation is generally faithful to the original. To take an important example, he writes "square on the side," not "square of the side," which can easily be confused with "side squared" and thus lead to the algebra of (2.1); even Heath's distinguished predecessor Robert Simson had used it in his influential edition (Euclid [Simson] 1756; for example, p. 51 for book 2, proposition 4 and (2.1)].[7] Yet in his commentaries Heath rewrote many of Euclid's propositions in common algebra without seeming to notice the variance from his own translation that inevitably follows (see his summary of geometric algebra in Euclid, *Elements*, vol. 1, 372–374): in a few cases his algebraic proofs differ from Euclid's originals (for example, book 6, proposition 28).

It is now much better understood that identity (2.1) belongs to the heritage from Euclid, especially among some Arabs with their word-based algebra (the phrase "completing the square" is Arabic in origin), and then in European mathematics, when symbols for quantities and operations were gradually introduced.[8] The actual version used in (2.1) corresponds more or less to the early seventeenth century, with figures such as Thomas Harriot and René Descartes: Euclid and the relevant Arabs are part of their history, they are part of the heritage from Euclid and those Arabs, and our use of (2.1) forms part of our heritage from both of them.[9] Here we have

6. Again, Euclid defined lines as "breadthless" (book 1, definition 2); often criticized by inheritors, he made clear an aspect of his own history, in replacing the Babylonian use of "lines" *as objects with width* (Høyrup 1995, 2002 *passim*).

7. Translations of mathematical texts often entail tricky questions of history and heritage, along with semantic and syntactic issues. These latter are especially marked when the languages involved belong to different families; in particular, Hoe (1978) translates Chinese into English or French character by character rather than by the word structure of the final language. See also ¶4.6 on general words.

8. There is of course another large history and heritage from Euclid, inspired by the alleged rigor of this proofs. It links in part to the modernization of his geometry, but I shall not discuss them here.

9. This last feature applies also, regrettably, to the supposed history (Rashed 1994) of Arabic algebra, where the Arabs seem already to have read Descartes.

various history and heritage statements, all in one sentence: fine, but do not muddle them up!

This advice seems to have been offered by E. J. Dijksterhuis (1892–1965) in his inaugural lecture as professor of the history of exact sciences at Utrecht University in 1953. He used the adjectives "genetic" or "evolutionary" to characterize heritage and "phenomenological" for history (Struik 1980, 12–13: the last adjective was perhaps not well chosen). Not coincidentally, his edition of Euclid was much more historically sensitive than Heath's, with the notations of geometric algebra avoided; for example, the square on side *a* was denoted *T(a)*, with *T* for "tetragon" (Euclid [Dijksterhuis] 1929, 1930).

In the rest of this chapter I shall concentrate on general historical and historiographical issues. In so doing no claim is made that history is superior to heritage, or superordinate upon it. A companion essay to this one dealing with good and bad practices in the prosecution of heritage is very desirable. History and heritage are twins, each profiting from practices used in the other. I only claim, *outside* of the discussion to follow, that it is often worthwhile to have some knowledge of the history of any context or subject in which one is interested.

3. Some Attendant Distinctions

3.1. Pre- and Posthistory

The distinction between history and the heritage of *N* clearly involves its relationships to its prehistory and posthistory. The historian may well try to spot the historical *foresight*—or maybe lack of foresight—of his historical figures, the ways in which they thought or hoped that the notions at hand may be developed. He should be aware of the merit as well as the difficulties of "not being wise after the event" (Agassi 1963, 48–67). By contrast, the inheritor may seek historical *perspective* and hindsight about the ways the notions actually seemed to have developed. This distinction, quite subtle, is often overlooked.

The distinction is emphatically *not* that between success and failure; history also records successes, but with the slips and delays exposed. A nice example is Hawkins (1970), a fine history of the application of point set topology to refine the integral from the Cauchy-Riemann version through content in the sense of Camille Jordan and Georg Cantor to the measure theory of Emile Borel and Henri Lebesgue. Hawkins not only records the progress achieved but also carefully recounts conceptual slips made en route: for example, the belief until its refutation that denumerable set, set of measure zero, and nowhere dense set were coextensive concepts.

The general situation may be expressed as follows. Let N_0, N_1, and N_2 form a sequence of, say, three notions holding some contextual (not necessarily logical) relationship, and lying in forward chronological order; then the heritage of N_1 for N_2 belongs also to the history of N_2 relative to N_0 and N_1. In both history and heritage it is worth finding out whether or not N_0 played an active role in the creation of N_1, N_2, . . . (as with the Euclid example for some Arabs), or if it is simply being used as a test case for them. However, more is involved than the difference between pre- and posthistory; *both* categories use posthistory, though in quite different ways. In the elaboration below are some further examples, though for reasons of space they are treated rather briefly; fuller historical accounts would take note of interactions with the development of other relevant notions.

3.2. *History Is Usually a Story of Heritages*

The historian records developments and events where normally a historical figure inherited knowledge from the past in order to make his own contributions heritage style. Conversely, heritage unavoidably involves various histories. Some attention to the broad features of history may well enrich the inheritance, and perhaps even suggest a research topic.

Sometimes tiers of history may be exposed. Work produced in, say, 1700 was historical in 1800 and in 1900 as well as in 2000. Thus the historian in 2000 may have needed to note how it was (mis-)understood by later figures, including historians as well as mathematicians, when it formed part of their heritage. If a mathematician really did treat a predecessor in a historical spirit, at least as he (mis-)understood it, then the (now meta-)historian should record accordingly (see, for example, Stedall 2001 on John Wallis's partly and dubiously historical *Algebra* of 1685).

3.3. *Types of Influence*

Types of influence raise important issues. Heritage is likely to focus only on positive influence, whereas history needs to take note also of negative influences, especially of a general kind, such as reaction against some notion or the practice of it or importance accorded some context. For example, one motive of A. L. Cauchy to found mathematical analysis in the 1820s upon a theory of limits (¶4.1) was his rejection of J. L. Lagrange's approach to the calculus using only notions from algebra. Further, as part of his new regime Cauchy stipulated that "a divergent series has no sum," regarding as illegitimate the results obtained by Leonhard Euler (Hofmann 1959) and various other contemporaries and successors; but in the 1890s Borel reacted against precisely this decree and became a major figure in the development of summability and formal power series (Tucciarone 1973). Note that we have here not only negative influence but also heritage from Euler and from Cauchy and history of Borel at the same time.

3.4. *The Role of Chronology*

The role that chronology plays differs greatly. In history it can form a major issue; for example, possible differences between the creations of a sequence of notions and those of their publication. Further, the details available may only give a crude or inexact time course, so that some questions of chronology remain unanswerable. It is particularly difficult or even impossible to determine for ancient mathematics and for ethnomathematics. In heritage chronology is much less significant; however, mathematicians often regard questions of the type "Who was the first mathematician to . . . ?" as the prime type of historical question to pose (May 1975b), whereas historians recognize these kinds of questions as often close to meaninglessness when the notion involved is very general or basic. For example, "to use a function?" could excite a large collection of candidates according to the state, generality, or abstractness of the function theory involved (Thiele 2000; compare ¶4.6). The only questions of this kind of genuine historical interest concern priority disputes, when intense parallel developments among rivals are under investigation and chronology is tight—and where again maybe no answer can be found.

3.5. *Use of Notions Later than* N

This is a major matter. Later notions are *not* to be ignored; the idea of forgetting the later past of a historical episode, often put forward as desirable historiography, is impossible to achieve, since the historian has to know which notions *are* later, and this requires the historical task already to have been accomplished (¶5.1). Instead, when studying the history of N_0, by all means recognize the place of later notions N_1, N_2, . . . but avoid feeding them back into N_0 itself. For if that does happen, the novelties that attended the emergence of N_1, N_2, . . . will not be registered. Instead time loops are created, with cause and effect over time becoming reversed: when N_2 and N_1 are shoved into N_0, then they seem to be involved in its creation, whereas the converse is (or may be) the case. In such situations not only is the history of N_0 messed up but also that of the intruding successors, since their absence before introduction is not registered.

For example, in the late eighteenth century Lagrange realized that the solvability of polynomial equations by algebraic operations alone was connected to properties of certain functions of their roots when the latter were permuted; and this achievement played a role in the development of group theory during the nineteenth century (Wussing 1984, 70–84). Now to describe his work in terms of group theory not only distorts Lagrange but also muddies the (later) emergence of that theory itself by failing to note its absence in him. Sometimes such modernizations are useful to save space on notations, say, or to summarize mathematical relationships, but the ahistorical character should be stressed: "in terms of group theory (which Lagrange did not have), his theorem on roots may be stated thus: . . ."

A valuable use of later notions when studying the history of N_0 is as sources for questions to ask about N_0 itself—but do not expect positive answers! However, negative answers need to be examined carefully; lack of evidence does not provide evidence of lack.

By contrast, when studying the heritage of N_0, by all means feed back N_1, N_2 . . . to create new versions: it may be clarified by such procedures; the chaos in the resulting history is not significant; maybe even a topic for mathematical research will emerge. But it is only negative feedback, unhelpful for both history and heritage, to attack a historical figure for having found only naïve or limited versions of a theory that, as his innovations, helped to lead to the later versions upon which the attack is based. To resume the case of summability from ¶3.3, it is not informative to regard Euler on that topic as an idiot; but he did not foresee the rich panoply of uses to which the name "divergent series" is now put.

3.6. A Schematic Representation of the Distinction

The difference is shown in figure 2.2, where time runs from left to right. For history the horizontal arrows do not impinge positively on the preceding notions whereas those for heritage do. That is, in history one should avoid feeding later notions back into N if they did not play roles there; by contrast, such practices are fine for the purposes of heritage and indeed constitute a common and fruitful way of conducting research (¶3.1).

Each N may be a collection of notions, with some or maybe all playing roles in the creation of successors in the next collection. Arrows pointing forward in time could be drawn to represent foresight, hopes for further progress.

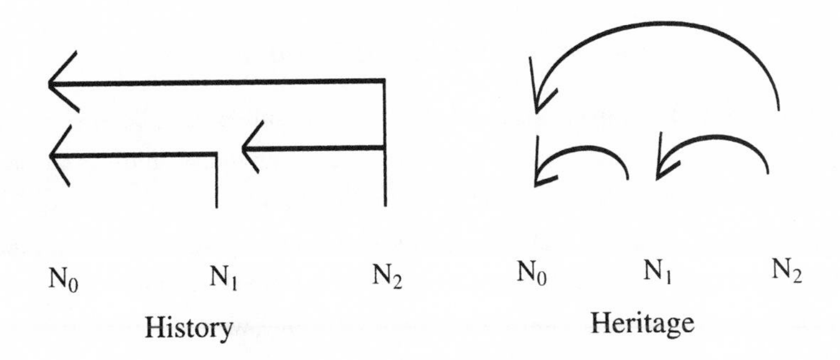

Figure 2.2

3.7. Foundations Up or Down?

The distinction can be extended when N is an axiomatized theory, which proceeds logically through concepts C_1, C_2, C_3 . . . ; for to some extent the respective historical origins move *backward* in time, thus broadly the reverse of the historical record. A related difference is thereby exposed: heritage suggests that the foundations of a mathematical theory are laid down as the platform upon which it is built, whereas history shows that foundations are dug down, and not necessarily onto firm territory. For example, the foundations of arithmetic may start with a 1900s version of mathematical logic (hopefully free from paradoxes!), use set theory as established mainly by Cantor in the 1880s and 1890s, define progressions via the Peano axioms of the later 1880s, and then lay out the main properties of integers as established long before that.

A figure important in that story is Richard Dedekind, with his book of 1888 on the foundations of arithmetic. The danger of making historical nonsense out of heritage is well shown in a new supposed translation. A typical example of the text is the following passage, where Dedekind's statement that (in literal translation) "All simply infinite systems are similar to the number-series N and consequently by (33) also to one another" comes out as *"All unary spaces are bijective*[1] to the unary space[2] N and consequently, by ¶33,[3] also to one another"; moreover, of the three editorial notes, the first one admits that "isomorphic" would be more appropriate for Dedekind but the second one informs that "*unary space* . . . is what he means" ("Dedekind" 1995, 63).

3.8. Indeterminism or Determinism?

Especially if history properly records missed opportunities, and delayed and late arrivals of conception and/or publication, it will carry an indeterministic character: the history did indeed pass through the sequence of notions N_0, N_1, N_2, . . . but it might have been otherwise. Everything in this chapter is proposed in an explicitly indeterministic spirit. The inheritor can take a hint from the historian here: in the past, many theories have developed slowly and/or fitfully, with long periods of sleep; so which theories are sleeping today?

By contrast, even if not explicitly stressed, a deterministic impression is likely to be conveyed by heritage: N_0 *had* to lead to N_1. Appraisal of historical figures as "progressive" or "modern," in any context, is normally of this kind: the appropriate features of their work are stressed, the others ignored. In this respect, and in some others such as the stress on hindsight and the flavor of determinism, heritage resembles Whig history, the seemingly inevitable success of the actual victors, with predecessors assessed primarily in terms of similarities with the dominant position. For scientists, Isaac Newton as a modern scientist gains a "yes," but Isaac Newton the major alchemist is a "no."[10] Again, the inheritor may read something by, say, Lagrange and exclaim, "My word, Lagrange here is very modern!"; but the historian should reply: "No, we are very Lagrangian."

A fine example of indeterminism is provided by the death of Bernhard Riemann in 1866. The world lost a very great mathematician, and early; had he lived longer, new theories might have come from him that arrived only later or maybe not at all. On the other hand, his friend Dedekind published in 1867 two manuscripts that Riemann had prepared in 1854 for his *Habilitation* but had left unpublished, seemingly indefinitely. While both manuscripts contained notions already present in the work of some other mathematicians, they made rapid and considerable impacts on their appearance. Had the one on mathematical analysis and especially trigonometric series not appeared then, there is no reason to assume that Cantor, a young number theorist in the early 1870s, would have tackled the problem of exceptional sets for Fourier series (to use the later name) that Riemann exposed, thereby inventing the first elements of his set theory (Dauben 1979, chs. 1–2); but then many parts of mathematical analysis would have developed differently. The other manuscript, on the foundations of geometry, is noted at the end of the next subsection.

10. Arnol'd (1990) is a supposedly historical assessment of Isaac Newton's remarkable theorem in the *Principia* on the class of closed convex curves expressible by algebraic formulae; apparently it was a theorem about the topology of Abelian integrals (ch. 5, including a fantasy on p. 85 about Cauchy's motivation to complex-variable analysis).

3.9. Revolutions or Convolutions?

When appraising heritage, interest lies mainly in notions in (fairly) finished form without special concern about the dynamics of their production. A deterministically construed heritage can convey the impression that the apparently inevitable progress shows mathematics to be a *cumulative* discipline.

But history suggests otherwise; some theories die away, or at least their status is reduced. The status or even occurrence of revolutions in mathematics is historically quite controversial (Gillies 1992); I have proposed the metanotion of convolution, where new and old notions wind around each other as a (partly) new theory is created (Grattan-Guinness 1992; also ch. 6). Convolution lies between, and can mix, three standard categories: revolution, in the sense of strict *replacement* of theory; innovation, where replacement is absent or plays a minor role (I do not know of a case where even a remarkably novel notion came from literally *no* predecessors); and evolution, similar to convolution in itself but carrying many specific connotations in the life sciences that are not necessarily relevant here.

One of the most common ways in which old and new mix is when a new notion is created by connecting two or more old notions in a novel way. Among very many cases, in 1593 François Viète connected Archimedes's algorithmic exhaustion of the circle using the square, regular octagon, and so forth, with the trigonometry of the associated angles, and obtained this beautiful infinite product:

$$2/\pi = \sqrt{1/2}\sqrt{1/2 + 1/2\sqrt{1/2}}\sqrt{1/2 + 1/2\sqrt{1/2 + 1/2\sqrt{1/2}}}\sqrt{}\ \ldots \qquad (2.2)$$

Again, in the 1820s Niels Henrik Abel and Carl Jacobi independently linked the notion of the inverse of a mathematical function with Adrien-Marie Legendre's theory of "elliptic functions" (to us, elliptic integrals) to produce their definitive theories of elliptic functions. Heritage may also lead to new connections being effected.

Sometimes convolutions, revolutions, and traditions can be evident together. A very nice case is found in the work of Joseph Fourier in the 1800s on heat diffusion (Grattan-Guinness and Ravetz 1972).

1. Apart from an unclear and limited anticipation by J.-B. Biot, he innovated the differential equation to represent the phenomenon.
2. The method that he used to obtain it was traditional, namely, Euler's version of the Leibnizian differential and integral calculus (which is noted in ¶4.1).
3. He refined the use of boundary conditions to adjoin to the internal diffusion equation for solid bodies.

4. He revolutionized understanding of the solution of the diffusion equation for finite bodies by using infinite trigonometric series; the solution had been known before him but was importantly misunderstood, especially about the manner in which a periodic series could represent a general function at all.
5. He innovated the Fourier integral solution, for infinite bodies.

Delays often arise from connections *not* being made. A well-known puzzle is the slowness to recognize non-Euclidean geometries when there was a long history of map making that surely exhibits one kind of such a geometry. J. H. Lambert is an especially striking figure, because he worked with some luster in both areas in the later eighteenth century. The answer seems to be that, like his predecessors and several successors, he understood the geometry problem as being just the status, especially provability, of the parallel axiom *within the Euclidean framework* rather than the more general issue of alternative geometr*ies*, which was fully grasped only by Riemann in his 1854/1867 manuscript (Gray 1989). Thus the link, which seems so clear in our heritage, was not obvious in the earlier times.

3.10. Description or Explanation?

Both history and heritage are concerned with description; but, as was mentioned in ¶1.1, history should also attempt explanations of the developments found and of the delays and missed opportunities that are noticed. These explanations can be of various kinds, not just of the technical insights that were gained but also the social background, such as the (lack of) educational opportunities for mathematics in the community or country involved. Especially in ancient and medieval times, and not only in the West, prevalent philosophical and religious stances could play important roles. One feature especially of the nineteenth century that needs explanation is the differences between nations of the *(un)popularity* of topics or branches of mathematics (France doing loads of mathematical analysis, England and Ireland with rather little of it but working hard at several new algebras, and so on).

Heritage studies need to consider explanation only from a formal or epistemological point of view. For example, it would explain the mystery of having to use complex numbers when finding the real roots of polynomials with real coefficients in terms of closure of operations over sets, an insight which has its own history (Sinaceur 1991, pt. 2).

3.11. Levels of (un)Importance

This last task relates to another difference, that a notion rises or falls in importance. Heritage does not need to give such changes much attention; the modern level of importance is taken for granted. But history should watch and ponder the changes carefully. A general class of cases is considered in ¶4.4.

A fine example is provided by trigonometry, which for a long time has been an obviously useful but rather minor topic in a course in algebra—and, correspondingly, there has been no detailed general history of it since (von Braunmühl 1900, 1903). By contrast, in the late Middle Ages it was a major branch of mathematics and was handled geometrically so that, for example, the sine was a length measured against the hypotenuse as unit, not as a ratio of lengths. In further contrast, spherical trigonometry was more important than planar trigonometry because of its use in astronomy and navigation.

As a converse example, probability theory and especially mathematical statistics had long and slow geneses; most of the principal notions in statistics are less than two centuries old, and the cluster of them that is associated with Karl Pearson and his school has celebrated their centenary only recently. The slowness of the arrival of this discipline, now one of the most massive parts of mathematics while often functioning separately from it, is one of the great mysteries of the history of mathematics; its modest place during most of the nineteenth century is especially astonishing. But this tardiness need not disturb a seeker of heritage within it.

3.12. Handling Muddles

One way in which knowledge of all kinds increases, especially the mathematical, is by the cleaning up of unclarities and ambiguities by bringing in new distinctions of sense; for example, the convergence of series of functions was split, largely by Karl Weierstrass and his followers from the 1870s onward, into various modes of uniform, nonuniform, and quasiuniform convergence (Hardy 1918). Such housework forms part of the heritage that the mathematician will deploy (unless he has reason to question it historically). The historian will also note the modern presence of such distinctions, but he should try to *reconstruct* the old unclarities, as clearly as possible, so that the history of the distinctions is itself studied. An important example is included in ¶4.1.

This historical procedure seems to contradict the claim of ¶3.5 that history usually stresses differences between notions while heritage highlights similarities; preserving muddles keeps things the same while cleaning them up brings out differences. However, there is no difficulty; to continue with the example of the various modes of convergence before the Weierstrassians, the historian will stress the difference between the ignorance of them among predecessors and our knowledge of them, while the inheritor will insert them into that earlier work and so make it more similar to the later version.

3.13. On Some Consequences for Mathematics Education

The issue of heuristics in mathematics, and the discovery and later justification of mathematical notions, are strongly present in this discussion, with obvious bearing on mathematics education. The tradition there, especially at university level or equivalent, is to teach a mathematical theory in a manner very much guided by heritage. But reactions of students—including myself, as I still vividly recall (§1.1)—are often distaste and bewilderment; not particularly that mathematics is very hard to understand and even to learn but mainly that it turns up in "perfect" dried-out forms, so that if there are any mistakes, then necessarily the student made them. Mathematical theories come over as all answers but no questions, all solutions but no problems—and only the cleverest students possess enough intelligence to understand it.

A significant part of the growth in interest in the history of mathematics has been inspired by a negative influence (¶3.3) of such situations, and there is now a strong international movement for making use of history in the teaching of mathematics, at all levels (Fauvel and van Maanen 2000). In a companion paper (Grattan-Guinness 2004) I consider the bearing of the distinction between history and heritage on mathematics education in some detail; the main points are rehearsed here, and another one in ¶5.4.

Long ago I proposed the metatheoretical notion of "history-satire," where the broad historical record is respected but many of the complications often contained in the messy details are omitted or elided (Grattan-Guinness 1973): if one stays solely within, say, Newton's historical context all the time, then one will stop where Newton stopped. Otto Toeplitz's "genetic approach" to the calculus (Toeplitz 1963) is close to a special case of this approach (Schubring 1978). (Note from ¶1.2 the use of "genetic" by Dijksterhuis to characterize heritage.) It is also very well deployed in Bressoud (1994), a textbook on real-variable mathematical analysis {and now also Bressoud (2008) on measure theory}.

Where does mathematical education lie in between history and heritage? My answer is, exactly there, and a very nice place it is. Educators can profitably use both history *and* heritage for their purposes. For example, the algebraic version of Euclid, so important in its heritage, is often and well used in this kind of teaching. But also available is the real Euclid of arithmetic and geometry, including the beautiful theory of ratios, for me the mathematical jewel of the work, both fine mathematics in its own right and an excellent route in to the notoriously difficult task of teaching (the different topic of) rational numbers. (To make another contrast between history and heritage, Euclid used only the reciprocals $1/m$ among the rational numbers, and no irrational numbers at all.) Following history-satire, the differences between the two Euclids should be stressed; indeed, they could start off lots of nice points about the relationships between these three branches of mathematics in element-ary contexts, such as the difference between lines (geometry without arithmetic) and lengths (geometry with arithmetic). A recent attractive study of the history of algebra, including the role of Euclid, is provided by Bashmakova and Smirnova (1999), though in my view the authors conflate history and heritage statements throughout (Grattan-Guinness 2004, sect. 8).

4. Prevailing Habits: Six Cases

> Anything that has become background, or context, or tradition is no longer salient, sometimes no longer represented symbolically at all.
>
> —*James Franklin (2001, 344)*

I consider six special cases of aspects of mathematics where the conflation of history and heritage seems to be especially acute, including among historians. The cause seems to be habitual use of the notions involved, which becomes so commonplace that it is not questioned. The examples come mostly from the nineteenth and early twentieth centuries, which not accidentally is my own main period of research; thus no claim of optimal importance or variety is made for them. Examples of the distinctions made in ¶3 are also included.

4.1. The Calculus and the Theory of Limits

There have been four main ways of developing the calculus (Grattan-Guinness 1987; also ch. 11): in chronological order,

1. Newton's "fluxions" and "fluents" (1660s onward), with the theory of limits deployed, though not convincingly;
2. G. W. Leibniz's "differential" and "integral" calculus, based on dx and $\int x$ (1670s onward), with infinitesimals central to and limits absent from all

the basic concepts: reformulated by Euler in the mid-1750s by adding in the "differential coefficient," the forerunner of the derivative;
3. Lagrange's algebraization of the theory, in an attempt to avoid both limits and infinitesimals, with a new basis sought in Taylor's power-series expansion (1770s onward), and the successive differential coefficients reconceived in terms of the coefficients of the series as the "derived functions"; and
4. Cauchy's approach based on a firm *theory* (and not just an intuition) of limits (1810s onward); from it he defined the basic notions of the calculus (including the derivative as the limiting value of the difference quotient) and of the theories of functions and of infinite series, to create "mathematical analysis."

Gradually the last tradition gained wide acceptance, with major refinements brought in with Karl Weierstrass and followers from the mid-century onward. In particular, they honed Cauchy's basically single-limit theory into one of multiple limits with a plethora of new distinctions (including the modes of convergence noted in ¶3.12). Thus it has long been the standard way of teaching the calculus; but historians should beware using it to rewrite the history of the calculus where any of the other three traditions, *especially* Newton's and Cauchy's, are being studied. It also contains an internal danger. The (post-)Weierstrassian refinements have become standard fare and are incorporated into the heritage of Cauchy; but it is mere feedback-style ahistory to read Cauchy (and contemporaries such as Bernard Bolzano) as if they had read Weierstrass already (Freudenthal 1971). On the contrary, their own pre-Weierstrassian muddles need clear historical reconstruction (¶3.12). Again by contrast, inheritors can acknowledge such anachronisms but ignore them, and just see whether or not the mathematics produced is interesting.

4.2. *Part-Whole Theory and Set Theory*

An important part of Cauchy's tradition by (some of) the Weierstrassians was the introduction from the early 1870s of set theory, principally by Cantor (¶3.8). Gradually it too gained a prominent place in mathematics and then in mathematics education; so again conflations lurk around its history. They can occur not only in putting set-theoretical notions into the prehistory, but in particular confusing that theory with the traditional way of handling collections from antiquity: namely, the theory of whole and parts, where a class of objects contains only parts (such as the class of Australian cathedrals as a part of the class of cathedrals), and membership was not distinguished from inclusion. Relative to set theory, parthood corresponds to improper inclusion, but the theory can differ philosophically from Cantor's doctrine, on matters such as the status of the empty class/set, and the class/set as one and as many; so care is needed. An interesting example occurs in avoiding the algebraization of Euclid mentioned in ¶2: Mueller (1981) proposed an algebra alternative to that in (2.1) in ¶2, but he deployed set theory in it whereas Euclid had followed the traditional theory, so that a different distortion arises. As in earlier points, inheritors need feel no discomfort.

4.3. Vectors and Matrices

In a somewhat disjointed way, vector and matrix algebras and vector analysis gradually developed during the nineteenth century and slowly became staple techniques during the twentieth century, including in mathematics education (Grattan-Guinness 1994, articles 6.2, 6.7, 6.8, 7.12). But then the danger just highlighted arises again; for earlier work was not thought out that way. The issue is *not* just one of notation; the key lies in the associated notions, especially the concept of laying out a vector as a row or column of quantities and a matrix as a square or rectangular array, and manipulating them separately or together according to stipulated rules and definitions. Similar remarks can be applied to tensor analysis.

A particularly influential example of these anachronisms is Truesdell; in very important pioneering historical work of the 1950s, he expounded achievements by especially Euler in continuum mathematics that previously had been largely ignored (see, for example, Truesdell 1954). However, in the spirit of heritage in his remark quoted at the head of this chapter, he treated Euler as already familiar with vector analysis and some matrix theory, as well as using derivatives as defined via the theory of limits, whereas in fact Euler had actually used his own elaboration of Leibniz's version of the calculus mentioned in ¶4.1. Therefore Truesdell's Euler was out of chronological location by at least a century. It is quite amusing to read Truesdell's editorial commentaries and then Euler's original texts in the same volumes (11 and 12 of the second series of the *Opera omnia*). Much historical reworking of Euler's mechanics is needed, not only to clarify what and how he had actually done and not done but also to eliminate the mess-ups of feedback.[11] The history of vectors and matrices needs to be clarified by noting the absence of these notions in Euler.

11. {The Euler centenary year 2007 led to some quite historically sympathetic writing; see, for example, Bradley and Sandifer (2007).}

4.4. The Status of Applied Mathematics

This case exemplifies the variation of levels of importance raised in ¶3.11, in a case where certain features of heritage have affected levels of historical interest. During the middle of the nineteenth century the professionalization of mathematics increased quite notably in Europe; many more universities and other institutions of higher education were created or expanded, so that the number of jobs increased. During that period, especially in the German states, then Germany as a whole, and later internationally, a rather snobbish preference for pure over applied or even applicable mathematics began to develop. Again this change has affected mathematics education, for the worse.[12]

The tendency has also influenced historical work in that the history of pur(ish) topics has been studied far more than that of applications; the history of military mathematics is especially ignored. But a mismatch of levels of importance arises, for prior to the change, applications and applicability were very much the governing motivation for mathematics, and the balance of historical research should better reflect it. Euler is a very good case; studies of his contributions to purish mathematics far exceed those of his applied mathematics (hence the importance of Truesdell's initiative in looking in detail at his mechanics). Some negative influence from current practice is required of historians to correct this imbalance.

4.5. The Place of Axiomatization

From the late nineteenth century onward David Hilbert encouraged the axiomatization of mathematical theories in order to make clearer the assumptions made and to study metaproperties of consistency, completeness, and independence. His advocacy, supported by various followers, has given axiomatization a high status in mathematics, and thence in mathematics education. But once again dangers of distortion of earlier work attend, for Hilbert's initiative was then part of a *new* level of concern with axiomatization (Cavaillès 1938); earlier work was rarely so preoccupied, although the desire to make clear basic assumptions was frequently evident (for example, in the calculus as reviewed in ¶4.1). Apart from Euclid, of the other figures named above only Dedekind can be regarded as an axiomatizer; it is out of line so to characterize the others, even Lagrange, Cauchy, Weierstrass, or Cantor.

12. Both history and heritage attach to the words "pure" and "applied" mathematics, and to cousins such as "mixed." The history of these adjectives is itself worth study.

4.6. Words of General Import

One aim of many mathematical theories is generality; and attendant to this aspiration is the use of correspondingly wide-ranging words or phrases, such as "arbitrary" or "in any manner," to characterize pertinent notions. The expressions may well still be used in many modern contexts; so again the danger of identification with their past manifestations needs to be watched.

A good example is the phrase "any function" in the calculus and the related theory of functions; it or some cognate (such as "functio quomodocumque") will be found with (at least) John Bernoulli in the early eighteenth century, Euler about 40 years later, Lagrange and S.-F. Lacroix around 1800, J. P. G. Dirichlet in the late 1820s, and Lebesgue and the French school of analysts in the early twentieth century. Nowadays it is usually taken to refer to a mapping (maybe with special conditions such as isomorphism), with set theory used to specify range and domain and no other details or conditions. But the universe of functions has not always been so vast; generality has always belonged to its period of assertion. In particular, Dirichlet (1829) mentioned the characteristic function of the irrational numbers (to use the modern name); but he quite clearly regarded it as a pathological case, for it did not possess an integral. The difference is great between his situation and that of Lebesgue's time, for the integrability of such a function was a good test case of the new theory of measure to which he was a major contributor; indeed, this detail is part of the heritage from Dirichlet.

5. History and Heritage as Metatheories

So far the concerns and examples treated in this chapter have centered on mathematics alone; but clearly the issue of history and heritage is more general. One can see the same kinds of issue arising in the histories of the other sciences and of technology, and indeed outside the sciences altogether; some nice examples arise in music, in connection with preferred practices in the execution of "authentic performance" of older works. Thus, while mathematics seems to provide by far the richest context and examples (at least to my knowledge), the issues themselves have a broader remit. In this section I state the four principles that inform the discussion above.

5.1. History Is Unavoidable

We work out the present from the past, whether we like it or not. Thus ignorance of history does not produce immunity from it any more than ignorance of food poisoning saves one from attacks of it. On the contrary, influence is all the more likely to be exerted.

This principle questions a basic issue in mathematics (and other sciences) and its teaching: namely, should one bother with the history or ignore it completely? Recognition of its unavoidability shows that *the question itself is falsely posed:* the issue is *not* history yes or no, but history how? A dried-out formulation of a theory of the kind mentioned at the end of ¶1.1, denuded of human names or background or heuristic, is still not immune from history; for example, it continues a historical tradition of presenting mathematical theories in a dried-out formulation, denuded of human names and background and heuristic. For the same reasons, heritage also is unavoidable. So it is better to be aware of both of them and the relationships that they excite and unavoidably impose.

5.2. *The Stratification and Self-Reference of Knowledge*

If history is unavoidable, then it has to be addressed somehow. We have some historical text before us, say, Euclid's *Elements*. How can we read it in a historical spirit? A popular answer, put forward for all kinds of history, goes as follows. When reading Euclid's work, forget all theories in the field that been developed since; step into his shoes (more likely sandals, in this case) and read his work with his eyes.

Unfortunately, as was noted briefly in ¶3.5, this method suffers from a difficulty; namely, it is *completely useless.* To ignore all knowledge produced since Euclid one needs to know what that knowledge is in the first place. But to know that we must be able distinguish it from the knowledge produced before and during Euclid's time. But to know that we need to know the history of Euclid's work—before studying the history of Euclid's work! Q.E.D.[13]

To avoid this contradiction it is necessary to realize that when historians study historical figures, they have to realize that they are thinking *about them,* not *with* them. It is claimed that the distinction between theory and metatheory is of central importance for knowledge, whether mathematical or any other kind. The positions of the horizontal arrows above the notions in the history part of figure 2.2 form an

13. In a posthumously published consideration of "History as Re-enactment of Past Experience," which has been much discussed by philosophers of history, Collingwood (1946, 282–289) took book 1, proposition 5 of Euclid's *Elements*, that "In isosceles triangles the angles at the base are equal to one another," and contrasted Euclid's own thoughts about the theorem from the thoughts about it made by a later historian. However, he tended to stress the similarities of the thoughts rather than the differences and did not explicate metatheory in the way advocated in this chapter.

image of this situation, in contrast to the feedback imaged in the heritage part of the figure.

The importance of this distinction lies in its generality.[14] This emerged from the 1920s onward, inspired principally by the logicians Kurt Gödel and Alfred Tarski after several partial anticipations, of which Hilbert's program of metamathematics as practiced during the 1930s was the most notable (Grattan-Guinness 2000, chs. 8–9). In most other disciplines the distinction is too obvious to require special emphasis; clearly a difference of category exists between, say, properties of light and laws of optics, or between a move in chess and a rule of chess. By contrast, in logic, a very general branch of knowledge, the distinction is uniquely subtle (and therefore desirable); for example, "and" features in both logic and metalogic, and failure to register the distinction led to much incoherence and even to paradoxes, such as one arising from "this proposition is false." Its importance and generality can be seen in Tarski's theory of truth (his own main way to this distinction): "snow is white" (in the "metalanguage," Tarski's word) if and only if snow is white (in the language). His theory is neutral with respect to most philosophies and sidesteps generations of philosophical anxiety about making true (or false) judgments or holding such beliefs.

Consider now a mathematical theory M. Its history is one kind of metatheory of it, its heritage is another, Hilbert-style metamathematics is a third if M is suitably axiomatized, questions about how to teach it is a fourth, and there may well be others. As with theory itself, metatheory requires its own metametatheory, and so on up as far as may be needed; thus theory becomes stratified. An example of metametatheory is the history of the history of mathematics, upon which a comprehensive book has recently been published (Dauben and Scriba 2002); the comments on Heath's translation of Euclid and his own algebraic rewritings (¶2), and the example of Wallis (¶3.2), also belong to the history of history. Another example, indeed a self-referring one (¶6), is this chapter; it belongs to the history of history (of mathematics), although whether it also will enjoy a heritage is another matter! If such a miracle were to occur, then the chapter would belong to another third-order theory: the history of philosophy of history, an interesting subject for which a good source book has recently been published (Burns and Rayment-Pickard 2000).

One great advantage of adopting stratification is that the assumptions chosen

14. Generality is not a necessary virtue. I agree with the maxim attributed to Saunders Mac Lane, "We do not need the greatest generality, but the right generality," {and have applied it to some philosophical questions in Grattan-Guinness (2008)}.

to underlie the theory do not have to be adopted also for its metatheory. An interesting and explicit case lies in L. E. J. Brouwer. Especially from the 1920s, he put forward a constructivist approach to mathematics, called "intuitionism," in which he rejected the law of excluded middle (van Dalen 1999). However, his metamathematics, which he called "mathematics of the second order," was classical, with that law in place; a proof was intuitionistically correct or not. No contradiction arises, since the levels are different.

The same freedom attends historians when they see themselves as metatheorist. For they do *not* have to defend or even like what they try to describe or to explain. Why should they? After all, they were not there. This point aligns with commonplace understandings; that historians of, say, Hinduism do not have to be Hindu, although they might be. Similarly, inheritors have to take what they can find, maybe without enthusiasm. Stratification also sidesteps the fashionable modern chatter about narratives and discourses (Windschuttle 1997) and the relativism and such waffle that often accompanies it.

5.3. *Knowledge Is Based on Ignorance*

Knowledge is based on ignorance in the important sense that theories explain knowns in terms of unknowns. To take Euclid again, the primitives in his geometry include the "common notions" and axioms given in book 1 (and indeed more axioms than he realized, as has been understood for over a century); but these primitives cannot be known in terms of other notions, for then they would no longer be primitive. To take another case, one of the bases of parts of Newton's mechanics is his inverse square law of central attraction, which is unknown, maybe unknowable, and certainly mysterious.

This principle is worth stressing partly because it is often confused with an important but quite different way in which theories develop; namely, having been created in one context, they are then applied to new ones to see how they fare. To continue with the Newton case, Euler and others applied his theory to areas of continuum mechanics such as elasticity theory and fluid mechanics, where Newton had not said a great deal. Euler also took the second law of motion to apply in *any* direction, whereas Newton himself had restricted it to special directions such as tangents and normals to given curves (Truesdell 1968, chs. 3 and 5). Such developments tempt one to say that Newton's theory explained the unknown in terms of the known; but such claims are *methodological,* concerning the important process of changing from contexts already known to contexts currently unknown. The principle put forward in this subsection is *epistemological,* concerning the structure of theories as such.

5.4. Knowledge and Ignorance Go Together

Knowledge and ignorance go together at the metatheoretic level in a profound way. For we have knowledge (of a fact, or theorem, or whatever) and maybe knowledge of that knowledge, such as a proof of a theorem. But we also have *knowledge of ignorance,* especially when forming a problem or conjecture: when asking whether or not some property does or does not obtain in a theory, the poser *knows that he does not know* the answer. There is also *unawareness,* that is, *ignorance of ignorance,* when one can see that some workers on a topic did not know that they did not know the substance of the problem because the properties and connections were not known at the time in question. The emphasis on ignorance, especially the granting to it of a status metatheoretically equal to that of knowledge, is the principal novelty of the approach advocated here.

The posing of problems enjoys high prestige in mathematics. To recall a historically famous case, Hilbert (1901) posed in 1900 a string of them (some in rather sketched form) for mathematicians to tackle. In each case he knew that answers were not yet known. One of them was Cantor's continuum hypothesis about the number of real numbers, which claimed that two particular infinite numbers were equal to each other. Speculations on the infinite go back earlier than Cantor, but his predecessors did not know that they did not know whether or not his continuum hypothesis was true or false because none of them knew of two different ways of constructing infinite numbers in the first place.

The same kind of relationship obtains also in history (of mathematics as just one special case). Historians know various facts, say, and can even prove some, for example by finding authoritative documents. But they too can pose problems, concerning matters that they know that they do not know; and they can be unaware of other problems until new connections come to light. Layers of history as exemplified in ¶3.2 concerning the history of history can also involve knowledge and ignorance; what historians did (not) know at intermediate periods.

This scheme works also for (mathematics) education. One important task there is laying out a syllabus, and the planning could focus much on deciding how long the students will be kept unaware of some topic or theorem, when an associated problem should be posed, and when solutions should be given to it. To use common algebra again, a school course hoping to advance as far as the formula for the root of a cubic equation will surely spare the youngsters knowledge of the horrible cube roots to come, but the problem could be posed when the roots of the quadratic equation have been dealt with; and when the formula has been obtained, further new questions posed, such as formulae for all three roots, and the possibility of going further with formulae for the roots of the quartic, the quintic (big shock to come!), the sextic, and so forth. At every main stage in the teaching, the interplay of knowledge and ignorance could play a major role in the teaching, though preferably not muddled together (¶3.13).

6. Philosophical Prospects

As logicians have long known, generality skirts self-reference, which sometimes generates paradoxes. Here is a hopefully virtuous example. The discussion in the last section was a sketch of a theory of relationships between knowledge and ignorance. I know that it constitutes a problem (and, I believe, an important one), but I am ignorant of a general solution of it, which would be a detailed account of the main relationships and their own metarelations. Elaboration could be well guided by consideration of the many ways in which changes take place in notions, especially in theorems and theories. These include extending known notions or generalizing them or abstracting them or all three; making new classifications of the mathematical objects involved; reacting to counterexamples by seeking the defective components of the refuted theory; exposing hitherto unnoticed assumptions; in cases where foundations are significant, interchanging some theorems, axioms, and definitions; devising new algorithms, or modifying old ones; making new applications or extending known ones, both within mathematics and to other disciplines; and forging new connections between branches of mathematics, or eliminating or avoiding some (recall from ¶2, for example, that in his geometry Euclid did not arithmeticize his quantities).

A general theory of history and heritage would not be restricted to mathematics, which, however, is a particularly rich source of examples and issues. What are the prospects for further philosophical progress?

There are other intellectual contexts in which ignorance plays an active role. In mathematics it can come into economics, where the actors in an economic situation are ignorant of the intentions of the other actors. In probability theory, values are sometimes interpreted in terms of degrees of ignorance. Some nonclassical logics are relevant, such as the logic of asking questions (Wisniewski 1995). Ignorance has been aired occasionally in science; in particular, much interest was aroused by two "encyclopedias" on it for science and for medicine, collections of articles in which specialists posed then-unsolved problems for their fields (Duncan and Weston-Smith 1977, 1984).

All these cases are perfectly respectable, though inevitably oriented around specific contexts.[15] To find the generality that informs ours, we must move to philosophy proper, especially theories of knowledge. But there a kind of converse scenario emerges: the generality is indeed present, but ignorance is treated like a disease, to be cured by the acquisition of knowledge, however the philosophy at hand claims this should be done. The same attitude seems to inform those philosophies of history that address ignorance at all.[16]

The tradition in which this approach has been developed most systematically is skepticism, in which Descartes was a major figure. It is a highly dystopian philosophy, a disenchantment inspired by the fact that one does not know things for certain.[17] Well, that is true, and for certain (note the use of metatheory here); but skepticism can degenerate into unwelcomely negative positions, such as pure relativism or nihilism (nobody can know anything, at least not better than anyone else).

The insight lacking is the positive one that it is nice to be ignorant, for that is where the problems come from. The only philosophy of which I am aware that both exhibits this insight and carries the required generality is to be found in some writings of a philosopher who was deeply influenced by Tarski from an early stage in his own career: Karl Popper. I have in mind his concern with "the sources of knowledge and of ignorance" (Popper 1963, intro.), and with the tricky self-referential problem of rationally criticizing rationality itself (Watkins 1969). Also relevant is his detailed metaphilosophical arguments for indeterminism and against determinism (Popper 1982), which he also applied to historiography itself.[18]

15. Here is an unexpected example from the time of writing of this paper: "He was describing how, in a world of asymmetrical terrorist threats, there is no way to predict the future, so NATO must plan accordingly. 'There are "known knowns,"—there are things we know we know,' Mr. Rumsfeld said. 'There are "known unknowns," that is to say there are things we now know we don't know. But there are also "unknown unknowns"—things we do not know we don't know.' Mr. Rumsfeld summarized: 'The absence of evidence is not evidence of absence.'" Press release of 2002, kindly supplied to me by the British Broadcasting Corporation.

16. A wide-ranging survey of other philosophical approaches to history in general is provided by Stanford (1997). There is no explicit discussion of the historiography of mathematics, though some space is given to that for science: his references to mathematics concern either its place in the history of science or its use in mathematical history.

17. See Unger (1975) for a nice elaboration of skeptical positions, some linked to (lack of) facts and others related to (possibly false) beliefs. However, chapter 7 on "the impossibility of truth" is a very disappointing monistic treatment based for some reason on relating proposed truth to "the whole truth about the world," a notion that is indeed impossible to handle (but then why invoke it?), so that truth as a notion is rejected. No use is made of stratification, not even in the (brief) treatment of Tarski's theory.

18. See Popper (1945, ch. 25; 1957). Note carefully his rather nonstandard use of the word "historicism." On the historiography of science, in a somewhat Popperian spirit, see also Agassi (1963). Stanford (1997) considers other aspects of Popper's philosophy than those mooted in this paper and seriously misdescribes him as a philosopher "of positivist inclinations" (p. 39). The excellent index does not have entries for "ignorance," "(in)determinism," or "self-reference."

But even from Popper the hints are limited. Like most philosophers, he said little about the formation of scientific (including mathematical) theories; he was mainly concerned instead with the ways in which they could be tested. Further, despite his strong emphasis on theories, he was dismissive of questions of ontology, that is, doctrines concerning existence and being in both the physical world and in commitments of these kinds made in theories (Grattan-Guinness 1986).[19] He also did not write much on the philosophy of mathematics and was disinclined to enter into discussion of it (personal experience, on several occasions); ironic, then, that mathematics is such a rich source! However, maybe his insights can be elaborated; if so, the outcome would corroborate one of his maxims: "All life is problem-solving."

ACKNOWLEDGMENTS

For their reactions to lectures on this topic delivered between 2000 and 2003, I express gratitude to the audiences of the meetings held in Spain (Bilbao and Grand Canaries), Germany (Ingst, with the Deutsche Mathematiker-Vereinigung; and Bonn, with the Foundation for the Formal Sciences), Portugal (Braga), India (Delhi, with the Indian National Science Academy), Canada (Toronto, with the Canadian Society for the History of Mathematics), and the United States (Mathematical Association of America, joint annual meeting with the American Mathematical Society).

BIBLIOGRAPHY

Agassi, J. 1963. *Towards a historiography of science*, 's Gravenhage, Netherlands: Mouton.

Arnol'd, V. I. 1990. *Huygens and Barrow, Newton and Hooke. Pioneers in mathematical analysis and catastrophe theory from evoluents to quasi-crystals*, Basel, Switzerland: Birkhäuser.

Atwood, M. 1985. *The handmaid's tale*, London: Bloomsbury.

Bashmakova, I. G., and Smirnova, G. 1999. *The beginning and evolution of algebra* (trans. A. Schenitzer), Washington: Mathematical Association of America.

Bourbaki, N. 1974. *Eléments d'histoire des mathématiques*, 2nd ed., Paris: Hermann.

19. The place of ignorance in this scheme makes questions of ontology and ontological commitments particularly important but tricky to handle, especially because we have to consider mathematics with science, philosophy with logics, and history with its methods. Ontology has to cover not only the actual physical world but also possible ones asserted in theories (Jacquette 2002), including reconstructions of the past, where the issues are extremely perplexing (Hacking 2002, ch. 1). For me, a key notion is reification, an as-if approach in which the referents of a theory (of any kind) are assumed to exist in a liberal spirit, but where any of them are readily thrown away or reduced to others if deemed necessary, worthwhile, or just convenient (Grattan-Guinness 1987). Preferences over ontology are often made between objects, properties, and propositions as candidate categories for existence; but in many contexts, this may not be necessary, since the associated ontologies are intertranslatable (Brink and Rewitzky 2002).

Bradley, R. E., and Sandifer, C. E. 2007. (Eds.), *Leonhard Euler: life, work and legacy*, Amsterdam: Elsevier.

Bressoud, D. M. 1994. *A radical approach to real analysis*, Washington, D.C.: Mathematical Association of America.

Bressoud, D. M. 2008. *A radical approach to Lebesgue's theory of integration*, Washington, D.C.: Mathematical Association of America.

Brink, C., and Rewitzky, I. 2002. "Three dual ontologies," *Journal of philosophical logic, 31*, 543–568.

Burns, R. M., and Rayment-Pickard, H. 2000. (Eds.), *Philosophies of history: From Enlightenment to postmodernity*, Oxford: Blackwells.

Cavaillès, J. 1938. *Méthode axiomatique et formalisme*, 3 pts., Paris: Hermann. [Also in *Oeuvres complètes de philosophie des sciences*, Paris: Hermann, 1994, 221–374.]

Collingwood, R. G. 1946. *The idea of history*, Oxford: Oxford University Press.

Dauben, J. W. 1979. *Georg Cantor*, Cambridge, Mass.: Harvard University Press. [Repr. Princeton, N.J.: Princeton University Press, 1990.]

Dauben, J. W. 1999. "*Historia mathematica*: 25 years/context and content," *Historia mathematica, 26*, 1–28.

Dauben, J. W., and Scriba, C. J. 2002. (Eds.), *Writing the history of mathematics: its historical development*, Basel, Switzerland: Birkhäuser.

"Dedekind, J. W. R." 1995. *What are numbers and what should they be?* Orono, Maine: RIM Press. [German original 1888.]

Dieudonné, J. 1989. *A history of algebraic and differential topology 1900–1960*, Basel, Switzerland: Birkhäuser.

Dirichlet, J. P. G. Lejeune-. 1829. "Sur la convergence des séries trigonométriques," *Journal für die reine und angewandte Mathematik, 4*, 157–169. [Also in *Gesammelte Werke*, vol. 1, 1889, Berlin: Reimer (repr. New York: Chelsea, 1969), 117–132.]

Duncan, R., and Weston-Smith, M. 1977. (Eds.), *The encyclopaedia of ignorance*, Oxford: Pergamon.

Duncan, R., and Weston-Smith, M. 1984. (Eds.), *The encyclopaedia of medical ignorance*, Oxford: Pergamon.

Dyson, F. 1972. "Missed opportunities," *Bulletin of the American Mathematical Society, 78*, 635–652.

Euclid, *Elements*. Edition used: Ed. and trans. T. L. Heath, *The Thirteen Books of Euclid's Elements*, 2nd ed., 3 vols., 1926, Cambridge: Cambridge University Press. [Repr. New York: Dover, 1956. First ed. 1908.]

Euclid (Simson), 1756. *The elements of Euclid* (ed. and notes by R. Simson), Edinburgh: Foulis. [Contains books 1–6, 11 and 12.]

Euclid (Dijksterhuis), 1929, 1930. *De Elementen Van Euclides* (ed. E. J. Dijksterhuis), 2 vols., Groningen: Noordhoff.

Fauvel, J., and van Maanen, J. 2000. (Eds.), *History in mathematics education: The ICME study*, Dordrecht, Netherlands: Kluwer.

Franklin, J. 2001. *The science of conjecture: Evidence and probability before Pascal*, Baltimore: Johns Hopkins University Press.

Freudenthal, H. 1971. "Did Cauchy plagiarise Bolzano?," *Archive for history of exact sciences, 7*, 375–392.

Gillies, D. 1992. (Ed.), *Revolutions in mathematics*, Oxford: Clarendon Press.

Gispert, H. 1999. "Les débuts de d'histoire des mathématiques sur les scènes internationales et le cas de l'entreprise encyclopédique de Felix Klein et Jules Molk," *Historia mathematica, 26*, 344–360.

Grattan-Guinness, I. 1973. "Not from nowhere: History and philosophy behind mathematical education," *International journal of mathematics education in science and technology, 4*, 421–453.

Grattan-Guinness, I. 1986. "What do theories talk about? A critique of Popperian fallibilism, with especial reference to ontology," *Fundamenta scientiae, 7*, 177–221.

Grattan-Guinness, I. 1987. "What was and what should be the calculus?," in I. Grattan-Guinness (ed.), *History in mathematics education*, Paris: Belin, 116–135. [Also ch. 13 here.]

Grattan-Guinness, I. 1992. "Scientific revolutions as convolutions? A sceptical enquiry," in S. S. Demidov, M. Folkerts, D. E. Rowe, and C. J. Scriba (eds.), *Amphora: Festschrift für Hans Wussing zu seinem 65. Geburtstag*, Basel, Switzerland: Birkhäuser, 279–287.

Grattan-Guinness, I. 1994. (Ed.), *Companion encyclopedia of the history and philosophy of the mathematical sciences*, 2 vols., London: Routledge. [Repr. Baltimore: Johns Hopkins University Press, 2003.]

Grattan-Guinness, I. 1996. "Numbers, magnitudes, ratios and proportions in Euclid's *Elements:* how did he handle them?" *Historia mathematica, 23*, 355–375. [Printing correction: 24 (1997), 213. Also ch. 11 here.]

Grattan-Guinness, I. 1997. *The Fontana history of the mathematical sciences: The rainbow of mathematics*, London: Fontana. [Also as *The Norton history of the mathematical sciences: The rainbow of mathematics*, New York: Norton, 1998.]

Grattan-Guinness, I. 2000. *The search for mathematical roots, 1870–1940: Logics, set theories and the foundations of mathematics from Cantor through Russell to Gödel*, Princeton, N.J.: Princeton University Press.

Grattan-Guinness, I. 2004. "History or heritage? An important distinction in mathematics and for mathematics education," *American mathematical monthly, 111*, 1–12. [Also in G. van Brummelen and M. Kinyon (eds.), *Mathematics and the historian's craft: The Kenneth O. May lectures*, 2005, New York: Springer, 7–21.]

Grattan-Guinness, I. 2008. "Levels of criticism: Handling Popperian problems in a Popperian way," *Axiomathes, 18*, 37–48.

Grattan-Guinness, I. 1972. In collaboration with J. R. Ravetz; *Joseph Fourier 1768–1830. A survey of his life and work, based on a critical edition of his monograph on the propagation of heat, presented to the Institut de France in 1807*, Cambridge, Mass.: MIT Press.

Gray, J. J. 1989. *Ideas of space*, 2nd ed., Oxford and New York: Clarendon Press.

Hacking, I. 2002. *Historical ontology*, Cambridge, Mass.: Harvard University Press

Hardy, G. H. 1918. "Sir George Stokes and the concept of uniform convergence," *Proceedings of the Cambridge Philosophical Society, 19*, 148–156. [Also in *Collected papers*, vol. 7, Oxford: Clarendon Press, 1979, 505–513.]

Hawkins, T. W. 1970. *Lebesgue's theory of integration*, Madison: University of Wisconsin Press. [Repr. New York: Chelsea, 1975.]

Hilbert, D. 1901. "Mathematische Probleme," *Archiv für Mathematik und Physik, (3)1,* 44–63, 213–237. [Various reprints. English trans.: "Mathematical problems," *Bulletin of the American Mathematical Society, 8* (1902), 437–479, itself variously reprinted.]

Hoe, J. 1978. "The Jade Mirror of the Four Unknowns—some reflections," *New Zealand mathematical chronicle, 7,* 125–156.

Hofmann, J. E. 1959. "Um Eulers erste Reihestudien," in K. Schröder (ed.), *Sammelband der zu Ehren des 250. Geburtstages Leonhard Eulers,* Berlin: Akademie Verlag, 139–208.

Høyrup, J. 1995. "Linee larghe: Un'ambiguità geometrica dimenticata," *Bollettino di storia delle scienze matematiche, 15,* 3–14.

Høyrup, J. 2002. *Lengths, widths, surfaces: A portrait of old Babylonian algebra and its kin,* New York: Springer.

Jacquette, D. 2002. *Ontology,* Chesham, U.K.: Acumen.

May, K. O. 1973. *Bibliography and research manual in the history of mathematics,* Toronto: University of Toronto Press.

May, K. O. 1975a. "Historiographic vices: I. Logical attribution," *Historia mathematica, 2,* 185–187.

May, K. O. 1975b. "Historiographic vices: II. Priority chasing," *Historia mathematica, 2,* 315–317.

May, K. O. 1976. "What is good history and who should do it?" *Historia mathematica, 3,* 449–455.

Mueller, I. 1981. *Philosophy of mathematics and deductive structure in Euclid's Elements,* Cambridge, Mass.: MIT Press.

Pickstone, J. V. 1995. "Past and present knowledge in the practice of the history of science," *History of science, 33,* 203–224.

Popper, K. R. 1945. *The open society and its enemies,* London: Routledge and Kegan Paul.

Popper, K. R. 1957. *The poverty of historicism,* London: Routledge and Kegan Paul.

Popper, K. R. 1963. *Conjectures and refutations,* London: Routledge and Kegan Paul.

Popper, K. R. 1982. *The open universe: An argument for indeterminism,* London: Hutchinson.

Rashed, R. 1994. *The development of Arabic mathematics: between arithmetic and algebra,* Dordrecht, Boston, and London: Kluwer.

Russell, B. A. W. 1956. *Portraits from memory and other essays,* London: Allen & Unwin.

Schubring, G. 1978. *Das genetische Prinzip in der Mathematik-Didaktik,* Stuttgart, Germany: Klett-Cotta.

Sinaceur, H. 1991. *Corps et modèles,* Paris: Vrin.

Stanford, M. 1997. *An introduction to the philosophy of history,* Oxford: Blackwells.

Stedall, J. A. 2001. "Of our own nation: John Wallis's account of mathematical learning in medieval England," *Historia mathematica, 28,* 73–122.

Struik, D. J. 1980. "Why study the history of mathematics?" *Undergraduate mathematics and its applications, 1,* 3–28.

Thiele, R. 2000. "Frühe Variationsrechnung und Funktionsbegriff," in R. Thiele (ed.), *Mathesis: Festschrift zum siebzigsten Geburtstag von Matthias Schramm,* Berlin: Diepholz, 128–181.

Toeplitz, O. 1963. *The calculus: A genetic approach,* Chicago: University of Chicago Press.

Truesdell, C. A., III. 1954. "Prologue," in L. Euler, *Opera omnia*, ser. 2, vol. 12, Basel, Switzerland: Orell Füssli, ix–cxxv. [On fluid mechanics.]

Truesdell, C. A., III. 1968. *Essays in the history of mechanics*, Berlin: Springer.

Tucciarone, J. 1973. "The development of the theory of summable divergent series from 1880 to 1925," *Archive for history of exact sciences, 10*, 1–40.

Unger, P. 1975. *Ignorance: A case for skepticism*, Oxford: Clarendon Press.

van Dalen, D. 1999. *Mystic, geometer, and intuitionist: The life of L. E. J. Brouwer*, vol. 1, *The dawning revolution*, Oxford: Clarendon Press.

von Braunmühl, A. 1900, 1903. *Vorlesungen über Geschichte der Trigonometrie*, 2 vols., Leipzig, Germany: Teubner.

Watkins, J. M. W. 1969. "Comprehensively critical rationalism," *Philosophy, 44*, 57–62. [See also the ensuing discussion by various authors in 46 (1971), 43–61.]

Weil, A. 1980. "History of mathematics: Why and how," in O. Lehto (ed.), *Proceedings of the International Congress of Mathematicians, Helsinki 1978*, Helsinki, Finland: Academia Scientarum Fennica, vol. 1, 227–236. [Also in *Collected papers*, vol. 3, 1980, New York: Springer, 434–443.]

Windschuttle, K. 1997. *The killing of history: How literary critics and social theorists are murdering our past*, New York: The Free Press.

Wisniewski, A. 1995. *The posing of questions: Logical foundations of erotetic sciences*, Dordrecht, Netherlands: Kluwer.

Wussing, H. 1984. *The genesis of the abstract group concept*, Cambridge, Mass: MIT Press.

CHAPTER THREE

Decline, Then Recovery

An Overview of Activity in the History of Mathematics during the Twentieth Century

To the affectionate memory of Roy Porter (1946–2002)

A graph of the level of activity in the history of mathematics during the twentieth century resembles a U-shaped curve. Around 1900 and until the Great War the field was very active, building on a considerable body of work from the 1870s onward. Since the first half of the 1970s also it has been in a pretty lively state. But in between, there was a substantial drop in level, though some figures kept alive a modest flame of learning.

1. Introduction

The first part (¶2–¶10) of the unusual story sketched here is a largely factual summary of activity from 1900 to around the 1990s. The second part (¶11–¶16) reflects on the story in various ways: imbalances in the measures of study of the various branches and aspects of the history of mathematics, changes in interpretations of some topics and developments, causes of the decline, relationships with mathematics education, and attitudes positive and negative to our subject among non-practitioners. The account is usually divided by countries, since differences between them are marked; however, work in each one does not necessarily focus on its own developments, for many historians have not displayed any national preference in their choice of studies. The treatment of work is deliberately selective, especially in recent decades: many of the items cited exemplify some particular category or unusual feature, and silence on the others reflects only the desire to avoid just recording a long list of doubtless excellent achievements.

Recently, some historians of mathematics became interested in the history of

The original version of this essay was first published in *History of science*, 42 (2004), 279–312; reprinted by permission of Science History Publications. The original section 11, on my 1994 encyclopedia, has been merged with a similar discussion in §5.1.

their subject to the extent of that they published a large collective book on it in 2002 (Scriba and Dauben 2002).[1] I refer to this book for more ample accounts of all aspects of this article, and for the history of the history of mathematics before 1900; some features of the book are considered in ¶15.

PART I: REPORTAGE

2. The Enterprises of Felix Klein

Let us start in Germany, whose language was by far the most important one for our subject around 1900. Then in his early fifties, the mathematician Felix Klein (1849–1925) at Göttingen University was heavily involved in various large-scale projects concerning the development of mathematics and mathematical education; the Leipzig publishing house of Teubner was the chief outlet (Schulze 1911, esp. pp. 266–374). Of those projects that bore upon historical work, the most important was the *Encyklopädie der mathematischen Wissenschaften mit Einschluss ihrer Anwendungen* (EMW), which had been launched in 1894 as a detailed account of all the main areas of mathematics, pure and applied, at the time. The founder president was Klein's former student, the algebraist Franz Meyer (1856–1934).

The EMW was divided into six parts, each with its own editor(s) and publishing schedules:

1. arithmetic and algebra (editor Meyer), 1,250 pages, 1898–1904;
2. analysis (coeditor Heinrich Burkhardt), 4,020 pages, 1900–1927;
3. geometry (coeditor Meyer), 5,200 pages, 1903–1935;
4. mechanics (coeditor Klein), 3,000 pages, 1901–1935;
5. physics (editor Arnold Sommerfeld), 3,280 pages, 1903–1926;

6, section 1. geodesy and geophysics (coeditor Philipp Furtwängler), 970 pages, 1905–1922;

6, section 2. astronomy (coeditor Karl Schwarzschild), 2,300 pages, 1905–1934.

1. It is important to distinguish the history of mathematics from mathematical history. The latter is a branch of applied mathematics, in which the mathematics involved could be modern or even new: for example, some kind of statistics used to assess prosopographically, say, priests in eighteenth-century Mexico, or methods such as seriation for retrieving data from incomplete sources such as archaeological remains (see, for example, Rashevsky 1968). While a small field, mathematical history, sometimes known as "cliometrics," has been practiced long enough to have its own history (see footnote 10). A related subject is computing history, where software methods are developed specially for historical needs; the Association for Computing and History furthers the cause internationally.

In total it filled around 20,000 pages. Many articles were the first of their kind on their topic, and several are still the last or best. Some of them have excellent information on the deeper historical background. This is especially true of articles on applied mathematics, including engineering, which was stressed in its title.

The German Mathematicians' Union, which had been founded in 1890, was closely involved. It also published several important survey-historical articles in its *Jahresbericht*. An outstanding historical example was the 1,804-page review of mathematical methods in analysis by Heinrich Burkhardt (1861–1914), an editor of the second part of the EMW; he placed there a 540-page snippet of this longer piece.[2]

The edition had been planned to include a part 7 on history, philosophy, and education; unfortunately not a line was to appear, but Klein's other major activity lay in education, for at the International Congress of Mathematicians in Rome in 1908 he helped launch an International Commission for Mathematics Education. Up to 1920 nearly 200 books and pamphlets were produced, and over 300 reports. They included not only many national reports but also broader topics that are still of interest today, such as teaching mathematics to girls (Schröder 1913) and the use of the history of mathematics in mathematics education (Gebhardt 1912). Wilhelm Lorey (1873–1955) was a distinguished author here and elsewhere, especially on institutional history (see especially Lorey 1916). For its "organ" of regular communication to the outside world, the commission appointed the Swiss journal *L'enseignement mathématique*, which had been founded in 1899.[3]

These projects were not just German but very Göttingen, because of Klein's strong emphases on applications and on education. The mathematicians at Berlin, the other main mathematical pole in Germany and a citadel for pure mathematics, were not invited to collaborate on the EMW and are reputed to have sneered at it. But Klein's commitment to the diffusion of mathematics extended to giving pioneering lecture courses on the history of nineteenth-century mathematics and on "elementary mathematics from an advanced standpoint," viewed with a deft use of both history and heuristics; both works have been reprinted or translated (Klein 1926–1927, 1925–1928).

2. Burkhardt (1901–1908); the snippet occupies pp. 819–1354 of part 2, section 1 of the EMW.

3. This commission deserves a full-length study; meanwhile, see Coray and others (2003), especially the articles by F. Furinghetti and G. Schubring on pp. 19–66; Schubring (1988); and Grattan-Guinness (1993).

3. Other Work Largely in German

The *Vorlesungen über die Geschichte der Mathematik* of Moritz Cantor (1829–1920) was the major general history of mathematics of the time. Underlying its massive scope was his strong belief in the unity of knowledge. Appearing from Teubner from 1880 onward, the first three volumes all received second editions in the early 1900s, about 2,800 pages in total. A fourth and last volume, covering 1759–1799, appeared as Cantor 1908, a collective 1,100-page effort by nine colleagues. It included a short survey of the history of history of mathematics by Siegmund Günther (1848–1923). The article on trigonometry was by Anton von Braunmühl (1853–1908), whose own two-volume *Vorlesungen über die Geschichte der Trigonometrie* (1902–1903, also Teubner) imitated Cantor not only in title but also in thoroughness; it remains uneclipsed. Cantor also continued to edit until 1913 a series of Abhandlungen on history that he had founded with Teubner in 1877; many valuable pieces appeared there, some of book length.

Denmark had been contributing importantly to the history of ancient mathematics since the late nineteenth century, sometimes to Cantor's discomfort. The mathematician Hieronymus Zeuthen (1839–1920) wrote extensively on Greek mathematics, though in a rather modernizing manner (¶12) that led him into disputes with Cantor. His compatriot J. L. Heiberg (1854–1928) made the most sensational discovery of the time; in connection with a new edition of Archimedes, he identified in 1906 an important manuscript that had been found some years earlier.[4] Axel Bjørnbo (1874–1911) also wrote on ancient mathematics but focused attention on the transmission of Greek and Arabic mathematics into Europe in the Middle Ages, especially in manuscripts. This latter interest was also pursued by the German scholar Heinrich Suter (1848–1922).

The most important journal was *Bibliotheca mathematica*, which Gustav Eneström (1852–1923) had founded in 1884 in his native Sweden, initially as an appendix to the mathematical journal *Acta mathematica*. It reached a peak with its third series (1900–1915) of 14 volumes, published by Teubner; for example, Heiberg's edition of the Archimedes text appeared there in 1907. However, Eneström somewhat spoiled the enterprise with long lists of criticisms of Cantor's *Vorlesungen*, of greatly varying quality. (As a critic of Cantor he followed Zeuthen.) Among other sources,

4. This manuscript, a palimpsest, came into private hands seemingly around 1920 and re-emerged only at auction in the late 1990s. Its anonymous purchaser is financing the cleaning of the manuscript at the Walters Art Gallery, Baltimore, and the transcription of the texts (an extremely difficult task, because the mathematical ones are the originals). {Important discoveries are still being made as cleaning and reading progresses: see www.archimedespalimpsest.org.}

the mathematical abstracting journal of the time, the *Jahrbuch über die Fortschritte der Mathematik* (1867–1942), maintained a reasonable level of commentary on historical writings until its closure, though its production faltered much in its last decade (Siegmund-Schultze 1993).

4. Major Editions

Several editions of collected or selected works of mathematicians were prepared or at least started during the late nineteenth century, and very many more during the twentieth century (Higgins 1944): the total number seems to be *far* greater than those for the other types of scientist. The majority have been just photoreprints of the mathematician's writings, some requiring many volumes, but with no special editorial involvement. However, some editions made much greater demands, with introductions, annotations, bibliography, translations, and indexes. For example, the edition of C. F. Gauss (1777–1855), published and unpublished, had begun to appear already in 1865 but was far from completion by 1900: Klein not only was involved in continuing the edition but also helped to inspire colleagues to produce substantial historical essays on aspects of Gauss's work, towards an "intellectual biography." Several of the essays appeared as supplements to the edition.[5]

Another major edition with a long and interesting history is that for Christiaan Huygens (1629–1695), prepared in the Netherlands by an anonymous committee of the Dutch Society of Sciences. Most unusually, they started with the correspondence (10 volumes, 1882–1905) before tackling the publications and manuscripts (12 volumes, 1908–1950). Each volume was meticulously indexed, and its sheets are watermarked alternately "Christiaan" and "Huygens."

The edition that has made perhaps the greatest impact on historians of mathematics is the one for Leonhard Euler (1707–1783). A commission was formed under the auspices of the Swiss Scientific Society in 1907, the bicentenary of Euler's birth. Eneström produced a bibliography (Eneström 1910–1913) published by Teubner for the German Mathematicians' Union, which was also closely involved. From the list Paul Stäckel (1862–1919) sketched out the edition in three series (Stäckel 1909): the final plan, announced by the president of the Euler Commission, Ferdinand Rudio (1856–1929), proposed 18 volumes for the first series (pure mathematics), 16 for the second (mechanics and astronomy), and 11 for the last (physics, the philosophical

5. See especially the ten essays composing the second parts of Gauss, *Werke*, vols. 10–11 (Leipzig, 1922–1933; repr. Hildesheim, 1973), 1,300 pages in all. Some essays had appeared earlier as papers with the Göttingen Academy.

Lettres à une princesse d'Allemagne [1768–1772], and unpublished correspondence; Rudio 1911, xxxiii–xxxiv). Work started in 1910, and the first two volumes were published by Teubner in the following year; ten more appeared until the Great War, and two during it. The next stages are recorded in ¶6.

A valuable edition project of a different kind was the *Ostwalds Klassiker der exakten Wissenschaften* (1889–), where many major scientific texts were reprinted (or, where necessary, translated into German) and furnished with notes. Mathematical writings were prominent from the start.

5. Work in Other Languages

French mathematicians soon began to prepare their own translation and elaboration of the EMW, as the *Encyclopédie des sciences mathématiques pures et appliquées*, with Teubner working with Gauthier-Villars. All six parts were started, and some of the revisions were very remarkable: in particular, the historical content of several articles was increased, thanks especially to additional notes and material furnished by Eneström and Jules Tannery (1848–1910). Some of the articles on set theory and functions were so good that their German colleagues retranslated them back in the mid 1920s as additional pieces for their own second part. But much greater enthusiasm was shown for translating articles on pure than on applied mathematics; Klein's philosophy did not fully diffuse into France. Then the death in 1914 of the general editor of the *Encyclopédie*, Jules Molk, and the general circumstances of the war, led to Gauthier-Villars withdrawing from the project in 1916; some articles stop in midsentence at the end of a 32-page signature.[6]

Efforts were made to launch part 7, on the history, philosophy, and education of mathematics. In particular, the First Congress on Mathematical Philosophy was held in Paris in April 1914, closely following one on mathematics education held there under the auspices of the International Commission for Mathematics Education.[7] But Heinrich Timerding (1873–1945), who was responsible for the part, recorded at the 1914 congress unexplained "very great difficulties" in its preparation,

6. On the history of the relationship between the two encyclopaedias, see Gispert (1999) and further historical items cited there. In 1995 the published parts of the *Encyclopédie* were photoreprinted with new editorial material by Gispert; a collective history of it is in preparation. An example of an incomplete article, important because not obviously so, is Vilfredo Pareto's account of mathematical economics, which happened to end with a sentence on the last line of page 640 (= the twentieth signature) of its part.

7. The philosophers published their papers in the *Revue de métaphysique et de morale* from 1914 to 1916; a report Reymond (1914) was provided in *L'enseignement mathématique;* the educators quickly published their proceedings in this same volume on pp. 167–226, 245–534 (a part

and after the war it seems to have been abandoned, in both the French and the German versions (Reymond 1914, 370–371).

Other French work in our subject was characterized by that concern with history mixed with philosophy that has long marked their scholarship. A striking example is Louis Couturat (1868–1914), for in advocating (the history of) symbolic logics, he isolated himself from his compatriot mathematicians; his historical studies of logic in G. W. Leibniz (1646–1716) were better received. Other figures include Gaston Milhaud (1858–1918) and Abel Rey (1874–1940) on Greek science and philosophy; mathematician Pierre Duhem (1861–1916), especially on medieval mechanics and the philosophy of science; and Leon Brunschvicg (1869–1944) and Pierre Boutroux (1880–1922) on progress in mathematical thought, and as important contributors to an edition of the works of Pascal (14 volumes, 1904–1914). The editions of the works of René Descartes (13 volumes, 1897–1913) and Pierre de Fermat (5 volumes, 1897–1912, 1922) excited admiration for the level of detailed scholarship deployed; Paul Tannery (1843–1904) was prominent in their preparation, in addition to his own major studies of Greek mathematics.

In Italy Gino Loria (1862–1954) launched in 1898 a *Bollettino di bibliografia e di storia delle scienze mathematiche*, imitating the title of a remarkable *Bullettino* that had been edited by Baldassare Boncompagni (1821–1894) between 1868 and 1887. Loria's journal also lasted for twenty years, and then continued as an appendix to the *Bollettino di matematica;* but it never reached the same level of importance as its predecessor. However, he also published the first "guide" to our subject (Loria 1916), where he reflected on the practice of history and its own history as well as providing bibliographies. His main research specialty lay in the history of geometry, an interest shared by Roberto Bonola (1874–1911), whose book on *Non-Euclidean Geometry* (English translation 1912) is still well known.

Ettore Bortolotti (1866–1947) specialized in Italian mathematics during the Renaissance and afterward, following a tradition from Boncompagni. Bonola's doctoral supervisor, Federigo Enriques (1871–1956), stands out among his countrymen as not only an excellent mathematician but also a French-style *storico-filosofo*, viewing a mathematical theory as dynamically intertwined with its history. As well as writing for part 3 of the EMW, he tried to promote part 7 on history, philosophy, and education. He also applied his approach to (the history of) logic; a school in Turin around Giuseppe Peano (1858–1932) concentrated on the formalization of

comprising national reports on teaching the calculus). The meeting is appraised by J.-P. Kahane and P. Nabonnand in Coray and others (2003, 167–178, 229–249).

logic and of mathematical theories, but also adorned their accounts with extensive historical notes.[8]

The most important Italian edition of the period was that for Galileo Galilei (1564–1642), prepared by Antonio Favaro (1847–1922) in 20 volumes (1890–1909). However, although officially "national," it met with little response at first and has become known mainly through a reprint with additions (1929–1939).

While a few British and American authors had contributed to the EMW, no English edition was prepared, and efforts to start one in Britain met with apathy. The main historians of the time were Sir Thomas Heath (1861–1940), with his various editions and commentaries on Greek mathematics; Sir Thomas Muir (1844–1934), with his mammoth four-volume *The Theory of Determinants in the Historical Order of Development* (1906–1923); and Sir Edmund Whittaker (1873–1956), with several very fine obituaries and articles on applied mathematics (including one in the EMW), and also the well-known *History of the Theories of Aether and Electricity* (1911, second edition in two volumes 1951–1953). These men received knighthoods for other activities. Philip Jourdain (1879–1919) is also worth recalling for some fine pioneering work in the history of set theory and symbolic logics (Jourdain 1991). At Cambridge W. W. Rouse Ball (1850–1925) wrote extensively on the history of the local mathematicians, and produced a remarkable and unique history, *Mathematical Recreations and Problems of Past and Present Times* (1892), which took its seventh edition in 1917 and is still in print in its thirteenth edition (1987).[9]

In addition, a separate tradition developed in Britain, concerning probability and mathematical statistics. The principal modern founder of mathematical statistics was Karl Pearson (1857–1936); he also took an active interest in its history, especially from the 1920s onward, when historical pieces by him and others appeared in the journal *Biometrika*, which he had started in 1901.[10]

Many valuable long survey-historical articles appeared in the 10th (1902–1903) and especially the 11th (1910–1911) editions of the *Encyclopaedia Britannica*. These vast and glorious quartos contain the equivalent of little books on many topics,

8. The notes are found principally in the five editions of the Peanists' compendia *Formulario di matematica* (Turin, 1895–1908). One member of the group was the historian Giovanni Vacca (1872–1953), who drew Couturat's attention to the logic manuscripts of Leibniz at the International Congress of Mathematicians in Paris in 1900.

9. With the third edition (1896) Rouse Ball changed the title of his book to *Mathematical Recreations Past and Present*. The posthumous editions (11th–13th) were edited by H. S. M. Coxeter.

10. For collections of reprints of historical articles largely from *Biometrika*, see Kendall (1970, 1977). Pearson's own historical lectures were published posthumously, thanks to his son, as Pearson (1978). Some of his other work belongs to the history of mathematical history mentioned in footnote 1.

mathematics included; for example, to my knowledge the 120-page pair on "ship" and "shipbuilding" (Watts 1911), surpasses anything of that time on recent developments, the EMW included. Earlier editions of the *Encyclopaedia Britannica*, and some contemporary encyclopedias in other languages, also contain valuable but neglected articles; historians should be more aware of such sources.

The United States was emerging as a significant mathematical country around 1900, partly under the influence of Klein (Parshall and Rowe 1994). Some benefits for historical work flowed, especially the relationship between history and mathematics education. The Swiss-born Florian Cajori (1859–1930) had already reported on *The Teaching of the History of Mathematics in the United States* in 1890. At the Columbia University Teachers' College in New York from 1901, D. E. Smith (1860–1944) was pursuing the same line; indeed, it was a suggestion by him in *L'enseignement mathématique* in 1905 that led to the creation of the International Commission for Mathematics Education (¶2). G. B. Halsted (1853–1922) was a pioneer advocate of non-Euclidian geometry in the United States, to the extent of translating some of the original major texts.

For publishing, one unusual venue was the Open Court Publishing Company (McCoy 1987), which had started in the late 1880s under direction of the German mathematician and philosopher Paul Carus (1852–1919) as a channel for importing German culture, especially philosophy, into the United States (Henderson 1993). Carus's journal *The Monist* (1890–) was an important site for papers (including translations) on logic and on the philosophy of mathematics and of science. As well as publishing new books, Open Court issued translations, including reprinting some of Halsted's. Among several German authors, Ernst Mach (1838–1916) was quite a hero, and others included the mathematicians Richard Dedekind (1831–1916) and Georg Cantor (1845–1918). The latter was handled by Jourdain in 1915, soon after he became the company's European editor: he succeeded Carus as general editor in 1919 but died himself later that year. The archives of the company, held at Southern Illinois University at Carbondale, are amazingly rich.

Our subject was pursued also in the Far East, though the West may have known only of two books in English by the Japanese historian Yoshio Mikami (1875–1940), one of them written with Smith (Mikami 1913; Mikami and Smith 1914). The respective publishers were Teubner (in Cantor's Abhandlungen series) and Open Court.

6. A Fairly Quiet Time: To the Second World War

After 1920 the level of activity fell off quite noticeably. One continuing line was that of the French *historien-philosophe*: new figures of this genre include the Russian-born Alexandre Koyré (1892–1964), who deployed his religiosity and Platonism to convey a rather purist vision of the development of astronomy and mechanics from Copernicus to Galileo; Gaston Bachelard (1868–1944), who wrote on some aspects of the history of applied mathematics and of physics; and Jean Cavaillès (1903–1944), who specialized in the history of foundational questions in mathematics (Sinaceur 1987) and contributed to the 1930s editions of the works of Dedekind and Cantor. His death during the Second World War is no accident of chronology, for he sacrificed himself for the cause of the French Resistance and is now buried in the chapel of the Sorbonne in Paris, near Descartes.

In the Netherlands E. J. Dijksterhuis (1892–1965) produced some notable work on ancient mathematics, for example, studies of mechanics, and an excellent partial translation of Euclid's *Elements* (1929–1930). This edition launched a book series under his coeditorship to encourage the links between history and education among schoolteachers (van Berkel 1996).

In Switzerland (and elsewhere) the Euler edition continued in various hands, especially those of Otto Spiess (1878–1966) and Andreas Speiser (1885–1970): 15 more volumes were published between 1920 and 1944, though none in the second series for mechanics. From 1933 Teubner shared the publication with the Zurich house Orell Füssli. Spiess also began planning an edition of the works of Euler's friends and mentors, the Bernoulli family. Read on in ¶7.

Several articles in most parts of the EMW continued to appear. Those for geometry and mechanics were the last to be completed, in 1935, when the president was Constantin Carathéodory (1873–1950), a major mathematician with considerable historical interests. A second edition of the EMW was being worked out, with foundations of mathematics, algebra, and number theory as the initial topics; but the Second World War took it as a victim, and it seemed to stop around 1950 after only a few articles were published.

A notable German figure was Heinrich Wieleitner (1874–1931). He continued the effort to update Cantor's *Vorlesungen*, and wrote on the history of various mathematical topics, especially during the Middle Ages. Contemporary with him was Johannes Tropfke (1866–1939), whose work was more closely linked to educational purposes. His *Geschichte der Elementarmathematik in systematischer Darstellung* (1902–1903) was greatly enriched in its multivolume second (1921–1924) and third (1930–1937) editions by studies of original texts. A further posthumous volume was produced in 1940 by Kurt Vogel (1888–1985); 40 years later he coedited a large new edition of the first three volumes as one big book, covering arithmetic and algebra (Tropfke 1979). Vogel's own research lay initially in ancient mathematics, and later with the medieval European period, including the transmission of Greek and Arabic mathematics there. Several compatriots, such as Carl Schoy (1877–1925) and the philologist Julius Ruska (1867–1949), specialized in Arabic mathematics, including astronomy. Some mathematicians formed seminars on our subject in their universities, especially at Kiel, Frankfurt (led by Lorey), and Bonn.

Vogel's interest in ancient mathematics was shared by the most internationally significant German-born figure of this period. Otto Neugebauer (1899–1990) specialized in ancient mathematics, in ways interestingly different from those of the French philologist François Thureau-Dangin (1872–1944) (Høyrup 1991). For most of the 1930s he also ran a series, *Quellen und Studien zur Geschichte der Mathematik*, a successor to Cantor's series.

After five years at Copenhagen University, Neugebauer decided in 1939 to immigrate to the United States. There the interest in history and education had continued. Smith produced the influential *History of Mathematics* in 1923 and *A Source Book in Mathematics* six years later. The journal *Scripta mathematica* was launched in 1932 by Jekuthiel Ginsburg (1889–1957) partly to pursue this cause, and to treat educational and philosophical topics in mathematics (after his death, this policy was quickly dropped). The practice of teaching history courses spread to many educational institutions in the United States, probably far more than in any other country.

Among activities related to research, in 1934 Smith and Ginsburg put out *A History of Mathematics in America before 1900;* the publisher was the Mathematical Association of America, in a series funded by a legacy from the widow of Carus, the founding editor of *The Monist*. Smith also bought many mathematical books and masses of manuscripts, collections of major importance now housed in Columbia University. Cajori published with Open Court an important *History of Mathematical Notations* (1928–1929); in Colorado, isolated from sources, he used to pay scholars to transcribe texts from holdings in major libraries.[11] In 1934 he put out a revised edition of Andrew Motte's 1729 translation of Newton's *Principia mathematica*, which has become very well known.

Perhaps the best American research historian was R. C. Archibald (1875–1955); he wrote on several topics in nineteenth-century mathematics, which then was still much unstudied. It is a pity that he produced no substantial tome: he lacked the prolificacy of the Belgian-born George Sarton (1884–1956), who had moved to the United States in 1915. Best known as the founding editor of the history of science journals *Isis* in 1912 and *Osiris* in 1936, Sarton is one of very few historians of sci-

11. I owe this information to the late John Fauvel.

ence who have taken our subject seriously. He produced in 1936 a valuable handbook, *The Study of the History of Mathematics*, as a companion to a contemporary review of *The Study of the History of Science* (Sarton 1984).

The best-known American figure is the Scottish-born E. T. Bell (1883–1960). He produced books such as *Men of Mathematics* (1937) and *The Development of Mathematics* (1940)—unfortunately, for especially the first book is a hit-or-miss attempt to chronicle the lives or supposed lives of major mathematicians, with no references given. Apparently the style of work behooved the man himself, who also wrote science fiction (Reid 1993). His historical books are still reprinted quite regularly.

Special factors attended the development of our subject in the Soviet Union. Before the Revolution of 1917, work was modest, though V. V. Bobynin (1849–1919) and A. V. Vasiliev (1853–1929) were known internationally for their work on Russian mathematics, and I. Y. Timchenko (1863–1929) in Odessa for excellent contributions to the history of mathematical analysis. After the political change, the subject was encouraged among both historians and leading mathematicians as part of the general conception of history of knowledge as a Good Thing; but of course it had to be presented in a Good Way, and on Good Topics. Thus the history of the calculus, the one branch of mathematics that Karl Marx had studied intensively, became a preferred branch; and writers on all topics often cited the Marx-Engels correspondence in their prefaces, though often giving only factual accounts thereafter. In the same vein, history courses were given in several universities. Leading figures include Sof'ya Yanovskaya (1896–1966), with a special concern for the history of logic and of foundational topics in mathematics.

In the Far East, activity seems to have fallen away: Mikami wrote a great deal about Japan and China but published relatively little. But India came into the picture at this time. B. Datta (1888–1958) and A. N. Singh (1901–1954) produced *Hindu Mathematics: A Source Book* in English (1935–1938); and some Western achievements were well captured by G. Prasad (1876–1935), especially in his survey of 14 figures entitled *Some Great Mathematicians of the Nineteenth Century: Their Lives and Their Works* (1933–1934).

7. A Fairly Quiet Time: The Postwar Period up to the 1970s

In Germany Wieleitner's most important student was J. E. Hofmann (1900–1973). Fascinated by genius in mathematics, he gave much attention to major figures of the early modern period, normally up to 1800 (Hofmann 1990). In 1954 he founded a seminar for the history of mathematics at the Mathematical Research Centre at

Oberwolfach, a series of week-long meetings that has became of major importance for our subject (¶8). Among his contemporaries Helmut Gericke (1909–2007) was the first historian of mathematics to gain a full chair in the history of science in Germany (Munich, 1963); his main contributions lay in the history of algebra and of foundational questions, and he wrote some more general histories. The most prominent of Hofmann's students is Christoph Scriba (1929–), who gained a chair in Berlin in 1969 and then in Hamburg in 1975. In research he has focused especially on geometry, number theory, the early development of the calculus, and historiographical questions.

Of all historical figures who fell under Hofmann's scrutiny, the most significant for him was Leibniz. So he worked on the *Sämtliche Briefe und Schriften* of Leibniz, which is possibly the most massive edition of all for a single historical figure. Conceived in seven series under the auspices of the German Academy of Sciences in Berlin, volumes had begun to appear in 1923, but slowly: the first for the mathematical ones (letters in the third series, publications and manuscripts in the seventh) came out only in 1976 when Hofmann completed the first volume of letters. The complications of the edition involve not only its sheer scale (for example, the Paris period 1672–1676 alone will need eight volumes) and the extreme difficulty of transcribing many manuscripts, but also the postwar fact that the direction lay with the Academy of Sciences of the German Democratic Republic while the documents and many of the scholars were in Hannover and elsewhere in the Federal Republic.

Another major German edition was a *Gesamtausgabe* of publications and manuscripts for Johannes Kepler (1571–1630). It had begun to appear in 1937 under the Bavarian Academy of Sciences, and volumes appeared steadily even during the war and until 1959, in the hands of von Dyck and Franz Hammar and especially Max Caspar (1880–1956). Then there was a long pause, until further volumes started to come out again in 1983, under the leadership of Volker Bialas; most of the edition is now done.

The Gauss edition had been completed in 1933, but continued interest in his life and work and its context led to the founding in the 1960s of Gauss-Gesellschaft, which publishes the journal *Mitteilungen*. Some biographies appeared, but none yet matches the massive range of his achievements; so Klein's hope of "intellectual biography" (¶4) remains unfulfilled.

This last judgment can be made also for Euler. He was the third subject in a series of useful booklet-length biographies of major mathematicians that was published from 1947 to 1980 by the Swiss journal *Elemente der Mathematik*. But the Euler edition continued well. The original scheme of three series for his publications and correspondence in 18 + 16 + 11 volumes had become 29 + 31 + 12 volumes for the publications alone,[12] and 33 more volumes appeared from Orell Füssli between 1945 and 1970. The story is continued in ¶10.

As usual, most activity in France was conducted in Paris. In addition to more *historiens-philosophes*—most notably Georges Canguilhem (1904–1995), as well as François Rostand with original but much neglected meditations on inexactness in mathematics (Rostand 1960, 1962)—two distinct *équipes* were formed. One was led by René Taton (1915–2004) and Pierre Costabel (1912–1989), with the *Revue d'histoire des sciences* (1947–) as a main outlet; the other operated under Gaston Bachelard's daughter Suzanne (1919–2007). Much work of both groups focused on French mathematics from the time of the founding of the Académie des Sciences in the 1660s up to the early twentieth century. Schoolteacher Jean Itard (1902–1979) became involved in the interface between history and education as well as researching Greek mathematics and the seventeenth century.

Separate from both groups was the Bourbaki *collectif* of prominent French mathematicians, who presented central parts of mathematics for mathematicians in a highly formalized manner in their multivolume treatise *Eléments des mathématiques* {(Mashaal 2006)}. But they included short historical essays on various topics, which they collected as the volume *Eléments d'histoire des mathématiques* (1969 and later editions); it is one of the historical sources best known to mathematicians for relatively modern developments and has received several translations.

Some of the Italian work was driven by the powerful academic place there of the history of philosophy, so it focused on foundational questions. An important example is Ludovico Geymonat (1908–1991), especially after the Second World War, when the demise of the Fascists allowed him to take university chairs, at Turin and then at Milan. Ettore Carruccio (1908–1980) wrote on questions of method and proof in general and also on Italian mathematics; Attilio Frajese (1902–1986) specialized in Greek mathematics, including Italian translations; and Ugo Cassina (1897–1964) analyzed the achievements of his old master Peano and the followers, and prepared an edition of Peano's works (three volumes, 1957–1961).

In Britain the principal efforts centered on Isaac Newton (1642–1727). A seven-volume edition of his correspondence was produced between 1959 and 1977, under various editorial hands. More directly significant for our subject was *Mathematical*

12. For a listing of the contents of the three series of Euler's publications, see Speiser (1948). A summary history of the edition to date, followed by his massive secondary bibliography on Euler, is provided in Burckhardt (1983) and preceded by Biermann (1983) on the prehistory of the edition.

Papers, the edition of his numerous unpublished papers, which were edited by D. T. Whiteside (1932–2008) with the assistance of Adolf Prag (1906–2004) and Michael Hoskin (1930–), and appeared between 1967 and 1981 as eight large quartos from Cambridge University Press. The medieval and early modern periods attracted some fans, including Alistair Crombie (1915–1996), who from 1967 chaired a Thomas Harriot Seminar to study the achievements of that remarkable but shadowy polymath.

In the Soviet Union the leading historian was Adolf Pavlovich Yushkevich (1906–1993). His extensive studies of the medieval period, the development of the calculus, and of the work of Euler were complemented by a massive study of the development of mathematics in Russia prior to the revolution, one of the first modern histories of the mathematics of a country (Yushkevich 1968).[13] Among many other contributions, in 1948 he cofounded an annual for our subject, entitled *Istoriko-matematicheskogo issledovaniya* (historico-mathematical writings); it was the first new journal (or equivalent) for decades. Of his colleagues, Isabella Bashmakova (1921–2005) stands out for her studies of ancient and medieval mathematics and of algebraic number theory. Collective Soviet productions included several good editions of the works of Russian mathematicians: for example, those of N. I. Lobachevsky and P. L. Chebyshev are far more complete and scholarly than the earlier French-language ones.[14] Several biographies of mathematicians, Russian/Soviet or otherwise, were written and given print runs in the tens of thousands.

Among the Soviet colonies the German Democratic Republic became the most active, thanks especially to Hans Wussing (1927–), who built up the important Institute for the History of Science at Leipzig University. His main specialties were the histories of group theory and medieval German arithmetic, but he and many colleagues also prepared a wide range of more general works, especially a valuable volume of short biographies of many mathematicians of which there is no counterpart in any other language (Gottwald and others 1990). In Berlin K.-R. Biermann (1919–2002) became a leading authority on institutional history and on Gauss (and on Alexander von Humboldt). In a typical postwar German situation, *Ostwalds Klassiker der exakten Wissenschaften* (¶4) continued in both countries, in West Germany with a "new series."

In the United States the links with education or at least with popularization were continued by new figures such as Carl Boyer (1906–1976), with a general

13. A few more (partial) histories of mathematics in a country have been produced since, for example, Finland, Sweden, Poland, Hungary, Portugal, and the United States (ref. 65).

14. Contrast the five-volume *Polnoe sobranie sochinenii* (Lobachevsky 1946–1951; Chebyshev 1944–1951) with the two-volume editions of *Oeuvres* (respectively 1883–1886, 1899–1907).

history (1968) and histories of some specific branches; Philip Jones and Howard Eves, very active in the Mathematical Association of America; and Morris Kline (1908–1992), whose *Mathematics in Western Culture* (1953) brought mathematics to a remarkably wide audience. His 1,200-page *Mathematical Thought from Ancient to Modern Times* (1972) was then novel in the large proportion of space given to the nineteenth and early twentieth centuries, as was Edna Kramer's *The Nature and Growth of Modern Mathematics* (1970). The Dutch-born Dirk Struik (1894–2000), resident since the mid-1920s, pioneered efforts from the 1930s to bring social factors to bear on the history of mathematics—in his case, influenced by Marxism (Struik 1980). His *A Concise History of Mathematics* was soon recognized as the best of its compass upon its first appearance in 1948, and has been translated in 18 languages. Among many other activities, he was heavily involved in an excellent edition of "the principal works" of Simon Stevin (1548–1620), including English translations, that was initiated in the Netherlands by Dijksterhuis and others (five volumes, 1955–1966).

The mathematical component of the work of Bernard Cohen (1914–2003) included the history of economics and of computing, but it was dominated by studies of Newton. In particular, in 1972 he was the coeditor of a new edition of the third edition (1726) of Newton's *Principia mathematica*, and in 1999 of a new English translation of the work that should come to eclipse the Motte-Cajori translation of 1934 noted in ¶6. Koyré was the other editor of the 1972 edition just mentioned; he came to enjoy quite an influence in the United States because of his periods of academic residence there, especially at the Institute for Advanced Study at Princeton, where Marshall Clagett (1916–2005) pursued the history of medieval mathematics, and then Egyptian science.

Quite separate from these workers, during the mid-1950s C. Truesdell III (1919–2000) began to make notable inroads into the neglected histories of elasticity theory and hydrodynamics; in work undertaken largely in connection with the Euler edition, he elevated Euler to a very high place.[15] In 1960 he also launched with Springer the *Archive for History of Exact Sciences*, where our subject features frequently.

A department for the history of mathematics was established under Neugebauer at Brown University in 1948. But it was a Marxist self-exilist from the United States who was to have the greatest impact on the practice of the history of mathematics, as we shall now see.

15. Truesdell's main historical writings are found in the Euler *Opera omnia*, ser. 2, vols. 11–13 (Zurich, 1954–1960).

8. Revival from the 1970s: The Role of the International Commission on the History of Mathematics

One of the main interests in mathematics for the American Kenneth O. May (1916–1977) was information retrieval, and this led to him to study its history seriously. In 1973 he published with the University of Toronto Press (his university) an 800-page bibliography for our subject that, while worryingly dense with errors of reference, provides a comprehensive classification of historical literature (May 1973). The year following, the Canadian Society for the History and Philosophy of Mathematics was formed.

At that time May, with support from Taton and Yushkevich, also set up the International Commission on the History of Mathematics within the framework of the International Union for the History and Philosophy of Science. Under his chairmanship, and that of his successors (initially Scriba and then Joseph Dauben [1944–]), it has been one of the most active commissions, with national representatives across the world. In 1985 it was also recognized by the International Mathematical Union. Since 1989, and forthwith at each International Congress for the History of Science, the commission awards May medals, in memory of its founder, to historians who have enriched the subject not only by their scholarship but also for developing the community of historians. The first awards went to Struik and Yushkevich, followed in 1993 by Scriba and Wussing and in 1997 by Taton.

One of the main tasks of the commission is to administrate *Historia mathematica*, a journal created by May in 1974. Originally published by the University of Toronto Press, it was transferred three years later to Academic Press. Intended as more than a repository of articles and reviews, it included departments on projects and on archival sources, abstracts of relevant publications, news of meetings and appointments, and other news. One of May's main decisions was to incorporate an abstracts section: at first about 250 items per year were listed; by the mid-1980s the number had grown to 300–400; and for some years now at least 700 have appeared. Under May's successor editors (1985–1993), initially Dauben (1977–1985) and Eberhard Knobloch (1943–), it has played an important role in the growth in activity.

A very significant stimulus to our subject was the series of week-long meetings at Oberwolfach that was founded by Hofmann (¶7). After his death in 1973, the series was taken over by Scriba, and until 2000 it met annually or biennially. It allowed 50 historians of mathematics from around the world not only to hold a meeting but also to plan other events: in particular, the commission met there. Without this opportunity some of its and others' initiatives might well not have occurred, or at least would have been much harder to effect.

One of Dauben's main achievements as chairman was to edit a substantial annotated bibliography of historical literature in our subject, which appeared in 1985. Smaller in scale than May's, it was more focused in organization and information (Dauben 1985). An extended edition was prepared under the direction of Albert Lewis (1943–) and published as a CD-ROM by the American Mathematical Society (Lewis 2000).

9. Revival from the 1970s: By Country or Region

Only a few features of the last 30 years in the most active countries can be recorded here; some preference is given to collective rather than individual enterprises. Italy became particularly active (Barbieri and Pepe 1992); the depth of work is well exemplified by the volume on Italian mathematics between the two world wars ([Guerraggio] 1987). The Italian Mathematical Society has sponsored since 1980 the *Bollettino di storia delle scienze mathematiche*, edited by Enrico Giusti (1940–). In addition, a general magazine on mathematics and its history, *Lettera Pristem* (the latter word is the acronym of a study group) started in 1991, partly modeled on the American general journal *Mathematical Intelligencer*. The history of mathematics also appears sometimes in the similar publication *Informazione filosofiche*. In 1994 the Circolo Matematico di Palermo added to its renowned *Rendiconti* a biennial historical supplement, edited by U. Bottazzini (1947–). A national society was formed in 2001, with Silvia Roero (1952–) as founder-president.

A long-term project that continues an Italian tradition from Bortolotti and others is the Centre for Medieval Mathematics, founded in 1981 at Siena University by Laura Toti Rigatelli (1941–) and Raffaella Franci (1940–). Two years later it launched a series of editions of original Italian sources, Quaderni del Centro Studi della Matematica Mediovale; it has also provided a bibliography of the many articles on Renaissance mathematics written by Gino Arrighi (1906–2001) and often published in obscure locations (Simonetti 1996).

Another major center is Paris, where various équipes function, with much work on the important French educational institutions and societies. In 1980 the *Cahiers du séminaire d'histoire des mathématiques* was launched, with some specialty for transcribing manuscripts and correspondence. Conferences take place occasionally at Luminy (Marseille) in the mathematics center of the French Mathematical Society, which also supported in 1995 the founding of the journal *Revue d'histoire des mathématiques* to succeed the *Cahiers*.

The German Mathematicians' Union launched in 1985 a book series of German documentary sources and editions. It also supports a history section, inspired largely by its absorption in the early 1990s of the sister society of the former German Democratic Republic. A book series with Vandenhoeck and Ruprecht contained some excellent doctoral theses. A project led by Menso Folkerts (1943–) tracks down medieval manuscripts, including those of the German Rechenbuch tradition of early algebra. In 1988 he also launched a book series, Algorismus, for the history of science (often mathematics) and became the editor of another one, Boethius (1963–).

Soviet work continued in the same style and quantity as previously, especially in Moscow and Leningrad. Yushkevich edited a three-volume general history of mathematics, up to the nineteenth century (Yushkevich 1970–1972), followed by a trio edited with A. N. Kolmogorov (1903–1987) on the nineteenth century itself; the latter has been translated into English (Kolmogorov and Yushkevich 1992–1998). The change of the Soviet Union into Russia does not seem to have had major effects on the work done there. With the death of Yushkevich in 1993, his mantle has fallen upon Sergei Demidov (1942–), who maintains the annual *Istoriko-matematicheskogo issledovaniya.*

Czechoslovakia has produced a respectable body of work, led by Lubos Novy (1929–) and Jaroslav Folta (1933–); some of it was concerned with the Czech specialty of geometry. Rather less has come out of Poland, but in compensation the history of logic is active there, especially concerning the country's great importance in that subject since the 1920s. Thanks to organizer Christa Binder (Vienna), week-long meetings take place biennially in Austria, serving as a valuable focus for historians, especially in Central Europe and Germany.

The British Society for the History of Mathematics was formed in 1971 to provide a forum for adherents to the subject. Since the mid-1980s the society has substantially increased both its membership and range of meetings and developed a triannual newsletter from 3–4 pages of typescript in 1986 to 60 or so pages of print nowadays; it was elevated to a bulletin in 2004. Its founder and president was Gerald Whitrow (1912–2000), whose historical work focused on cosmology and theories of time, an area of interest shared by C. W. Kilmister and J. D. North (1936–2008). The latter has also worked much on the medieval and early modern periods, including in the Thomas Harriot Seminar (¶5), which has continued since the early 1980s under the direction of Gordon Batho. In the United States large special sessions take place regularly at the annual joint meetings of the American Mathematical Society and the Mathematical Association of America, and are well attended. In 1988 the society started a book series, which is now cosponsored by the London Mathematical Society. Among large-scale projects, the *Dictionary of Scientific Biography* gave appropriate attention to mathematics, both in its original 16 volumes (1970–1980) and in the two supplementaries (1990); the editor Charles Gillispie (1918–) had Boyer and Clagett on his editorial board. Many researchers are to be found in the United States, with the links to education particularly strong.

In Mexico City a group around Alejandro Garciadiego (1953–) started in 1985 the journal *Mathesis*, which has been especially concerned with the history of foundational subjects. Interest has grown substantially in Latin America: in 1992 the Association for History, Philosophy and Pedagogy of Mathematics was established, and nine years later there was founded a national society in Brazil, together with a journal. Part of the stimulus has come from policies in the Iberian Peninsula to encourage the history of science as an academic discipline. Spain has become quite active, especially with a group under Mariano Hormigon (1946–2003) at Zaragoza; in Portugal a seminar for our subject was formed in the late 1980s, with an especial concern for meetings suitable for students.

Another region to develop its own interests is Africa. Under the inspiration of Paulus Gerdes (Mozambique), the Commission on the History of Mathematics in Africa was established in 1987. A good deal of revisionist history is emerging from its meetings and newsletter, especially concerning the content of African mathematics and developments after the ancient glory of Egypt, which hitherto were usually ignored (Gerdes 1994). Partly in this connection, considerable interest has developed in ethnomathematics; an astonishing range of theories created worldwide in all sorts of contexts has been exposed. In addition to Gerdes, leading practitioners include Ubiratan d'Ambrosio (1926–) in Brazil (a May medalist in 2001) and Marcia Ascher (1935–) in the United States (for example, Ascher 2002).

Further north on the African continent, Arabic and Islamic sciences have gained a new level of interest and subsidized support, with mathematics featuring prominently. An institute has been established at Aleppo, and meetings occur regularly. For mathematics, one main task is to locate unknown manuscripts; in addition to many indigenous sources, important versions of lost works by Apollonios and by Diophantos have been found. Several editions of the works of important figures have appeared, some with translations into a Western language. (A few texts have been edited more than once, with some disturbance of the scholarly atmosphere [Hogendijk 2002].) Several manuscripts are in India, whose own rich history has received a considerable impulse: a national society was formed in 1978, and their bulletin was launched the following year under the title *Ganita-Bharati*. The annual meetings of the society have become more international recently, and proceedings have appeared for the 2001 gathering (Grattan-Guinness 2003).

Similarly, the history of mathematics in the Far East has increased both in the countries involved and elsewhere—in China after a period of isolation reinforced by the "cultural revolution." An important international example is a project between Dauben and various Chinese and Taiwanese colleagues to produce an English edition of the ancient classic *Nine Chapters of the Mathematical Art*, in enhanced echo of the collaboration decades earlier between Smith and Mikami (¶5). In the same international spirit, several historians in Far Eastern countries study aspects of Western mathematics, even effecting some translations of primary sources.[16]

One major general task is to convey to nonreaders of Chinese the *sense* as well as the content of the mathematics involved, for six characters may have been rendered as, say, fifteen words in English, and in another order of expression. To cope with this problem Jock Hoe introduced a semisymbolic language corresponding as closely as possible to the order of the characters, thereby bringing the reader closer to Chinese thought (Hoe 1978). His proposal has not been adopted to the extent that it deserves.

10. Revival from the 1970s: Old and New Editions

Among the editions mentioned earlier,[17] the mathematical series of the Leibniz edition (¶7) is now supervised by Knobloch. Reorganized during the 1980s with a new eighth series to cover some science and technology, the scale of the enterprise has forced planning of future volumes into long periods (Knobloch 1988), and editions of individual manuscripts or groups of them are appearing elsewhere. Since its founding in 1979 the journal *Studia Leibnitiana* has been keeping the reader abreast.

In the Euler edition, the first and third series were completed in 1956 and 2003 respectively, and two astronomy volumes in the second one are now left.[18] However, the fourth series, comprising two subseries of his correspondence and his notebooks, is far from complete; only four of the ten thick volumes of letters are out, and none for the notebooks. From 1952 Orell Füssli was the sole publisher of the edition; this responsibility was taken over in 1981 by Birkhäuser (Basel, Switzerland), who are also issuing the edition of the Bernoulli family conceived by Spiess, helped by funds that he left. Currently under the general editorship of P. de Radelet-Grave, the latter includes the contributions of the Bernoullis' close colleague Jacob Hermann (1678–

16. For example Joseph Fourier's *Théorie analytique de la chaleur* (1822) appeared in Japanese in 1993 and in Chinese in 1994—shortly after a Spanish translation of 1992.

17. For a survey of editions old, new, and pending, see Giusti and Pepe (1986).

18. These volumes were to have been edited by Eric Aiton (1920–1991), the editor of the volume on cosmic physics and the only Briton among the various editors. {The second series is to be completed in 2009, just over a century after the start of the edition.}

1733; Mathúna 2000). Birkhäuser also started in 1989 (initially with a house in the German Democratic Republic) the book series Science Networks for the history of mathematics and the physical sciences, parallel to one for both, Studies and Sources, already launched by their parent company Springer in the 1980s as a successor to Neugebauer's series of the 1930s (¶6).

Among newer editions, some large ones involve historical figures in the interaction between mathematics, logic, and philosophy. The *Gesamtausgabe* for the Bohemian polymath Bernard Bolzano (1781–1848; 1969–, Frommann) began with a beautiful biography by the Austrian historian Eduard Winter (1896–1982; Winter 1969). Now run largely in German hands based around Jan Berg, the edition will contain in its more than 50 volumes not only Bolzano's many publications but also transcriptions of his mathematical notebooks and unpublished essays, as well as correspondence. It succeeds some partial editions produced earlier in the century in Bohemia and in the Czechoslovak Republic.

For scholarly editions of more recent figures, there is the 30-volume chronological edition of the *Writings* of C. S. Peirce (1839–1914; 1982–, Indiana University Press) in the tradition of algebraic logic (among his many accomplishments), and one of comparable compass for the papers and manuscripts of Bertrand Russell (1872–1970; 1983–, now Routledge) in the very different line of mathematical logic (and again many other things). The publications and manuscripts of Felix Hausdorff (1868–1942), important in set theory and several other mathematical topics, are in hand (eight volumes, 2003–, Springer); those of Kurt Gödel (1906–1978) have recently been completed (five volumes, 1986–2003, Oxford University Press).

In addition, interest in the history of the allied area of computing grew rapidly during the 1980s, especially with the launch of the *Annals for the History of Computing* in 1979 and reprints of several original books and unpublished reports. The mathematician to benefit most from this activity was Charles Babbage (1792–1871), whose collected works appeared in 1989 (11 volumes, Pickering), edited by Martin Campbell-Kelly.

Other major editions in progress include Pierre Crépel directing a full edition of Jean d'Alembert (1717–1783), to be published in around 40 volumes by the National Council for Scientific Research; and various hands toiling over the *Collected Papers* of Albert Einstein (1879–1955; 1987–, Princeton University Press). Among editions solely of manuscripts, an outstanding example is Clerk Maxwell's (three volumes, 1990–2003, Cambridge University Press), carried out by Peter Harman; and Scriba is coediting the correspondence of John Wallis (1616–1703; 2003–, Oxford University Press).

The photoreprint industry has been very busy. Much of the mathematical output reproduces several older editions of works, an extraordinary number of old books, some older books in the history of mathematics (including several cited earlier), many mathematical journals, and at least one historical journal (Boncompagni's). Some volumes even appeared in paperback, and several with new introductions or other editorial material (in particular, by Hofmann for Olms Verlag). Most of this reprinting has happened since the Second World War: one motive was library losses created by bombing, while two others were the creation of new universities and the expansion of old ones, many with empty or rather bare libraries. Historians, the historically minded, and librarians have benefited enormously, especially when the item has been made available in paperback.

PART 2: APPRAISALS

11. Changes in Approach and in Balance

We are in the midst, it would appear, of the springtime of the history of 19th-century mathematics.

—*M. Bernkpof (1971)*

The growth in activity since the 1970s has led to changes not only in quantity; different (im)balances between branches, and some new approaches, are also evident. The examples about to be noted are all significant, but not exhaustively so.

The largest change in period concerns the attention now given to the nineteenth and early twentieth centuries. When I started working on that period nearly 40 years ago, it had become largely deserted: most topics had received no detailed histories at all; general histories (of which there had been quite a few) rarely went beyond familiar stories about the discovery of non-Euclidean geometries and a few bits on group and set theory (all mistaken or oversimplified accounts, historical research would reveal); and the Oberwolfach meetings had few lectures on it. Now it seems to gain more attention than any other—not surprisingly, for once entered, an incomparably vast panorama is made visible. As an adjunct to *Historia mathematica*, an occasional book series of articles on the "history of modern mathematics" deals exclusively with developments since 1800 (Rowe and McCleary 1989; Knobloch and Rowe 1994).

One aspect of mathematics since 1800 to benefit is institutional history, where the origins of national differences in modern mathematics can often be traced. After the exciting work of Klein's commission early in the century (¶2), institutional history suffered an especially severe drop. However, we now have, for example, the admirable examinations by Biermann of mathematics in the Berlin Academy (Biermann 1973), and by Hélène Gispert of the French Mathematical Society (Gispert 1992). The remarkable rise of the United States in mathematics from the late nineteenth century onward, tardy but then fast, has also been well chronicled.[19] Scientific institutions running before 1800 and involving mathematics have been newly studied. The social history of mathematics has also gained attention, not only from Struik's approach (¶7) and concern with institutional history and education but also using approaches pursued in the history of other sciences.[20]

Another important change concerns probability and mathematical statistics. It achieved professionalization only in the twentieth century, and so constituted a third grouping, alongside but rather separate from pure and applied mathematicians. However, in imitation of its past, historians of mathematics often ignored probability and mathematical statistics, even the historical work available such as that fostered by Pearson in Britain (¶5). For example, the treatment in part 1 of the EMW was very modest. But the situation has now changed considerably, for much historical work has appeared since 1980. Of especial note is the two-volume study, *The Probabilistic Revolution* (1987), edited by Lorenz Krüger and others, the principal result of a large research project run at the University of Bielefeld in Germany in the early 1980s. Several other substantial monographs, and many papers, have appeared from individual authors, and one fervently hopes that the tradition of historians ignoring the history of probability and statistics has ceased.

Similar judgments can be passed for some other topics. The history of mathematical economics is largely studied by historians of economics, but its presence is gradually being noticed by historians of mathematics; notable recent contributions include Ingrao and Israel (1990) and Mirowski (1994). The same kind of situation holds for links between mathematics and the arts (including architecture). Recent noteworthy works include Radelet-de Grave and Benvenuto (1995), Field (1997), and Gouk (1999). One sees also welcome work on scientific instruments of all kinds, whether mathematical as such (such as compasses or calculators) or scientific and technological with a significant mathematical element (such as telescopes and winches). But relationships between mathematics and religions, which go *both* ways in influence, have not yet been deeply studied, least of all by the many his-

19. See especially Parshall and Rowe (1994); Duren (1988–1998), which launched the American Mathematical Society book series mentioned in ¶9; and Shell-Gellasch (2002).

20. Two pioneering sources are Otte and Jahnke (1980) and Mehrtens, Bos, and Schneider (1981); more recent successors include Goldstein and others (1996) and Bottazzini and Dahan (2001). An early deployment of postmodernism in our subject was made in Mehrtens (1990), in the context of foundational studies in mathematics around 1900.

torians of science working on the influence of religions; see further in chapters 14 and 15.[21]

The rise of feminist history in general has made due impact internationally. Some biographies or studies have been written recently, as well as collective biographical volumes (for example, Grinstein and Campbell 1987). A fair proportion of historians today are female, whereas before the 1960s there were hardly any.

Biographies of men also still attract attention, but the form of treatment has changed from the normal style of factual reportage of a century ago. A remarkable pioneer was Pearson, for his three-volume *Life and Labours of Francis Galton* (1914–1930) even applied Galton's own biometric methods! Nowadays, the personal and the technical are often mixed together in a particularly attractive kind of biography, corresponding to the way that the historical figure lived his life; for example among several, Hankins (1980) on W. R. Hamilton, Hodges (1983) on Alan Turing, Gillispie (1997) on P. S. Laplace, and Dauben (1979) on Georg Cantor.

Cantor's work lay largely in set theory, which formed a major part of foundational studies in mathematics around 1900. Such research lies at the interface of mathematics with logics and with philosophy—perfect circumstances for a ghetto (as evidenced, for example, by the absence of logic from the first edition of the EMW). I tried to improve the situation by founding in 1980 the journal *History and Philosophy of Logic* to cover all periods. Also available is *Philosophia mathematica* (1964–), which came under the auspices of the Canadian Society for the History and Philosophy of Mathematics in 1991. Large editions for three figures in this area were described in ¶10. It is becoming better recognized that the old story about formalism versus logicism versus intuitionism is only part of the rich picture of foundational studies pursued between 1870 and the 1930s.

After these various positive signs, let us consider a few negative ones. The dichotomy between pure and applied mathematics, with pure as superior, is itself a historical development, occurring largely from the mid-nineteenth century onward and tied in fairly closely with the increase in professionalization of mathematics. Along with it grew a rather snobbish mentality of fortress mathematics, presented as the queen of the sciences and with a hierarchy of preference among its topics: pure first, led off by arithmetic and number theory, and then analysis, geometries, algebras, topology; applied mathematics (and also numerical methods) second best; and probability and especially mathematical statistics and mathematical economics

21. On Christianity, Chase and Jongsma (1983) lists publications on a wide range of themes, including philosophical, logical, and educational issues, but rather few specifically on history. {See now also Koetsier and Bergmans (2005) on mathematics and religions.}

quite likely out of sight altogether. Much modern historical writing tends to adopt the same ranking. The Bourbakists (¶7) are a prominent example, especially in their *Eléments d'histoire des mathématiques* (1969 and later editions). In the same spirit one of the fraternity, Jean Dieudonné (1906–1992) edited a general history of mathematics during the eighteenth and nineteenth centuries; although probability and logic were included, applications were omitted, and in silence (Dieudonné 1978). The great difference between the mathematics profession in the two centuries is also passed over.

This attitude is disappointing because it is unhistorical, heritage much more than history (ch. 2); many mathematicians did *not* always subscribe to such rankings, especially before the mid-nineteenth century but also afterward. Thus general histories of mathematics written fortress-style are simply *not* general; large areas or periods often disappear partly or totally. (One of principal aims of the *Companion Encyclopedia* was to attack the fortress mentality.) For example, the balance of historical research on Euler is quite out of line with that in his own work. There has been no satisfactory general history of mechanics since the early years of this century;[22] the histories of engineering mathematics, and also of military mathematics, are especially neglected, though Booß-Bavnbek and Høyrup (2003) is a recent pioneering volume on military mathematics across the two world wars. Similarly, mathematical physics is not granted the importance comparable to its status from the early nineteenth century onward; historians of physics have written much of the best work. Many of the accounts of applied mathematics written early in the century, such as in the EMW, have no modern counterparts.

Finally, apart from recreational mathematics, much less attention has been paid to the history of amateur mathematics. Developments in school-level mathematics are also rather unfashionable, in contrast to Klein's commission (¶2). Another overly neglected topic is mathematical activity in society. The best studies to date are the bibliographical and prosopographical investigations by Peter Wallis (1918–1992) and his wife, Ruth, of British "philomaths," subscription lists and publications (such as Newtoniana) in the eighteenth century (see especially Wallis and Wallis 1986); other countries, please imitate.

22. On the historiography of mechanics see Grattan-Guinness (1990).

12. Differences of Interpretation

Many of these changes discussed above involve innovations and novelties, and revisions of historical understanding. One important example of the latter concerns Greek mathematics. It is discussed in more detail in chapter 11.

From the 1880s Paul Tannery and Zeuthen proposed to read much of it as "geometric(al) algebra": their phrase, which denoted the common algebra of constants and variables, roughly like Descartes's. In this view the more simple books of Euclid's *Elements* consist of identities, usually of quadratic or bilinear forms, and many of the later constructions correspond to the extraction of roots from equations. This interpretation soon became standard; one influential example among many is Heath's editions and histories mentioned in ¶5. In the 1920s it was adopted by figures such as Neugebauer, who then looked for the algebraic origins in Babylonian mathematics, and by his follower B. L. van der Waerden (1903–1996), whose articles and especially the book *Science Awakening* (1954) have been influential sources.

However, Dijksterhuis avoided geometric algebra in his Dutch translation of Euclid of 1929–1930 noted in ¶5; and during the last thirty years, substantial reservations have been voiced by historians Sabatei Unguru, Ian Mueller, and others. No direct historical evidence backs up this algebraization of the geometry; and if it had happened, why was it never made explicit in the centuries after Euclid? Further, algebra with symbols is quite different from geometry with diagrams; and common algebra is the wrong algebra anyway, for Euclid never multiplied geometrical magnitudes together (§11.3.1). Other historians also avoided the tradition of geometric algebra. For example, Wilbur Knorr (1944–1997) saw *The Ancient Tradition of Geometric Problems* (1986) as much preoccupied by geometric problem solving, relating its repertoire of curves to objects of nature such as ivy leaves; his account of *The Evolution of Euclid's Elements* (1975), and David Fowler's reading of *The Mathematics of Plato's Academy* (1987, 1999), has opened up questions about the Euclidean algorithm and its mathematical richness, and also about papyrology. Similarly, Jens Høyrup has been rethinking the context and content of their Babylonian and Mesopotamian predecessors (Høyrup 2002). From suggestions such as these, ancient mathematics should have an interesting and different future.

This rejection of a standard historical interpretation in a particular area exemplifies *a general* methodological change; historians now see that they should adopt a more interventionist attitude. Prior to the 1960s (and long before 1900) most writing was empirical: that is, the facts were assembled and maybe classified, with the accounts largely confined to technical details. General questions of historical method were not normally raised (for example, not in the EMW; maybe part 7 would have considered them). In addition, the notations and terms were often imported from current mathematics or at least from a later time than that supposedly under study. One sense of the phrase "Whig history" applies to such approaches. They are deeply *deterministic* in character, and thereby unhistorical: to continue with the above example, it asserts that since Euclid certainly influenced the development of algebra in the West (and earlier among the Arabs), then he *must* have been trying to create what *they* produced partly in inspiration from him.

By contrast, to some extent under influence from the history of science, such methodologies are now seriously in doubt. The admission of indeterministic approaches has also led historians to doubt or even reject the view that mathematical knowledge grows cumulatively; the possibility of major changes, even revolutions in mathematics, has been discussed (Gillies 1992), partly in connection with Thomas Kuhn's well-known ideas in *The Structure of Scientific Revolutions* (1962). This increased sense of historical period also encourages (some) historians to distinguish Descartes from the Greeks, and not to shower Euler Truesdell-style with vectors and matrices, Lagrange on algebra with abstract group theory, or all sorts of people with set theory and axioms. Then historical readings of a text can be more clearly distinguished from modern analyses of it—which itself is a legitimate form of mathematical research, of course.

One central issue here is the distinction between the *history of* some mathematical notion, its creation and development at some period in past time, and the *heritage from* it during later periods and maybe even up to now (ch. 2). While the distinction is quite general, it is particularly pressing in the history of mathematics because many mathematical ideas have a duration of relevance and reinterpretation that does not apply in other sciences: for example, Euclid can be modernized from first line to last, which cannot possibly be done with, say, Ptolemy and the history of astronomy or Galen and the history of medicine. In my view the measure to which this distinction is recognized could have a notable bearing on future developments of our subject. Or maybe it will just deteriorate into another orthodoxy, to be rejected by a later generation!

13. Educational Motives, Positive and Negative

In 1977 I published a bibliographical survey article on our subject (Grattan-Guinness 1977); now it looks very quaint. Why should our subject have grown so much in the last thirty years or so? Of the various motives to be found, the largest single one is *negative reaction to practices in mathematical education.* This usually occurs at one of two career stages: (1) as a reaction against being bottle-fed as an undergraduate loads of clever theories that show no connection with human action, have no motivation whatsoever, and so lead from nowhere to nothing except more complicated versions of themselves; or (2) as a reaction to bottle-feeding undergraduates as a teacher loads of clever theories that show no connection with human action, have no motivation whatsoever, and so seem to lead only to more complicated versions of themselves.

This reaction has highlighted the question of using history for educational purposes, both for the teaching of pupils and students and the training of teachers (ch. 9). Two nonexclusive uses are possible, for all clients: (1) *informative,* giving instruction, such as *special* courses, in the history of mathematics as part of an educational program in(volving) mathematics; and (2) *integrated,* incorporating historical material *into* the design and execution of mathematics instruction, such as mathematics courses. The activity early in the twentieth century seems usually to have followed the informative kind. Moritz Cantor's *Vorlesungen* are a prominent case: they are typical historical books in that they give accounts about the past, in this case concerning mathematics, but their educational utility is left to the teacher-reader. Presumably his teaching (at Heidelberg University) of some parts of those large volumes carried the same aim.[23]

This informative use is also common today; but the integrated use has now also come into play, especially at school level. Between them they form the largest single concern of members of (at least) the British and Brazilian Societies for the History of Mathematics. A French group known by the acronym IREM takes history as one of its concerns and has won admiration at home and abroad, especially for the integrated use (Fauvel 1990). Various publications of the Mathematical Association of America give this use some prominence (for example, Calinger 1996), whereas in the days of Cajori and Smith early in the twentieth century, the informative use seems to have been more favored. In Denmark in 1987 the government required that historical material be included in the mathematics curriculum in secondary schools, and later made it mandatory also in the training of school teachers.[24]

When history is deployed at university level, it seems to be used informatively, in special courses for undergraduates. Thanks to an initiative taken by H. J. M. Bos (1940–) at Utrecht, between 1995 and 1997 the Socrates program of the European

23. The same point about selective teaching from published Vorlesungen applies not only to Cantor but also possibly to von Braunmühl's large history of trigonometry (¶3). The publisher for both authors was Teubner, who regularly published substantial Vorlesungen on mathematical subjects; hence there was some tradition in that form of title.

24. I am indebted for this information to Kirsti Andersen.

Union supported three three-week summer schools in our subject for (prospective) doctoral students that were held in European universities. Courses of this type, or for schoolteachers, are given in various countries.

Several of the dozens of journals in mathematical education are aware of the bearing of both uses of history and feature historical material, for example, in the *Zentralblatt für Didaktik der Mathematik*, which also runs an abstracting department for the whole field. The strength of this interest was substantial enough to inspire the founding in 1976 under UNESCO of the International Study Group on the Relations between History and Pedagogy of Mathematics, which organizes meetings and distributes a newsletter; one substantial volume of proceedings has appeared (Fauvel and van Maanen 2000).

The principal educational benefit from historical immersion is this obvious but profound truth: *mathematics is there because somebody thought it up, and for a reason—and moreover his, not ours.* Further, the interaction between research and education seems to have been richer in mathematics than in other sciences: sometimes, especially in analysis and mechanics, a mathematician's research was even stimulated by his teaching experience. But *how* does one get from the old text to the modern classroom? I have proposed the name "history-satire" for the idea that the teacher learns the history to some extent and then imitates the story (ponder on this word) in the classroom but without getting bogged down in the complicated historical details. Many advocates of history in education follow a similar line, which is sometimes aired in the journals for mathematics education.

An important consequence is that one may teach a theory in different historico-satirical versions over the years; for example, some basic theorems of the calculus in the ancient Greek and medieval techniques of exhaustion, the precalculus forms of the early to mid-seventeenth century, and the very different versions of Newton, Leibniz, and Lagrange before the more rigorous but less intuitive ones inspired by A. L. Cauchy and Karl Weierstrass in the nineteenth century (ch. 13). The German mathematician Otto Toeplitz (1881–1940) explored this kind of approach already in the late 1920s in the Bonn seminar (¶6) and was published posthumously (Toeplitz 1949).

14. Cottage Industry or Ghetto?

> The history of mathematics is essentially different from the history of other sciences in its relationship with the history of science, because it was never an integral part of the latter . . . mathematics being far more esoteric than the other sciences, its history can only be told to a select group of initiates . . .
>
> It is a pity that this should be so, for the history of mathematics should really be the kernel of the history of culture.
>
> —*George Sarton (1936, 4)*

This quandary captures well the professional situation for historians of mathematics. On the positive side, the two abstracting journals *Zentralblatt für Mathematik* and *Mathematical Reviews* (launched in 1931 and 1940 respectively much under the influence of Neugebauer, incidentally) greatly improved their departments for history during the mid-1970s. Some of the more popular mathematical journals imposed or improved their referring procedures for historical articles. Further, as part of the heritage from Sarton, the historical bibliographies of *Isis* have always given mathematics normal treatment.

But the sample of professional mathematicians and statisticians who were or are generally sympathetic to history must be small among the total population of tens of thousands (as in other sciences, no doubt). For most of the rest our subject is far too historical: it counts only as a sideshow, where the accuracy of a historical text is of little importance as long as it is Bell-style "fun"; some examples are given in chapter 6. One may even have to be "revolutionary" in declaring one's interest in history: a friend confessed to me that he "proved his silly little theorems" in his chosen branch of mathematical research in order to gain tenure, and *then* went open about his tendencies.

Similarly, the value of history for education is still not *widely* recognized among mathematical educators. For example, a workshop on the topic was held at the University of Toronto in 1983, in memory of May. Money was available to publish the proceedings, but they were rejected by 13 publishers, one for our failure to satisfy the standards normally found in mathematical education but by the others for fear of poor sales. (However, one referee wanted to keep such an "interesting" manuscript.) The proceedings eventually appeared, thanks to an initiative taken by the French Society for the History of Science.[25]

25. Grattan-Guinness (1987). This old tome is not mentioned in the recent volume of the same title (Fauvel and van Maanen 2000); such is history. A moderate amount of historical mate-

A converse predicament arises relative to the historians of science; to them our subject is too mathematical (ch. 6). While several of the journals in that field will take mathematical articles, it is well-nigh impossible for historians of mathematics to lecture before historians of science and expect their material to be discussed with the seriousness granted to the history of chemistry, say, or medicine; indeed, in my experience the reception has sometimes been negative. As for professional historians in general, contact normally obtains only for ancient and some medieval and Renaissance mathematics, where the mathematics as such may be elementary (although not always; and the history can well be very tricky). Historians will recognize from (ironically, mathematical) words like "1753" and "1876" that history is at hand; but the strength of their disinclination to study it is, shall we say, impressive.

Reasons for distaste here are simple to determine: mathsphobia is widespread among the population, and historians furnish no exception, even when it compromises their professional concerns. But then history suffers, for the ignorance of historians takes precedence over proper scholarship about the past. Further, a vicious circle begins; for this mathsphobia usually arises from the historian's unpleasant childhood experiences of mathematics, which was governed by policies in mathematical education, which will have ignored history, so that. . . .

15. "Writing the History of Mathematics": A Look at the Community

Metahistorical concerns have been very prominent around the International Commission on the History of Mathematics. As well as the Dauben and Lewis bibliographies mentioned in ¶8, it mounted symposia on historiography at the International Congresses for the History of Science: the 1989 occasion resulted in proceedings (Demidov and Folkerts 1992), and the meditations offered in 1993 led to the formation of a working group to edit a book under the editorship of Dauben and Scriba. With the cooperation of dozens of colleagues around the world and some special gatherings of the working group at Oberwolfach, a 700-page book appeared in 2002 from Birkhäuser, under the main title *Writing the History of Mathematics* (¶1; Scriba and Dauben 2002).

The first major decision concerned the manner by which the corpus of historical work should be divided: by branches of mathematics? or by countries in which mathematics was practiced? or into periods? Quickly it became clear to the

---

rial, mostly for informative use, appears in the recent educational encyclopedia (Grinstein and Lipsey 2001).

working group that the governing criterion was *the country in which the historian worked*, whatever topic he studied and whether that developed in his homeland or elsewhere. As has emerged in some examples above, the country has usually had a greater bearing on the historians than on the subject matter that they studied. For example, German historians seem to have shown a philosophical concern with knowledge as such, which is evident also, though in different ways, among French historiens-philosophes and some Italian colleagues. British and American historians have rarely exhibited such concern; on the other hand, their interest in connections with education are rather more marked, as is now also the case in Central and Latin America. (Finer analyses could consider the differences in practice in different regions of a country.) Historians seem rarely to have advocated some philosophical stance, though they may have held one privately; some have expressed belief in the unity of mathematics (M. Cantor, perhaps Klein), and some have favored Marxism (Struik, May, Wussing).

These differences are manifest to some extent also in the posts and careers that historians (have been forced to) adopt. Some French historians had associations with philosophical groups or institutes (for example among those named earlier, Couturat, Koyré, and Bachelard, as well as Geymonat in Italy) that seem to be rather rare elsewhere. Some historians came to ancient mathematics from a philological background: Heiberg, Ruska, Thureau-Dangin. Links to departments of the history of science have been rather infrequent, even when departments in that subject were formed: they are most evident in the Germanies (Gericke, Scriba, Wussing, Folkerts), France (Taton, Costabel), and the Soviet Union (Yushekvich, Demidov, and several other colleagues not named above). A few historians worked substantially in the histories of other sciences: Sarton, Struik, Cohen, Clagett, Folta.

Connections have always been strongest with mathematics departments in universities, at two levels. Several historians have practiced principally as professional, even professorial, mathematicians: for example, Klein, Zeuthen, Enriques, Timchenko, Peano, Burkhardt, Stäckel, Spiess, Prasad, Bell, Carathéodory, Truesdell, Kline, Struik, Dieudonné, van der Waerden. Others held a more minor post, even an honorary one, simultaneously with a paid position as a schoolteacher: Suter, Tropke, Wieleitner, Loria, Dijksterhuis, Vogel, Hofmann, Itard. A few, such as Smith and May, were employed in mathematics education, and Muir was an educational administrator in South Africa. Late in his life Cajori became professor in the history of mathematics at Berkeley. A few historians have survived on special funding: Sarton at Harvard, and Whiteside until 1975 when he gained a post at Cambridge University while working on the penultimate volume of his Newton edition. Some historians made a living by other means: Paul Tannery as a business manager,

Eneström and Bjørnbo as librarians, Heath in the British Treasury, Thureau-Dangin in the Louvre Museum in Paris, Whittaker as an astronomer, Jourdain by writing and editing, Mikami in the Japanese Imperial Academy.

The historians of today show roughly the same proportions of professional appointment and connection, with perhaps a somewhat larger percentage of schoolteachers. There have been very few institutions for our subject; however, the group at the Open University in Britain was recently reformed into a center under the direction of J. J. Gray (1947–).

On the decline after the Great War, here are some tentative thoughts. The deaths around that time of Eneström, Molk, Burkhardt, Duhem, Stäckel, Jourdain, Zeuthen, and Cantor; the retirement of Klein (who died in 1925); and of course economic and social difficulties in general both reduced activity and harmed some major sources and projects: the loss of *Bibliotheca mathematica* and the abandonment of the French *Encyclopédie* were particularly serious. Despite some of the volumes in the Euler and Gauss editions, the continuation of the series of *Ostwalds Klassiker*, the essays on Gauss, the final articles for the EMW, and Neugebauer's series, the reduction of writing in German is quite noticeable. There was also a rise and spread of philosophical positivism that deprecated history (of science) in general, either implicitly or explicitly. The reduced interest in Europe in the links between history and education may have deterred some potential new recruits. On the other hand, the level of activity in the United States increased somewhat, in line with its growing importance as a mathematical country. There must be more processes involved, which would be worth exploring in the context of the development of the history of other sciences at that time.[26]

After writing the history of mathematics, it should be published. Some publishers have demonstrated a strong commitment to our subject, including the many editions of the works of mathematicians of which only a few have been mentioned above. The output was and is largely books, but some houses have sustained journals for our subject or are friendly to it (Dauben 1998, 1999). In Germany Teubner stands out, especially through its links with Klein: unfortunately, soon after the Great War and seemingly as a consequence of it, Teubner greatly reduced its publication of mathematics in all forms (although it maintained the EMW and the Euler edition). This decision left a publishing vacuum that was quickly filled

26. The critical bibliographies in *Isis* for the period would provide many basic data, using the appropriate parts of the cumulative version for 1913–1965 edited by Magda Whitrow (5 vols., London, 1971–1982). Also useful are the lists of various kinds furnished in Sarton (1952). {Many bibliographies are now available online.}

by Julius Springer Verlag, which is the parent of today's Springer. In Switzerland Birkhäuser was and is prominent; in France Gauthier-Villars has a good record, as do the Oxford and Cambridge University Presses in Britain and the United States. Other houses in the United States include several more university presses as well as commercial firms, among which the Open Court Publishing Company occupies a special place; and the American Mathematical Society and the Mathematical Association of America are now significant.

In recent years the means of publishing have extended, especially with the development of the Internet. Dozens of sites are now available, some associated with the societies named earlier; others are due to departmental or even individual initiatives and often focus on biographies, countries or regions, particular theorems, or educational possibilities (Mohan 2003). As usual, the benefits of access are compromised by greatly variable quality control.

16. Concluding Remarks

The history of mathematics shows that since antiquity, mathematics has developed as a wide and ever-widening rainbow of ideas and theories (ch. 5). But the rainbow has negative as well as positive aspects. It delights those who look at its many colors; but it is also *distant* from us, and seemingly irrelevant to our cultural lives—including those of most mathematicians, philosophers, and historians.

This chapter was written to bring information about an academic field that is little noticed by outsiders; and its own history is not well known to many of its practitioners. The account belongs to metahistory, or the history of history, a part of history about which historians in general are curiously incurious. But all of us who worked on the Writings book found the experience enlightening and unusual in many ways; we would encourage comparable attention to be paid to the histories of the histories of other sciences.

BIBLIOGRAPHY

Ascher, M. 2002. *Mathematics elsewhere*, Princeton, N.J.: Princeton University Press.

Barbieri F., and Pepe, L. 1992. "Bibliografia Italiana di storia delle mathematiche 1961–1990," *Bollettino di storia delle scienze mathematiche, 12*, 1–181.

Bernkopf, M. 1971. Book review, *Isis, 62*, 532.

Biermann, K.-R. 1973. *Die Mathematik und ihre Dozenten an der Berliner Universität 1810–1933*, 1st ed., Berlin (DDR): Akademie Verlag. [2nd ed. 1988.]

Biermann, K.-R. 1983. "Aus der Vorgeschichte der Euler-Ausgabe 1783–1907," in Burckhardt and others 1983, 489–500.

Booß-Bavnbek, B., and Høyrup, J. 2003. (Eds.), *Mathematics and war*, Basel, Switzerland: Birkhäuser.

Bottazzini, U., and Dahan, A. 2001. (Eds.), *Changing images in mathematics: From the French Revolution to the new millennium*, London: Routledge.

Burckhardt, J. J. 1983, "Die Euler-Kommission der Schweizerischen Gesellschaft—ein Beitrag zur Editionsgeschichte," in Burckhardt and others 1983, 501–510, 511–552.

Burckhardt, J. J., and others 1983. (Eds.), *Leonhard Euler: Beiträge zu Leben und Werk*, Basel, Switzerland: Birkhäuser.

Burkhardt, H. F. K. L. 1901–1908. "Entwicklungen nach oscillirenden Functionen und Integration der Differentialgleichungen der mathematischen Physik," *Jahresbericht der Deutschen Mathematiker-Vereinigung, 10*, pt. 2, xii + 1804 pp.

Calinger, R. 1996. (Ed.), *Vita mathematica: Historical research and integration with teaching*, Washington, D.C.: Mathematical Association of America.

Cantor, M. 1908. (Ed.), *Vorlesungen über die Geschichte der Mathematik*, vol. 4, Leipzig, Germany: Teubner.

Chase, G. B., and Jongsma, C. 1983. (Eds.), *Bibliography of Christianity and mathematics: 1910–1983*, Sioux Center, Iowa: Messiah College. [Available on www.messiah.edu/departments/mathsci/acms/bibliog.htm.]

Coray, D., and others 2003. (Eds.), *One hundred years of* L'enseignement mathématique*: Moments of mathematics education in the 20th century*, Geneva: L'enseignement mathématique.

Dauben, J. W. 1979. *Georg Cantor*, Cambridge, Mass.: Harvard University Press.

Dauben, J. W. 1985. (Ed.), *The history of mathematics from antiquity to the present: A selective bibliography*, New York: Garland.

Dauben, J. W. 1998. "*Historia mathematicae*: journals of the history of mathematics," in M. Beretta and others (eds.), *Journals and the history of science*, Florence: Olschki, 1–30.

Dauben, J. W. 1999. "*Historia mathematica*: 25 years / context and content," *Historia mathematica, 26*, 1–28.

Demidov S. S., and Folkerts, M. 1992. (Eds.), "History and the historiography of mathematics," *Archives internationales d'histoire des sciences, 42*, 5–144.

Dieudonné, J. 1978. (Ed.), *Abrégé d'histoire des mathématiques*, 2 vols., Paris: Hermann.

Duren, P. 1988–1989. (Ed.), *A century of mathematics in America*, 3 vols., Providence, R.I.: American Mathematical Society.

Eneström, G. 1910–1913. *Verzeichniss der Schriften Leonhard Eulers*, Leipzig, Germany: Teubner.

Fauvel, J. 1990. (Ed.), *History in the mathematics classroom: The IREM papers*, Leicester, U.K.: The Mathematical Association.

Fauvel, J. and van Maanen, J. 2000. (Eds.), *History in mathematics education: The ICMI study*, Dordrecht, Netherlands: Kluwer.

Field, J. V. 1997 *The invention of infinity: Mathematics and art in the Renaissance*, Oxford: Oxford University Press.

Gebhardt, M. 1912. *Die Geschichte der Mathematik im mathematischen Unterrichte der höheren Schulen Deutschlands*, Leipzig and Berlin: Teubner.

Gerdes, P. 1994. "On mathematics in the history of sub-Saharan Africa," *Historia mathematica, 21*, 345–376.

Gillies, D. 1992. (Ed.), *Revolutions in mathematics*, Oxford: Oxford University Press.

Gillispie, C. C. (main author) 1997. *Pierre Simon Laplace: A life in exact science*, Princeton, N.J.: Princeton University Press.

Gispert, H. 1992. *La France mathématique: La Société Mathématique de France (1870–1914)*, Paris: Belin.

Gispert, H. 1999. "Les débuts de l'histoire des mathématiques sur les scènes internationales et le cas de l'entreprise encyclopédique de Felix Klein et Jules Molk," *Historia mathematica, 26*, 344–360.

Giusti, E., and Pepe, L. 1986. (Eds.), *Edizioni critiche e storia della matematica*, Pisa: ETS.

Goldstein, C., and others 1996. (Eds.), *L'Europe mathématique: Mathematical Europe*, Paris: Maison des Sciences de l'Homme.

Gottwald S., and others 1990. (Eds.), *Lexikon bedeutender Mathematiker*, Leipzig, Germany: Bibliographisches Institut.

Gouk, P. 1999. *Music, science and natural magic in seventeenth-century England*, New Haven, Conn., and London: Yale University Press.

Grattan-Guinness, I. 1977. "History of mathematics," in A. Dorling (ed.), *Use of mathematics literature*, London: Butterworth, 60–77.

Grattan-Guinness, I. 1987. (Ed.), *History in mathematics education*, Paris: Belin.

Grattan-Guinness, I. 1990. "The varieties of mechanics by 1800," *Historia mathematica, 17*, 313–338.

Grattan-Guinness, I. 1993. "European mathematical education in the 1900s and 1910s: Some published and unpublished surveys," in E. Ausejo and M. Hormigon (eds.), *Messengers of mathematics: European mathematical journals (1800–1946)*, Madrid: Siglo XXI, 117–130.

Grattan-Guinness, I. 2003. (Chief ed.), *History of the mathematical sciences*, Delhi: Hindustani Book Agency.

Grinstein, L. S., and Campbell, P. J. 1987. (Eds.), *Women of mathematics: A biobibliographical sourcebook*, Westport, Conn.: Greenwood Press.

Grinstein, L. S., and Lipsey, S. I. 2001. (Eds.), *Encyclopedia of mathematics education*, New York and London: RoutledgeFalmer.

[Guerraggio, A.] 1987. (Ed.), *La matematica Italiana tra le due guerre mondiale*, Bologna: Pitagora.

Hankins, T. L. 1980. *Sir William Rowan Hamilton*, Baltimore: Johns Hopkins University Press.

Henderson, H. 1993. *Catalyst for controversy: Paul Carus of Open Court*, Carbondale, Ill.: Open Court.

Higgins, T. J. 1944. "Biographies and collected works of mathematicians," *American mathematical monthly, 51*, 433–445. [Addenda in 56 (1949), 310–312.]

Hodges, A. 1983. *Alan Turing: The enigma*, London: Burnett Books.

Hoe, J. 1978. "The Jade Mirror of the Four Unknowns—some reflections," *New Zealand mathematical chronicle, 7*, 125–156.

Hofmann, J. E. 1990. *Ausgewählte Schriften zur Geschichte der Mathematik*, 2 vols. (ed. C. J. Scriba), Hildesheim: Olms.

Hogendijk, J. P. 2002. "Two editions of Ibn al-Haytham's *Completion of the conics*," *Historia mathematica, 29*, 247–265.

Høyrup, J. 1991. "Changing trends in the historiography of Mesopotamian mathematics," *History of science, 34*, 1–32.

Høyrup, J. 2002. *Lengths, widths, surfaces: A portrait of old Babylonian algebra and its kin*, New York: Springer.

Ingrao, B., and Israel, G. 1990. *The invisible hand: Economic equilibrium in the history of science*, Cambridge, Mass.: MIT Press.

Jourdain, P. E. B. 1991. *Selected essays on the history of set theory and logics (1906–1918)*, (ed. I. Grattan-Guinness), Bologna: CLUEB.

Kendall, M. G. 1970, 1977. (Ed.), *Studies in the history of statistics and probability*, 2 vols. (vol. 1 ed. with E. S. Pearson; vol. 2 ed. with R. L. Plackett), London: Griffin.

Klein, F. 1925–1928. *Elementare Mathematik vom höheren Standpunkt aus*, 3rd ed., 3 vols., Berlin: Springer. [English trans. of vols. 1–2: *Elementary mathematics from an advanced standpoint*, London: Macmillans, 1932; repr. New York: Dover, 1939.]

Klein, F. 1926–1927. *Vorlesungen über die Entwicklung der Mathematik im 19. Jahrhundert*, 2 pts., Berlin; Springer. [Repr. New York: Chelsea, n.d.]

Knobloch, E. 1989. "Leibniz und die Herausgabe seines wissenschaftlichen Nachlasses," *Akademie der Wissenschaften zu Berlin. Jahrbuch* (1988), 475–483.

Knobloch, E., and Rowe, D. E. 1994. (Eds.), *History of modern mathematics*, vol. 3, New York: Academic Press.

Koetsier, T., and Bergmans, L. 2005. (Eds.), *Mathematics and the divine*, Amsterdam: Elsevier.

Kolmogorov, A. N., and Yushkevich, A. P. 1992–1998. (Eds.), *Mathematics of the nineteenth century*, 3 vols., Basel, Switzerland: Birkhäuser. [2nd. ed. of vol. 1 (2001). Original ed.: *Matematika XIX veka*, 3 vols., Moscow: Nauka, 1978–1987.]

Lewis, A. C. 2000. (Ed.), *The history of mathematics: A selective bibliography*, Providence, R.I.: American Mathematical Society [CD-ROM].

Lorey, W. 1916. *Das Studium der Mathematik an den deutschen Universitäten seit Anfang des 19. Jahrhunderts*, Leipzig and Berlin: Teubner.

Loria, G. 1916. *Guida allo studio della storia delle mathematiche*, 1st ed., Milan: Hoepli. [2nd ed. 1946.]

Mashaal, M. 2006. *Bourbaki: A secret society of mathematicians*, Providence, R.I.: American Mathematical Society.

Mathúna, D. Ó. 2000. *The Bernoulli project: Historic origins, development of mathematical works and the evolution of the Bernoulli edition*, Basel, Switzerland: Birkhäuser.

May, V. 1973. *Bibliography and research manual in the history of mathematics*, Toronto: University of Toronto Press.

McCoy, R. E. 1987. *Open Court: A centennial bibliography 1887–1987*, La Salle, Ill.; Open Court.

Mehrtens, H. 1990. *Moderne Sprache Mathematik*, Frankfurt/Main: Suhrkamp.

Mehrtens, H., Bos, H. and Schneider, I. 1981. (Eds.), *Social history of nineteenth century mathematics*, Basel, Switzerland: Birkhäuser.

Mikami, Y. 1913. *The development of mathematics in China and Japan*, Leipzig, Germany: Teubner. [Repr. New York: Chelsea, 1974.]

Mikami, Y., and Smith, D. E. 1914. *A history of Japanese mathematics*, Chicago: Open Court.

Mirowski, P. 1994. (Ed.), *Natural images in economic thought*, Cambridge: Cambridge University Press.

Mohan, M. 2003. "Useful web links on history of mathematical sciences," *Ganita-Bharati, 25*, 29–44.

Otte, M., and Jahnke, H. 1980. (Eds.), *Epistemological and social problems of the sciences in the early nineteenth century*, Dordrecht, Netherlands: Reidel.

Parshall, K. H., and Rowe, D. E. 1994. *The emergence of the American mathematical research community, 1876–1900: J. J. Sylvester, Felix Klein, and E. H. Moore*, [Providence, R.I.]: American Mathematical Society.

Pearson, K. 1978. *The history of statistics in the 17th and 18th centuries . . .* (ed. E. S. Pearson), London and High Wycombe: Griffin.

Radelet-de Grave, P., and Benvenuto, E. 1995. (Eds.), *Entre mécanique et architecture / Between mechanics and architecture*, Basel, Switzerland: Birkhäuser.

Rashevsky, N. 1968. *Looking at history through mathematics*, Cambridge, Mass.: MIT Press.

Reid, C. 1993. *The search for E. T. Bell, also known as John Taine*, Washington, D.C.: Mathematical Association of America.

Reymond, A.-F. 1914. "Le premier congrès de philosophie mathématique. Paris 6–8 avril 1914," *L'enseignement mathématique, (1)16*, 370–378.

Rostand, F. 1960. *Souci d'exactitude et scrupules des mathématiciens*, Paris: Vrin.

Rostand, F. 1962. *Sur la clarté des démonstrations mathématiques*, Paris: Vrin.

Rowe, D. E., and McCleary, J. 1989. (Eds.) *History of modern mathematics*, vols. 1–2, New York: Academic Press.

Rudio, F. 1911. "Vorwort zur der Gesamtausgabe der Werke Leonhard Eulers," in Euler, *Opera omnia*, ser. 1, vol. 1, Leipzig, Germany: Teubner, ix–lxi.

Sarton, G. 1936. *The study of the history of mathematics*, Cambridge, Mass.; Harvard University Press. [Repr. New York: Dover, 1957.]

Sarton, G. 1952. *Horus: A guide to the history of science*, New York: Ronald.

Sarton, G. 1984. [Various articles on the role of Sarton in the history of science in general], *Isis, 75*, 6–62.

Schröder, J. 1913. *Die neuzitliche Entwicklung des mathematischen Unterrichts an der höheren Mädchenschulen Deutschlands*, Leipzig and Berlin: Teubner.

Schubring, G. 1988. "The cross-cultural 'transmission' of concepts—The First International Mathematics Commission reform around 1900," Bielefeld University: Institute for Mathematics Education, occasional paper, no. 92.

Schulze, F. 1911. (Ed.), *B.G. Teubner. Geschichte der Firma*, Leipzig, Germany: Teubner.

Scriba, C. J., and Dauben, J. W. 2002. (Eds.), *Writing the history of mathematics: Its historical development*, Basel, Switzerland: Birkhäuser.

Shell-Gellasch, A. 2002. (Ed.), *History of undergraduate mathematics in America*, West Point, N.Y.: U.S. Military Academy.

Siegmund-Schultze, R. 1993. *Mathematische Berichterstattung in Hitlerdeutschland. Der Niedergang des "Jahrbuch über die Fortschritte der Mathematik,"* Göttingen, Germany: Vandenhoeck und Ruprecht.

Simonetti, C. 1996. "Scritti di Gino Arrighi," in R. Franci and others (eds.), *Itinera matematica*, Siena, Italy: University, 375–425.

Sinaceur, H. 1987. "Structure et concept dans l'épistemologie mathématique de Jean Cavaillès," *Revue d'histoire des sciences, 40*, 5–30. [See also pp. 117–129.]

Speiser, A. 1948. "Einleitung der sämtlichen Werke Leonhard Eulers," *Commentarii mathematici Helvetici, 20*, 288–318.

Stäckel, P. 1909. "Entwurf einer Einteilung der sämtlichen Werke Leonhard Eulers," *Vierteljahrsschrift der Naturforschenden Gesellschaft in Zürich, 54*, 261–88. [Also somewhat revised in *Jahresbericht der Deutschen Mathematiker-Vereinigung, 19* (1910), pt. 2, 104–116, 128–142.]

Struik, D. J. 1980. "The historiography of mathematics from Proklos [sic] to Cantor," *Schriftenreihe NTM, 17*, 1–22.

Toeplitz, O. 1949. *Die Entwicklung der Infinitesimalrechnung*, Darmstadt: Springer. [English trans.: *The calculus: A genetic approach*, Chicago: University of Chicago Press, 1963.]

Tropfke, J. 1979. *Geschichte der Elementarmathematik*, vol. 1, 4th ed. (ed. K. Vogel, K. Reich and H. Gericke), New York and Berlin: De Gruyter.

van Berkel, K. 1996. *Dijksterhuis. Een biografie*, Amsterdam: Bert Bakker.

Wallis, P. J., and Wallis, R. 1986. *Biobibliography of British mathematics and its applications*, pt. 2, *1701–1760*, Newcastle-upon-Tyne: University.

Watts, P. 1911. "Ship" and "Shipbuilding," in *Encyclopaedia Britannica*, 11th ed., vol. 24, London: Encyclopaedia Britannica, 860–922, 922–981.

Winter, E. 1969. *Bernard Bolzano*, Stuttgart, Germany: Frommann.

Yushkevich, A. P. 1968. *Istoriya mathematiki v Rossii do 1917 goda*, Moscow: Nauka.

Yushkevich, A. P. 1970–1972. (Ed.), *Istoriya matematiki s drevneisnikh vremen do nachala XIX veka*, 3 vols., Moscow: Nauka.

CHAPTER FOUR

On Certain Somewhat Neglected Features of the History of Mathematics

To the memory of a master of things abstract, Saunders Mac Lane (1909–2005)

After contemplating the relationship between different ways of handling past mathematics, I consider five features of the history of mathematics on which I would encourage more research or consideration: a classification of types of influence, the histories of applied mathematics and of the history of school-level mathematics, possible roles for anthropology, and the origins of names and attributions of mathematical notions.

1. Introduction

As we have just seen in the previous chapter, the history of mathematics received a remarkable degree of attention in from the 1870s to the Great War; it has been matched in the last 30 years or so, following an intermediate half century of relatively little activity. Some branches are now *much* more active than they ever were in the past, for example, the history of probability and statistics. But the measure of attention varies quite considerably among the diverse branches. I take here the question "Which parts or aspects of the history of mathematics are noticeably neglected at present and deserve a good deal more attention?" and argue for five cases.

The bibliography that could be associated with this paper is enormous; I cite items for specific points, and omissions do not necessarily entail criticism. The best general bibliographies for the history of mathematics are May (1974), Dauben (1985) and Lewis (2000). Current and earlier work is well covered by the abstracts section of the journal *Historia mathematica* (1974–), the appropriate sections and

The original version of this essay was first published in *Archives internationales d'histoire des sciences, 55* (2005: publ. 2006), 383–402, with two transplants in ¶6.1 from the article in chapter 5. Reprinted by permission of the editor of the *Archives* and Science History Publications.

subsections of the annual bibliography of the history of science journal *Isis* (1913–), and the history sections of the mathematics reviewing journals *Jahrbuch über die Fortschritte der Mathematik* (1867–1942), *Zentralblatt Math* (1931–), and *Mathematical Reviews* (1940–). Inspection of these sources helped me to formulate the cases of neglect.

Mathematics has not developed in isolation from other sciences, and neither has its history, although the extent to which historians of all *other* fields ignore *it* is disappointing. Some of the issues raised have analogies in the histories of other sciences, but I shall not explore them here.

2. Theory and Metatheory

Before proceeding to the cases, I need to examine the *kind* of question just posed. In recent years I have advocated a distinction between "history" and "heritage" as two *basically* different ways of handling the mathematics of the past (ch. 2). Briefly, history addresses the question "What happened in the past?" concerning some notion(s): branch, topic, theorem, proof method, algorithm, notation, or whatever. It should also consider the companion question "What did not happen in that past that one might expect to have happened?" By contrast, heritage tackles the different but quite legitimate question "How did we get here?" concerning some notion *N*, quite possibly a current one. It handles some modern version(s) of *N* and will update pertinent past mathematics to find out the arrival and development of *N*. The product may well be a chronological jumble, but will (or can) constitute perfectly decent modern(ish) mathematics. The situation is analogous to a building built in the past and still existing today as part of our cultural "heritage" in an updated form fitted with, say, running water, electrical wiring, telephones, computers, and central heating, and refurbished with new locks, furniture, carpets, paints and wallpapers; it may still be a perfectly good building.

The distinction itself deploys another one, between theory and metatheory; for history and heritage are both metatheoretic (that is, theoretical *about* the mathematical theories involved),[1] and neither is reducible or subordinate to the other. Many of the disputes about the legitimacy of a historical interpretation rest on *confusing the two.* In particular, quite reasonably mathematicians are often more interested in the heritage from a piece of past mathematics rather than its history; by

1. Note that this sense of metatheory is not of the formalized kind deployed in metamathematics, which indeed has a nice history of its own, often under the misleading name "formalism" (Mangione and Bozzi 1993, ch. 4).

contrast, historians are supposed to be making the opposite choice, although heritagelike modernizations may creep in unintentionally (such as by habit expressing in terms of set theory some past mathematics where it does not belong: a further example comes shortly). Thus two basic metaquestions to ask are "Is the contemplated historical task actually historical at all? And if so, with respect to which periods and figures is it historical?" Should the answer to the first question be "no," then heritage work will be undertaken instead. The same kind of question and answer obtains also to a contemplated heritage task.

Let us take an example of these considerations (and in more detail in ch. 9). Euclid's *Elements* (fourth century B.C.) presents a suite of postulates, axioms, definitions, and theorems about parts of plane and solid geometry and arithmetic; then from the ninth century onward the Arabs reworked them as part of their innovation of the subject that we call algebra. As one consequence, particularly from the mid-nineteenth century onward, the standard historical interpretation of Euclid's *Elements* was to treat it as a work of algebra dressed up in algebraic and arithmetical forms; fortunately, over the last 30 years or so this reading has been soundly and comprehensively criticized, especially the geometrical parts (ch. 11). In terms of the above distinction, Euclid had no algebraic notions in mind at all; but the algebraizing of the *Elements* is a very important part of the *heritage* drawn from Euclid by the Arabs, and it is also a very important part of the Arabs' own *history*.

As this example shows, the distinction between history and heritage partly rests on that between the pre- and the posthistory of the past mathematics involved; but many other consequences follow, some not temporal. One of them is the way that knowledge and ignorance intertwine around each other. We have not only knowledge but also knowledge of knowledge, for example, when one has not only a theorem but also a proof of it. We also have knowledge of ignorance, especially when posing a problem; for example, before the work of N. H. Abel and Everest Galois of the 1820s and 1830s, students of the question of whether or not the general quintic equation could be resolved by the standard algebraic operations *knew* that they did not know the answer (though a few thought that they had found out; Pesic 2003). Finally, there is also ignorance of ignorance, when a question cannot even be posed; to continue with this example, those students did not formulate any aspect of the theory of quintic equations in terms of ring theory, because it was not then available, and they did not happen to think it up in this context.

Knowledge interacts with ignorance also at the historical level. Historians too can know, know that they know (by having some kind of documentary evidence available, for example), know that they do not know (by the lack of such evidence), and not know that they do not know (by being unaware that the pertinent connections actually obtained, say). In heritage analyses, such manifestations of ignorance can be replaced quite legitimately by later knowledge.

One especially fruitful consequence of emphasizing the metatheoretical character of history and heritage is to offer an alternative to the historiography commonly advocated by historians of all kinds (in our area, recently by Leo Corry [2004, 5–6]): that to get back into the period and context of a historical figure, the historian should forget about all the notions that have developed since then. But this strategy has to fail, because *begs its own question;* for the historian will have to know in detail which notions are anachronistic and which are not, which requires the intended historical task *already* to have been fulfilled. Instead, bear later notions in mind by all means, but look for their possible *absence* from the period and context under study.

3. The Various Types of Influence

Let us now move to the first case of neglect. Take two mathematical notions from the past, *A* and a later *B*. How did *A* influence the formation of *B*? The theme of influence in general is not at all neglected; but I wish to point out some distinctions that often are not made, or at least are not brought out explicitly.

First, influence can be negative as well as positive: a mathematician may react against the established theories of his day and put forward an alternative. This is the correct sense of revolution in mathematics, or indeed in any science: a substantial measure of *replacement* occurs, in contrast to innovation, where the new theory is truly new and replacement is on a fairly minor scale.[2] M. J. Crowe (1975) put forward ten interesting theses about the "patterns of change" in mathematics. The last one denied the occurrence of revolutions in mathematics, a claim debated in Gillies (1992); the category of negative influence, and the distinction from innovation, might have been more strongly drawn out.

Second, influence of either polarity has at least the following four types:

1. *Reduction. A* not only actively plays a role in the formation and development of *B*, but the theorist also hopes to *reduce B* to the remit of *A*. In such situations analogies become special cases of *A* in *B*.
2. *Emulation. A* actively plays a role in the formation and development of *B*, but reduction is not asserted or maybe not even sought. Analogies are just similarities, not reductions.

2. In chapter 6 I offer a compromise between revolution and continuity in the development of theories by suggesting that maybe scientific revolutions are convolutions, in which new and old ideas wrap around each other in the formulation of new theories.

3. *Corroboration.* *A* plays little or no role in the formation of *B*, but the theorist draws on similarities to *A*, maybe including structural ones, to consolidate *B*, and maybe also to enhance the measure of analogy between *B* and *A*.
4. *Instantiation.* *A* and *B* have certain notions *C* in common, thereby creating analogies. *C* may have been imported into *A* and *B*, as a mathematical tool. But each occurrence of *C* is an example of its great generality, which surpasses the remits of *A* and *B*, which remain as examples or applications of *C*.

On occasion, lack of precise information may make it difficult to determine which type of influence is actually at work; the historical figure may not even have noticed while forming his theory. But frequently evidence is clear enough.

As sources of examples I review briefly two very well-known stories. First, Isaac Newton inaugurated the first full-blown calculus, the "fluxional" version using infinitesimals and limits (1660s onward); his fluxion was an innovation corresponding to the modern derivative *function*, and his "fluent" to the *in*definite integral. G. W. Leibniz formed a quite different version also using infinitesimals (but of a rather different kind) but deploying differentials instead of limits (1670s onward); his versions of the new notions were the differential and the integral, in a theory that Euler modified in the 1750s with the introduction of the differential coefficient. There is no evidence of positive or negative influence either way of Newton on Leibniz, although around 1700 Newton came sufficiently suspicious as to lead a charge of plagiarism against Leibniz that polarized the calculus community into two camps.

Then from the 1770s Lagrange, reacting negatively against both traditions, set up his own version of the calculus, reducing it to algebra and thereby avoiding both limits and differentials. Drawing on some informal background reasoning, he assumed that every function took a convergent expansion in a Taylor series, with the "derived functions" (his name) specifiable from the coefficients. Finally, from the 1810s onward A. L. Cauchy reacted negatively against especially Lagrange's version and set up his own. It was based on a *theory* of limits and of sequences of quantities (maybe) approaching a limiting value, and not just the intuitive idea of limits; he also emphasized the need to specify necessary and/or sufficient conditions on the truth of theorems.[3] Within his version of the calculus, limits were instantiated in

3. In 1822 Cauchy refuted Lagrange's assumption that functions always took a Taylor expansion with his counterexample of the function exp $(-1/x^2)$ at $x = 0$; but this had not been decisive for him, as he had already set aside Lagrange's faith in algebra (§11.7.6).

The history of the calculus and mathematical analysis has been well written up; see, for example, various articles in Grattan-Guinness (1994, esp. pts. 3 and 4).

the sense that he (and others) applied them elsewhere in mathematics: to the theory of functions and infinite series, thereby creating the essentials of mathematical analysis as we understand it; and to complex-variable analysis (his innovation). Again, both his and Lagrange's versions had to match the range of their two predecessors, for example, with their own renditions of major theorems; but these were passive corroborations rather than active emulations, since the foundations of the versions, and also sometimes the proof methods, were quite different.

Second, partly thanks to his new calculus Newton also inaugurated his theory of mechanics. Positive influence was claimed by Robert Hooke in certain respects, especially the inverse square law; the issue remains unclear (that is, we know that we do not know the actual situation). Negative influence is much clearer, especially from Descartes: any principal claim of that Frenchman, whether about algebra, mechanics, or optics, had to be replaced. Newton's rejection of vortex mechanics was well founded, especially on the evidence that Halley's comet traversed the planetary system in 1682 in the direction contrary to that of the planets (Kollerstrom 1999). His mechanics assumed central forces, his (only) three laws and the inverse square law applied to some aspect of our solar system (but not all; for example, forces of cohesion in extended bodies are not inverse square).[4] He verified Kepler's laws, it seems, as corroborations rather than emulations. His invocation of the ether was an instantiation of a very general notion that was also manifest in other sciences (Cantor and Hodge 1981).

4. The Balance between Pure and Applied Mathematics, Especially since 1800

From around the middle of the nineteenth century the professionalization of mathematics increased quite markedly in Europe (Parshall and Rice 2002); many new universities and engineering colleges opened while existing ones expanded, and all offered posts in mathematics or at least involving it. During this period, even though some technical universities or equivalents (such as civil and military engineering colleges) were created, a gradual preference among professional mathematicians of pure over applied mathematics (that is, over applications of mathematics to the physical world) seems to have developed: we mathematicians are now numerous enough to form a profession; therefore we should have our own subject. The history of this change in attitude among some (perhaps many) mathematicians would make

4. The Newton industry is of course vast. A good historical coverage of his mechanics is provided in Buchwald and Cohen (2001).

a nice study of a pretty complicated project; here are a few notes on it, leading to the second case of neglect.[5]

The preference for pure mathematics is especially discernible in Paris. The Ecole Polytechnique had quickly assumed great—though unintended—prestige after its founding in 1794 as a school for training civil and military engineers, but it "declined" to a more appropriate level of function during the later nineteenth century. Prominence in mathematics passed to the Ecole Normale Supérieure, where pure and maybe some applicable mathematics was the norm.[6] Italy also seems to show the change, while not so marked;[7] for example, it has long shown a good tradition in engineering mathematics and was to be distinguished in relativity theory.

The preference was also quite marked in Germany, with Berlin as a particularly influential center (Schubring 1981; Biermann 1988). However, there were some technical high schools in Germany, and a few of its universities kept up a high level in applications (for example, Leipzig [Schlote 2004]). Applied mathematics became most forthright at Göttingen University, especially from the 1900s onward when Felix Klein played an important role in the advance of applied mathematics, engineering, and physics there (Schubring 1989; Tobies 1981, 56–75). Nevertheless, despite Klein's considerable influence from the 1890s onward on the emergence of mathematics in the United States, pure and some applicable mathematics were developed far more markedly than applications (Parshall and Rowe 1994). Applications were also favored at Cambridge University, with the dominating place given to its Mathematical Tripos (Warwick 2003) and the Smith's essay prize (Barrow-Green 1999), both of which focused on applied mathematics; but even there pure mathematics improved notably from the 1890s onward.

These remarks are confined to the major countries of the period; it seems that more minor countries emulated them. In any event, this preference is still with us internationally; and most historians of mathematics today have had a mathematical training of some kind, and so must have experienced it. I still remember that as an undergraduate I was impressed by it, and mystified by the gulf that lay between the pure and applied worlds; for example, even different versions of the differential and integral calculus were used (respectively, Cauchy's and the Leibniz/Euler). The

5. For various countries, see the articles on various nations in Grattan-Guinness (1994, pt. 11). Examples of attitudes to pure and applied mathematics during our period occur in a few of the articles contained in Corry (2004), especially the account by S. Katz of the Hebrew University of Jerusalem in the early twentieth century.

6. The impression of preference in France emerges from Gispert (1991).

7. See Bottazzini (1981); and Guerraggio (1987), where the evidence comes mainly from the interwar years.

influence of such gulfs on my educational experience was strongly negative and led to my taking a serious interest in the (unmentioned) history of mathematics in the first place (ch. 1).

From this situation comes the second case of neglect. Whether because of exposure to this preference or other reasons, the revival in historical work since the 1970s has been much more marked on the pure than on the applied side, especially for mathematics since 1800, although the balance has improved somewhat in the last ten years. If and when the historian's previous educational experience has influenced his choice of research topic, this may be an example of the heritage of the preference for pure mathematics.

To illustrate this claim I use as benchmark the great German *Encyklopädie der mathematischen Wissenschaften mit Einschluss ihrer Anwendungen* (1898–1935), launched under the leadership of Klein (§3.2). As one would expect, it shows no sign of such preference; for in it he and his many co-workers described in detail the main parts of both pure and (as its title emphasizes) applied mathematics of that period and its recent past. While it was not conceived as a historical enterprise, many articles in it not only captured that past but also went further back.

The first three of its six parts covered arithmetic and number theory, algebras, the calculus and mathematical analysis (real and complex), and geometries with topology; and in recent decades historians have reworked a lot of the mathematics covered. The last three parts treated mechanics, mathematical physics, geodesy, geophysics, and astronomy; and, by contrast, many of the articles there are still the last detailed historical(like) statements on the topics involved.[8] Indeed, historians primarily concerned with applications themselves have written much of the best recent history of several parts of applied mathematics: physics and economics are good examples. The history of engineering is particularly spotty, especially the military sides, though some welcome recent redress of the latter is supplied by Booß-Bavnbek and Høyrup (2003). The same imbalance in the literature is evident for many historical periods and figures earlier than those covered by the *Encyclopädie*; for example, there is a good deal more to read about Leonhard Euler on, say, functions and on number theory than about Euler on hydrodynamics and celestial mechanics (see the bibliography in Euler 1983, 511–552).

8. I note especially an apparent paucity of recent material on the mathematical aspects of the histories of physiological mechanics, physical and engineering apparatus of many kinds, embankment theory, ship design and operation, geology, atmospheric science and meteorology, comets and shooting stars, and stellar astronomy. In addition and across the range, a general history of potential theory is particularly lacking. This list is not exhaustive; and conversely no criticism is leveled against the lonely contributors to the topics mentioned.

This imbalance is especially noticeable in general histories of mathematics. Applications are noted in the parts devoted to ancient and (to some extent) medieval mathematics; but after that their role is much reduced, especially for the nineteenth century onward (to the extent that that period is treated at all). Morris Kline's huge history (1972) is an especially puzzling case, since the treatment of the nineteenth century is quite extensive, and the author was himself an applied mathematician. When I was invited to edit a comprehensive encyclopedia for our subject, and to write my own general history of it, I decided for both projects to avoid this imbalance (and also to give the nineteenth century, and probability and statistics, better showings than usual in general histories). The resulting volumes (Grattan-Guinness 1994, 1997) carried the phrase "the mathematical sciences" in their titles to indicate the full range (§5.1–2).

Among branches of mathematics, trigonometry is an especially interesting instance. Enjoying a very long history, in the late medieval and early modern periods it was a major area of research, especially the spherical part; an offshoot of geometry driven especially by applications to astronomy, geodesy, and navigation, its functions were specified as recti- or curvilinear lengths relative to a master radius. Then, since around Euler's time trigonometry has usually been re-presented as a useful though hardly exciting offshoot of pure mathematics, mainly algebra, with the functions defined as ratios of lengths. This steep decline in prestige, and substantial conversion into pure though rather routine mathematics, which is unique among the branches of mathematics, may have caused the lack of a comprehensive history since (von Braunmühl 1900, 1903). Happily, Glen van Brummelen is now filling this gap with a general history in preparation, building on the important work that has been done during the intervening century on specific developments and figures in the Near East and Europe.

The imbalance in historical work between pure and applied mathematics is collectively unhistorical, for prior to the gradual change of emphasis recalled above, most mathematics was applied or at least applicable: trigonometry is a typical example. I hope for a better balance, not only on the topics that historians choose to study but also in those that they may have the chance to suggest to newcomers (such as doctoral students) to our subject.

5. Mathematics Education and Institutions, Especially at School Levels

A notable part of the history of mathematics concerns its educational and institutional sides, and not only for the period from the nineteenth century onward when the great institutional expansion took place. A reasonable body of work, not all of it recent, exists for most countries in Europe and North America, and for some (though not enough) countries elsewhere. But as far as mathematics education is concerned, usually the level studied has been that of the university or equivalent. The third case of neglect is the histories of *school-level* mathematics, which are far more patchily covered, for all countries—to combine with the last case, the history of school-level applied mathematics is very little studied.[9] Possibly the accounts would be rather similar between one country and another for all branches of mathematics, at least in terms of curricula. However, other aspects could well vary considerably; for example, the training of teachers, methods of teaching and of examining pupils, envisioned career prospects for them, financial and economic circumstances, and relationships to other sciences.

There is a potential source for encouraging researchers. A specially welcome feature of current activity in the history of mathematics is the large and international interest evident in using history *in* mathematics education in schools (and universities), both in separate historical courses and more integrally as parts of ordinary mathematical courses. In 1976 an International Study Group on the Relations between History and Pedagogy of Mathematics was founded under UNESCO. It distributes a newsletter and organizes meetings, from which at least one substantial volume of proceedings has appeared (Fauvel and van Maanen 2000). Some of those with this interest, especially mathematics teachers, want to enter into historical research but feel inhibited by their limited mathematical command; it would be worth encouraging them to work on the history *of* mathematics education at school levels, perhaps in the country or region in which they live and work. Again there is a legacy inspired by Klein: the International Commission for Mathematics Education, most active during the period 1908–1920, which has left a massive but little-known legacy of books and pamphlets on educational practices (and their own histories, in some cases) in many countries (§3.2).

9. {Gert Schubring (Bielefeld) launched in 2006 the *International Journal for the History of Mathematics Education* to cover school history.}

6. Influences from Anthropology?

I have often thought that an interesting essay might be written on the influence of race on the selection of mathematical methods.

—*W. W. Rouse Ball (1889, 123)*

6.1. Background

Some of the sociological approaches used in the history of science have been adapted to the history of mathematics (Mehrtens, Bos, and Schneider 1981). However, anthropology is little present: it seems to have been widely deployed only in ethnomathematics (see, for example, Ascher 2002) and in some aspects of ancient mathematics (Cuomo 2001).[10] This is the fourth case of neglect.

Among cultural manifestations, folkloric mathematics remains very patchily examined, despite the growth in ethnomathematics. Take this example from antiquity and continuously present ever since: clothes. That three-dimensional object—the human being—wears clothes, made out of shaping and cutting two-dimensional fabrics. *How* were the shapes determined? Was it entirely a story of empirical rules? Even if so, how about maximizing the use of the fabric to reduce waste and scraps? Some mathematics must have been used, even if implicitly. Cases of more sophisticated thought have been examined somewhat, such as the topology of knitting; but the area is largely unexamined.

As a discipline anthropology is itself part of Western culture, so that the differences between remote or "primitive" cultures and European-Western ones are more marked than those between cultures and nations within the West; but it is not thereby *confined* to non-Western cultures. Indeed, some topics in the history of mathematics are quite close to normal anthropological interests; for example, my student Abhilasha Aggarwal is studying the importation of Western mathematics into India by the British during the nineteenth century {(Aggarwal 2008)}.

Books such as Kroeber (1948) and Bohannan (1963) contain topics in social anthropology that might be adapted for the history of mathematics. These include, among others, modes of transmission between communities and diffusion within them, the formation and dissolution of institutions, the dominance of schools by

10. Roman mathematics is still curiously ignored; Romans are known for a peculiar number system, a famous architect (Vitruvius), and much measurement of land—and otherwise in general as a brutal culture, which has given them a poor reputation. But any race that built up an empire of that size must have been concerned with *far* more mathematics than that. A good study is awaited {see Cuomo (2001, chs. 5–6), and §12.9 on a detail}.

leaders and the place of organizers of groups, obtaining employment with tenure (the word comes from land and property ownership), and techniques used in fieldwork (these are relevant to oral history). I am particularly struck by the place given in anthropology to rankings and inequalities between and in social classes, over and above the classes themselves. These are evident both among mathematicians and among branches of mathematics (as construed by mathematicians): the snobbery of pure over applied mathematics among the growingly professionalized (¶4) is an important example of the latter. The modern emphasis on pluralities and lacks of centers could be fruitful as long it does not inspire doctrinaire relativism, the paradoxical claim that asserts relativism as an absolute truth.[11]

One merit of anthropology is to intertwine with social aspects the intellectual factors in life, such as theories, in a way that is often absent from sociology. Our subject includes mathematics education and institutions, and mathematicians lone or groupy, so why not try adapting anthropological theories, especially its social and cultural branches? For example, one might use or modify theories used in studying societal relationships in and between African tribes to examine the spread of a (mathematical) theory from one European country to several others. The recent collective volume (Parshall and Rice 2002) on the growth of the international mathematical community during the nineteenth century and later contains many examples of diffusion, but no author (myself included) used any techniques from anthropology.

The prospects for the utility of anthropology in mathematics education, at all levels, also seem attractive; ranking is a clear topic. In addition, some mathematicians as teachers built up schools that exhibited a remarkable degree of bonding among their students, positive influences that lasted throughout the students' later careers—a kind of kinship without rites. Prominent examples include Karl Weierstrass at undergraduate level at Berlin for 30 years from the mid-1850s (Biermann 1988, ch. 5)—a major source for the preferment of pure mathematics considered in ¶4—and R. L. Moore with postgraduates at the University of Texas for nearly half a century from the early 1910s (Lewis 2004).

Schools are to be distinguished from groups, which are collections of people with common aims that are more general than those of a school and with no dominating figure. Klein was prominent though not dominant in two groups alluded to earlier: around his encyclopedia and his project on mathematics education (¶4 and ¶5). Groups may also be susceptible to anthropological analysis.

11. On this point in the context of anthropology and social psychology see, for example, Shweder (2003, chs. 6 and 7); on its place within fallibilism as "comprehensively critical rationalism," see the suite of articles in CCR (1971).

Anthropology also deals with languages and linguistics. This topic bears on the history of science in general, including the history of translations of scientific texts from ancient times onward (Montgomery 2000). As ethnomathematicians already know well, mathematics (including symbolic logics) presents many special issues with its extensive use of symbols, even systems of them (Zellweger 1982), to represent not only objects but also properties, relationships, and operations. The algebraization of Euclid (¶2) is in part an example of translation. But I shall not pursue translations as such here, because they are not a neglected feature of our subject.

6.2. Modes of Diffusion

A pioneer in applying "cultural anthropology" to mathematics was R. L. Wilder (1974, 1981). Viewing theories as vectors striving for growth, he saw them as stimulating "stresses" of various kinds, all regarded as akin to forces. Taking some inspiration from the American anthropologist A. L. Kroeber, he was especially influenced by the anthropologist L. A. White, who mused anthropologically on the notion of mathematical reality and located it in "cultural tradition," where "Ideas interact with each other in the nervous systems of men" (White 1947).

Wilder was partly concerned with the ways in which theories (do not) diffuse across communities or cultures of all kinds and diffuse among them, topics important also in social anthropology. In particular, in mathematics just as in other fields, a theory, if diffused at all, is rarely taken over complete or unchanged. To take a whole class of possible examples, as a fairly attentive student of the nineteenth and early twentieth centuries, I am always struck by differences between countries (and between regions of the larger ones) in the topics and branches of mathematics that were or were not pursued; the latter form cases of national and cultural lag. For example, mathematical analysis was a staple topic in France and the German states and then Germany, joined by complex analysis from the late 1820s onward; but they took a minor place in England and Ireland, where in some recompense many of the new algebras (for example, matrices, determinants, quantics, quaternions, differential operators, functional equations, logic, and probability) were prosecuted with marked enthusiasm, and more diligently than in all other countries (which include Scotland). Again, to note a difference within a country, the difference between Berlin and Göttingen as centers for mathematics was noted in ¶4.

This emphasis on nations is historically fair only from about the late seventeenth to the mid-twentieth centuries, when the scale of mathematical activity was increasing to analyzable sizes. During the Renaissance and early modern periods, scholars used Latin as an international language and moved across Europe (and in some cases across Arab and Christian regions), crossing boundaries in ways that will be impossible in, say, Victorian Britain. Further, in the last 60 years or so, mathematics has gone international again, especially recently with the impulse from computing facilities. These features, which apply to other sciences also, are surely worthy of anthropological as well as sociological analysis.

Some lessons about diffusion and distortion (or, at least, modification) may be drawn also from social psychology, for there important differences seem to have been detected in the ways in which Westerners and Asians interpret experience in general, and these could affect the formation of theories, including emulation. Westerners focus on the individual actor more than the circumstances in which the person acts and the relationships formed there; Asians tend to reverse these priorities (Nisbett 2003); for example, Westerners see fish in a pond whereas Asians see a pond with fish in it. It would be worthwhile, though I suspect very tricky, trying to detect ways such differences have affected the development of Western and Asian mathematics. Another plane on which such differences may emerge is in historiography itself; that is, in the ways in which Western and Asian historians conceive, and have conceived, of historical developments in the first place.

A further approach to diffusion can be drawn from social network analysis. Here manners of networking in general are studied, including the formation of cliques and affiliations, and the growth of egos. Indeed, mathematics is deployed in it, especially graph theory, parts of matrix theory, some abstract algebras such as lattice theory for orderings, the logic of relations, and mathematical statistics. Wasserman and Faust (1994) shows this range well, though few applications to history are made. Should such techniques prove to be fruitful in the history of mathematics, then we have an example of mathematical history, that is, mathematics applied to historical questions.

6.3. Stimulus Diffusion

A particularly interesting form of diffusion was given the rather unhelpful name "stimulus diffusion" by Kroeber in his paper (1940). This is the situation when a culture is aware of some achievement in another culture and wishes to adopt it, but at first has no idea how to fulfill that desire. Kroeber's prime example was the exquisite Chinese tableware that Europe discovered, especially during the eighteenth century. Importation was possible, though very slow and expensive; but the technology was unknown in Europe. However, after much experimentation in various countries (in Britain Josiah Wedgwood was a major figure) emulation was achieved, and so the upper classes could have "china" in their dining rooms. This is an important example of this kind of diffusion: setting up a problem and thereby motivating the creation of theories.

Kroeber also mentions various examples of a culture trying to decipher the alphabet of another, which suggest a good analogy for mathematics concerning cryptography. He even gives a mathematical example, concerning the development of (common) algebra among the Chinese by the thirteenth century and its late diffusion into Japan four centuries later; he surmises that the delay may have been partly caused by difficulties of stimulus. I am not enough of a historian of Far Eastern mathematics to know whether his claim is well argued, but a range of similar questions for the early modern period surround the complicated sequences of diffusions of Greek and Arabic mathematics into early modern Europe, at both technical and more popular levels of theory. For example, was Fibonacci really as important as is normally thought? Montgomery (2000, pt. 1) exhibits very well not only the international and interlinguistic aspects of this history but also the development of languages themselves (for example, the positive influence of literary Persian on Arabic). Many examples of history versus heritage arise; for example, whether members of a culture were interested *historically* in the work of their preceding cultures on which they were drawing, or whether the earlier cultures were sources for their own—usually, it seems, the latter. The recent collective volume Dold-Samplonius and others (2002) has a wealth of information on transmissions during the early modern period, but again anthropology seems not to have been used.

For mathematics there are also intellectual examples of stimulus diffusion, where nations or communities tried hard to understand a strange or surprising theory and eventually succeeded: complex numbers and non-Euclidean geometries are two obvious examples. Wilder calls this kind of situation "conceptual stress." There are many cases where mastering a theory required of the mathematician considerable technical study; for example, elliptic functions from around 1840, and celestial mechanics at any time. Other cases of diffusion may not make such technical demands, although the conceptual difficulties can be substantial; several of the English-Irish algebras mentioned above are examples.

The converse process, apathy and indifference to new notions in a nation or community, are also legitimate matters for historical study, for they are cases of nondevelopment that I noted in ¶2. Let us finish this case with a somewhat related example.

6.4. Outsiders and Religions

A fascinating topic is outsiders and the communities to which they never came to belong:[12] Bernard Bolzano (Winter 1969) and Oliver Heaviside (Nahin 1988) are two examples of my acquaintance. Another type was mentioned in ¶1: that of the history of mathematics itself, usually separated from the history of science. Presumably detestation of mathematics from pupil and student times grants its history leper status among many "historians" of science, but the degree of enthusiasm with which it is often practiced is, shall we say, very impressive (ch. 6).

Maybe our understanding of outsider figures and communities could be enriched by anthropological and sociological studies of otherness in society in general. This topic, which is called "xenology," has a history of its own as long as that of mathematics, especially in connection with religions (Classen 2004). A major concern of anthropology is the creation and prosecution of religions; and here mathematics has an especially interesting place, because it seems to have influenced their formation and development, mainly through numerology and sacred geometry (see chapter 14 for the case of Christianity, and Koetsier and Bergmans 2005 more generally). The converse influence has played various roles, such as mathematics being dedicated to God, and the religious character of truths in applied mathematics and astronomy; these latter were evident down to Isaac Newton and Euler. But this influence declined steeply from around 1750, including during the time especially in the nineteenth century when it *increased* substantially in sciences such as geology and biology: Western mathematicians continued to be Christians, but even the fervent ones rarely involved their Christianity in their mathematics (ch. 15). This great difference of status among scientific disciplines has so far eluded my understanding; maybe anthropology will assist.

7. Well-Known Nonfacts

One of the manifestations of the diffusion of theories is the wide spread of certain names for mathematical notions as well as the notions themselves. This is the final case of neglect, which has arisen from derision: the general opinion is that the naming of notions is unimportant, and that the name that came to be assigned is often wrong anyway. Both views are true in themselves; however, the way in which a name spread widely (though perhaps slowly) can tell us something about the diffusion of

12. A paradox like Russell's is lurking here, concerning the International Society of Outsiders, from which nobody is excluded; perhaps a metasociety is needed!

the theory to which the notion belongs, and thereby is of some interest. The story is amusing, especially when the name of a Great Man is given to more than one notion without any historical evidence in any instance; for example, the "theorem of Thales," which was bestowed in the late nineteenth century to various results in the geometry of the triangle and the circle (Patsopoulos and Patronis 2004).

Tracking down the origin of these names can be very hard. Two that I have tackled recently are "the fundamental theorem of the calculus" and "the Navier-Stokes equations" (historically correct names this time). Neither name is as old as might be thought; {I have managed to locate the former to du Bois-Reymond (1880)}, and the second to around 1950 without, however, finding a definitive source.

When an innovation is made in mathematics, especially a new distinction of sense, then the use of names is *not* merely conventional. An example that has aroused my perplexity concerns the work of E. H. Heine and Georg Cantor around 1870 on the uniqueness of a Fourier series expansion following some suggestions of Bernhard Riemann. It led to the substantive difference between trigonometric series and Fourier series; however, while the distinction was well recognized, the *names* seem to have been used as synonyms (Lecat 1924) and rarely distinguished (Young [1910] is one example).

Popular attributions of theorems form a similar category, and they too can be wrong and hard to correct. One that surprised me recently was Euler's beautiful formula

$$\exp(i\pi) + 1 = 0, \tag{4.1}$$

which shows a connection between five of the principal numbers in mathematics. Euler must have known it, because it is a special case of his formula for $\exp(ix)$: the historical problem is that he never wrote it down, at least not in his publications or correspondence. The first printing of which I am aware (thanks to Christian Gilain) occurs in J. F. Français (1813), as part of a discussion about the geometric interpretation of complex numbers in which a certain Argand also participated.[13] However, neither author nor journal (Gergonne's) was major at the time, and Français made no special comment about the formula anyway; so surely this appearance cannot be the source of the later popularity. According to W. E. Byerly in Archibald (1925, 6), in the late 1860s Benjamin Peirce commented to students on the companion formula to (41) for $\pi/2$,

13. Hence the name "Argand diagram," as usual not tracked down, as well as being problematic, is attached to this representation. Further, Schubring (2001) casts great doubt on the identification of this Argand with a Swiss-born bookseller of that name living in Paris.

$$\exp(-\pi/2) = i^i, \tag{4.2}$$

which Euler *had* written down (see Archibald 1921 on [4.2]; [4.1] is not mentioned there). Peirce's words were repeated in a popular American book on mathematics, where they were tagged to (4.1; Kasner and Newman 1949, 103–104)! The rest of this diffusion story, a nice one in the instantiation of mathematical beauty, remains unknown, at least to me; knowledge of my ignorance, stimulus to be satisfied, . . .

ACKNOWLEDGMENTS

For comments on the draft I am grateful to Marcia and Bob Ascher, and to Niccolo Guicciardini.

BIBLIOGRAPHY

Aggarwal, A. 2008. "British higher education in mathematics for and in India, 1800–1880," doctoral dissertation, Middlesex University.

Archibald, R. C. 1921. "Historical notes on the relation $e^{-(\pi/2)} = i^i$," *American mathematical monthly, 28*, 116–121. [See also H. S. Uhler on pp. 114–116 on the "numerical value" of i^i.]

Archibald, R. C. 1925. (Ed.), "Benjamin Peirce, 1809–80," *American mathematical monthly, 32*, 1–3. [Repr. Oberlin, Ohio: Mathematical Association of America, 1925.]

Ascher, M. 2002. *Mathematics elsewhere: An exploration of ideas across cultures*, Princeton, N.J.: Princeton University Press.

Barrow-Green, J. 1999. "'A correction to the spirit of too exclusively pure mathematics': Robert Smith (1689–1768) and his prizes at Cambridge University," *Annals of science, 56*, 271–316.

Biermann, K.-R. 1988. *Die Mathematik und ihre Dozenten an der Berliner Universität 1810–1933*, rev. ed., Berlin: Akademie Verlag.

Bohannan, P. 1963. *Social anthropology*, New York: Holt, Rinehart, and Winston.

Booß-Bavnbek, B., and Høyrup, J. 2003. (Eds.), *Mathematics and war*, Basel, Switzerland: Birkhäuser.

Bottazzini, U. 1981. "Il diciannovesimo secolo in Italia," in D. J. Struik, *Matematica: un profilo storico*, Bologna: Il Mulino, 249–312.

Buchwald, J. Z., and Cohen, I. B. 2001. (Eds.), *Isaac Newton's natural philosophy*, Cambridge, Mass., and London: MIT Press.

Cantor, G., and Hodge M. J. S. 1981. (Eds.), *Conceptions of ether: Studies in the history of ether theories 1740–1900*, Cambridge: Cambridge University Press.

CCR 1971. [Suite of articles on comprehensively critical rationalism by several authors], *Philosophy, 46*, 43–61.

Classen, A. 2004. "Other, the, European views of," in M. C. Horowitz (ed.), *New dictionary of the history of ideas*, Detroit: Thomson, 1691–1698.

Corry, L. 2004. (Ed.), "The history of modern mathematics-writing and rewriting," *Science in context, 17*, 1–265.

Crowe, M. J. 1975. "Ten 'laws' concerning patterns of change in the history of mathematics," *Historia mathematica, 2*, 161–166.

Cuomo, S. 2001. *Ancient mathematics*, London: Routledge.

Dauben, J. W. 1985. (Ed.), *The history of mathematics from antiquity to the present. A selective bibliography*, New York: Garland.

Dold-Samplonius, Y., and others, 2002. (Eds.), *From China to Paris: 2000 years*['] *transmission of mathematical ideas*, Stuttgart, Germany: Franz Steiner Verlag.

du Bois-Reymond, P. 1880. "Der Beweis des Fundamentalsatzes der Integralrechnung: $\int_a^b F'(x)dx = F(b) - F(a)$," *Mathematische Annalen, 16*, 115–128.

Euler, L. 1983. *Leonhard Euler: Beiträge zu Leben und Werk*, Basel, Switzerland: Birkhäuser.

Fauvel, J., and van Maanen, J. 2000. (Eds.), *History in mathematics education: The ICMI study*, Dordrecht, Netherlands: Kluwer.

Français, J. F. 1813. "Nouveaux principes de géométrie de position, et interprétation géométrique des symbols imaginaires," *Ann. math. pures appl., 4* (1813–14), 61–71. [Repr. in J. R. Argand, *Essai sur une manière de représenter les quantités imaginaires dans les constructions géométriques*, 2nd ed. (ed. J. Houël), Paris: Gauthier-Villars, 1874 (repr. Paris: Blanchard, 1971), 63–74.]

Gillies, D. 1992. (Ed.), *Revolutions in mathematics*, Oxford: Clarendon Press.

Gispert, H. 1991. *La France mathématique: La Société Mathématique de France (1870–1914)*, Paris: Belin.

Grattan-Guinness, I. 1994. (Ed.), *Companion encyclopedia of the history and philosophy of the mathematical sciences*, 2 vols., London and New York: Routledge. [Repr. Baltimore: Johns Hopkins University Press, 2003.]

Grattan-Guinness, I. 1997. *The Fontana history of the mathematical sciences: The rainbow of mathematics*, London: Fontana. [Also publ. as *The Norton history . . .* New York: Norton, 1998.]

[Guerraggio, A.] 1987. [Ed.], *La matematica Italiana tra le due guerre mondiali*, Bologna: Pitagora.

Kasner, E., and Newman, J. R. 1949. *Mathematics and the imagination*, London: G. Bell.

Kline, M. 1972. *Mathematical thought from ancient to modern times*, New York: Oxford University Press.

Koetsier, T., and Bergmans, L. 2005. (Eds.), *Mathematics and the divine: A historical study*, Amsterdam: Elsevier.

Kollerstrom, N. 1999. "The path of Halley's comet, and Newton's late apprehension of the law of gravity," *Annals of science, 56*, 331–356.

Kroeber, A. L. 1940. "Stimulus diffusion," *The American anthropologist, new ser. 42*, 1–20.

Kroeber, A. L. 1948. *Anthropology: Race, language, culture, psychology, prehistory*, rev. ed., New York: Harcourt, Brace; London: Harrap.

Lecat, M. 1924. *Bibliographie des séries trigonométriques*, Louvain and Brussels: the author.

Lewis, A. C. 2000. (Ed.), *The history of mathematics: A selective bibliography*, Providence, R.I.: American Mathematical Society (CD-ROM).

Lewis, A. C. 2004. "The beginnings of the R.L. Moore school of topology," *Historia mathematica, 31*, 279–295.

Mangione, C., and Bozzi, S. 1993. *Storia della logica da Boole ai nostri giorni* [Milan]: Garzanti.

May, K. O. 1973. *Bibliography and research manual in the history of mathematics*, Toronto: University of Toronto Press.

Mehrtens, H., Bos, H., and Schneider, I. 1981. (Eds.), *Social history of nineteenth century mathematics*, Basel, Switzerland: Birkhäuser.

Montgomery, S. L. 2000. *Science in translation: Movements in knowledge through cultures and time*, Chicago: University of Chicago Press.

Nahin, P. J. 1988. *Oliver Heaviside*, Baltimore: Johns Hopkins University Press.

Nisbett, R. E. 2003. *The geography of thought: How Asians and Westerners think differently—and why*, London: Brealy.

Parshall, K. H., and Rice, A. 2002. (Eds.), *Mathematics unbound: The evolution of an international mathematical community*, Providence, R.I.: American Mathematical Society.

Parshall, K. H., and Rowe, D. 1994. *The emergence of the American mathematical research community*, Providence, R.I.: American Mathematical Society.

Patsopoulos, D., and Patronis, T. 2004. "An example of didactical 'use' of history of mathematics in textbooks at the end of nineteenth century: The name 'Theorem of Thales' as attributed to different theorems," in *HPM 2004: History and pedagogy of mathematics, Proceedings*, Uppsala, Sweden: University, 336–342.

Pesic, P. 2003. *Abel's proof: An essay on the sources and meaning of mathematical unsolvability*, Cambridge, Mass., and London: MIT Press.

Rouse Ball, W. W. 1889. *A history of the study of mathematics in Cambridge*, Cambridge: Cambridge University Press.

Schlote, K.-H. 2004. *Zu den Wechselbeziehungen zwischen Mathematik und Physik an der Universität Leipzig in der Zeit von 1830 bis 1904/05*, Stuttgart and Leipzig, Germany: Verlag der Sächsischen Akademie der Wissenschaften zu Leipzig in Kommission bei S. Hirzel. [*Abhandlungen der Sächsischen Akademie der Wissenschaften zu Leipzig, mathematisch-naturwissenschaftliche Klasse*, vol. 63, no. 1.]

Schubring, G. 1981. "The conception of pure mathematics as an instrument in the professionalization of mathematics," in Mehrtens, Bos, and Schneider 1981, 111–134.

Schubring, G. 1989. "Pure and applied mathematics in divergent institutional settings in Germany: The role and impact of Felix Klein," in D. Rowe and J. McCleary (eds.), *History of modern mathematics*, vol. 2, New York: Academic Press, 171–220.

Schubring, G. 2001. "Argand and the early work on graphical representation: New sources and intrepretations," in J. Lützen (ed.), *Around Caspar Wessel and the geometric representation of complex numbers*, Copenhagen: Reitzsel (Royal Danish Academy of Sciences and Letters, mathematical-physical section 46, no. 2), 125–145.

Shweder, R. A. 2003. *Why do men barbecue? Recipes for cultural anthropology*, Cambridge, Mass.: Harvard University Press.

Tobies, R. 1981. *Felix Klein*, Leipzig, Germany: Teubner.

von Braunmühl, A. 1900, 1903. *Vorlesungen über Geschichte der Trigonometrie*, 2 vols., Leipzig, Germany: Teubner.

Warwick, A. 2003. *Masters of theory: Cambridge and the rise of mathematical physics*, Chicago: University of Chicago Press.

Wasserman, S., and Faust, K. 1994. *Social network analysis: Methods and applications*, Cambridge: Cambridge University Press.

White, L. A. 1947. "The locus of mathematical reality: An anthropological footnote," *Philosophy of science, 14*, 289–303. [Repr. in J. R. Newman (ed.), *The world of mathematics*, vol. 4, New York: Simon and Schuster, 1956, 2348–2364.]

Wilder, R. L. 1974. "Hereditary stress as a cultural force in mathematics," *Historia mathematica, 1*, 29–46.

Wilder, R. L. 1981. *Mathematics as a cultural system*, Oxford: Pergamon.

Winter, E. 1969. *Bernard Bolzano: Ein Lebensbild*, Stuttgart-Bad Cannstatt, Germany: Frommann (Holzboog).

Young, W. H. 1910. "On the conditions that a trigonometrical series should have the Fourier form," *Proceedings of the London Mathematical Society, (2)9*, 421–433. [Repr. in *Selected papers*, Lausanne: Presses Polytechniques et Universitaires Romandes, 2000, 129–141.]

Zellweger, S. 1982. "Sign-creation and man-sign engineering," *Semiotica, 38*, 17–54.

CHAPTER FIVE

General Histories of Mathematics? Of Use? To Whom?

In honor of Wilbur Knorr (1946–1997)

I have written one general history and edited two larger ones. I describe their contents, and indicate the aims, which above all was to exhibit the great variety and ubiquity of mathematics.

1. How General Is General?

The three questions posed in this title were stimulated by my recent work on editing and writing general studies for a wide (academic) public. I edited the *Companion Encyclopedia of the History and Philosophy of the Mathematical Sciences* (CE), which was published in two volumes and 1,806 pages by Routledge early in 1994. A combined effort involving 132 other colleagues from 18 nationalities to produce 176 articles, it really did try to be general. The invitation from Routledge in 1988 to produce the book was very timely; for I doubt if it would have been possible even five years previously to find authors to cover all the topics needed. The reception has been positive enough for a paperback reprint to appear in 2003, from the Johns Hopkins University Press.

After an editorial introduction, the book was divided into the following parts:

1. Non-Western traditions up to Western superventions
2. Medieval and Renaissance, up to around 1600–1700

¶1 and ¶2 come from "General Histories of Mathematics? Of Use? To Whom?" in S. Nobre (ed.), *II encontro de historia da matematica: Anais,* Rio Claro (Brazil): University, 1997, 15–24. Included in ¶2 is a passage omitted from chapter 3. The rest of this article dealt with work, or lack of it, on specific topics; the comments on Roman and folklore mathematics are now in §4.6.1.

¶3 is new; its table is based on the one in my (ed.), *Landmark Writings in Western Mathematics 1640–1940,* Amsterdam: Elsevier, 2005, ch. 0, which is reprinted by permission of Elsevier Rightslink.

3. Calculus and analysis
4. Functions, series, and methods in analysis
5. Logics, set theories, foundations of mathematics
6. Algebras and number theory
7. Geometries and topology
8. Mechanics and mechanical engineering
9. Physics and mathematical physics and electrical engineering
10. Probability and statistics and the social sciences
11. Higher education and institutions
12. Mathematics and culture
13. Reference and information

The normal terminal point was in the 1930s. The range included in part 12 a survey by country of the development of universities and related institutions connected with higher-level mathematics; and not only (in Part 5) questions with which professional philosophers are concerned (including, in this case, modern work) but also the much wider issues as understood at various times in geometries, algebras, mechanics, and probability theory. The many other topics include

Korean mathematics
Medieval optics and instruments
Nomography
Polish logics
Lie groups
Operational research
Servomechanisms
Meteorology
Biomathematics
Probability and statistics in medicine
Mathematical economics
Mathematics in prose literature
History of history of mathematics
Ancient methods of doing fractions
Prehistory of fractals
Algebraic and mathematical logics
Calculating machines
Linear programming
Graph theory and combinatorics
Methods of navigation
Capillarity
Actuarial mathematics
Probability and statistics in engineering and technology
Tilings
Mathematics in poetry
Number theory

The applications to physical problems were strongly stressed, because they often inspired the mathematics in the first place: the preference for pure mathematics is a creation of the mid-nineteenth century (interestingly, the time when research and university-level mathematics itself became markedly more professional). Most general histories written since that time have followed the same snobbism, which apart from its intrinsic lack of merit is unhistorical. However, it is so well established that the word "mathematics" is often identified only with its pure parts; to combat this I had to use the phrase "the mathematical sciences" in the title of CE, and I repeated it in my second foray into general histories.

2. One Author This Time

I published a single volume entitled *The Fontana History of the Mathematical Sciences*, put out by Fontana Press (London) in May 1997 at 817 pages. Some months later it appeared in the United States from W.W. Norton (New York), called *The Norton History* . . . The structure was different from CE, in that chronology dominated topics rather than vice versa. The chapter titles were

1. Pre-viewing the Rainbow
2. Invisible Origins and Ancient Traditions
3. A Quiet Millennium: From the Early Middle Ages into the European Renaissance
4. The Age of Trigonometry: Europe, 1540–1660
5. The Calculus and Its Consequences, 1660–1750
6. Analysis and Mechanics at Center Stage, 1750–1800
7. Institutions and the Profession after the French Revolution
8. An Era of Continuity: Mathematical Analysis and Geometries, 1800–1860
9. The Expanding World of Algebras, 1800–1860
10. An Era of Linearity: Mechanics and Mathematical Physics, 1800–1860
11. International Mathematics, but the Rise of Germany
12. The Rise of Set Theory: Mathematical Analysis, 1860–1900
13. Algebras and Geometries: Their Relations and Axioms, 1860–1900
14. An Era of Stability: The Widening World of Mechanics, 1860–1900
15. An Era of Media: Mathematical Physics, 1860–1900
16. The New Century, to the Great War and Beyond
17. Re-viewing the Rainbow

The subtitle of the book, "The Rainbow of Mathematics," was chosen with two connotations in mind: a positive one, of the many topics and aspects of the subjects, like the colors of the rainbow; and a negative one, that the rainbow is always "over there," not part of our culture, with its proud and even aggressive nonnumeracy. (Note that even the word "numeracy" misidentifies mathematics with arithmetic, as if music were the same as sonatas.) Its main title was dictated by the fact that it belonged to a series of ten such histories that these houses were producing;[1] so was the format, a straightforward text (with bibliography, in my case) rather than a book specifically directed toward teaching, with examples and exercises. There is a tradition of such works, especially in the United States, with the emphasis laid on core branches and topics, such as arithmetic, algebra, Euclidean geometry, and the calculus.

3. Highlighting the Landmarks

In 2000 Arjen Sevenster, editor for mathematics at Elsevier, came to me with a nice idea. As a New Year greeting he once sent to his clients a suite of seven postcards containing the title pages of important mathematical books. Looking at them one day inspired the question, "Where could he read about the contents of these books?" Histories of Great Works are known in the humanities; for example, Ralph Hill edited successfully such books on the symphony and the concerto in the 1950s. Source books, selections of original texts, had been published for various sciences, mathematics included. But no science had ever been treated to a book of original commentaries on major works; it was time for a try.

The outcome was the book *Landmark Writings in Western Mathematics 1640–1940*, which appeared early in 2005 under my editorship. One boundary condition on its design was Elsevier's stipulated size of 1,000 pages (it came out at around 1,050 pages), which dictated the decision to restrict the period to those 300 years. Another decision was to grant around 10 pages to each writing, so that about 100 writings would be chosen. But how to pick out major writings from that time span? Outstanding quality was obviously a necessary condition, but the list of candidate writings was too long for it to be sufficient. Three criteria helped me, and my editorial board, to make the selection.

One criterion was that we were looking for publications that exercised influence, whether quickly or slowly, so fine manuscripts were excluded. The second criterion was guided by my previous general histories; reflect the ubiquity of mathematics as much as possible. It would have been all too easy to choose a long sequence of undoubtedly important monographs in mathematical analysis, algebra, and mechanics, as well as some other core branches of mathematics, but the rainbow would have been reduced to a very few shades. The last criterion was to embody subsequences of major writings in the major branches, so that the history contained subhistories.

1. {However, the series was stopped around 2000 after seven volumes, with one volume published outside the series and the other two abandoned.}

The outcome is summarized in table 5.1, which is organized by the principal subhistories; while no full references are given there, the writings in question should be easy to identify.

In the end 99 writings were covered in 77 articles, which were published in chronological order of the original first appearances. There were 62 authors of 16 nationalities. In a few cases a portrait of an author was included and the title page of a writing if it includes a nice design, say, or carries an interesting motto. In several cases we were able to take two or three writings in one article when the content and timing were both close, for example, A. L. Cauchy's two textbooks on mathematical analysis of the early 1820s. This case is also an exception from another point of view; all other writings were aimed at research, and a sequence of landmark writings in mathematics *education* would be a different one. Cauchy's books were chosen because they contained much new mathematics—to the distress of his colleagues and students (§13.7.3)!

Each article began with a bibliography of the writing(s), including all the reprints, editions, and translations that we could find. Then the article described the background, the content, and the influence—and the last proved to be the hardest to capture. For those writings of some length (and several were multivolume), a summary table of contents was added. One of the nicest parts of my task as editor was to add cross-references to other articles when appropriate, and thereby see the book binding itself together. No author had written a piece of precisely this kind before, and all seemed to enjoy the experience; and the reception of the book has been positive. It would be good to see general histories of this kind for other sciences.

4. Concluding Remark

A main role of general histories is to present the broader framework within which the history of the mathematical sciences has developed in all its parts, applications, and rainbow colors. A general history should be really general, unless it is oriented toward specified educational needs. Maybe one day mathematics will be granted a place in modern cultural life commensurate with its vastly varied and ubiquitous roles in the past and present. However, it is not obvious that this dream can become a reality. The resistance to historical knowledge, based on ignorance rather than exposure, has a long and influential history of its own; change is only piecemeal and gradual, though monotonically increasing for the last three decades.

TABLE 5.1
Writings grouped by principal branches of mathematics

Geometries	Algebras
1649 Descartes, *Geometria*	1649 Descartes, *Geometria*
1744 Euler on curves	1826 Abel on the quintic equation
1748 Euler, *Introductio* to analysis	1844 Grassmann, *Ausdehnungslehre*
1795 Monge, *Géométrie descriptive*	1853 Hamilton, *Lectures on Quaternions*
1822 Poncelet on projective geometry	1854 Boole, *Laws of Thought*
1844 Grassmann, *Ausdehnungslehre*	1863 Dirichlet, *Vorlesungen über Zahlentheorie*
1847 von Staudt, *Geometrie der Lage*	1872 Klein, Erlangen programme
1867 Riemann on geometries	1895–1896 Weber, *Lehrbuch der Algebra*
1872 Klein, Erlangen programme	1897 Hilbert on algebraic number fields
1899 Hilbert, *Grundlagen der Geometrie*	1930–1931 van der Waerden, *Moderne Algebra*
1905–1934 Enriques and Chisini on algebraic geometry	
Calculus	**Number theory**
1684–1693 Leibniz, first papers on the calculus	1801 Gauss, *Disquisitiones arithmeticae*
1734 Berkeley, *The Analyst*	1863 Dirichlet, *Vorlesungen über Zahlentheorie*
1742 MacLaurin, *Treatise on Fluxions*	1897 Hilbert on algebraic number fields
1744 Euler on curves	1919–1923 Dickson, *Number Theory*
1755 Euler, *Differentialis*	**Real and complex analysis**
1797 Lagrange, *Fonctions analytiques*	1823 Cauchy, *Résumé* of the calculus
1797–1800 Lacroix, *Traité du calcul*	1825, 1827 Cauchy, two main writings on complex analysis
Functions, series, differential equations	1851 Riemann on complex analysis
1655 Wallis, *Arithmetica infinitorum*	1867 Riemann on trigonometric series
1748 Euler, *Introductio* to analysis	1904 Lebesgue, *Intégration*
1797 Lagrange, *Fonctions analytiques*	1932 Bochner on Fourier integrals
1797–1800 Lacroix, *Traité du calcul*	**Set theory, foundations**
1799–1805 Laplace, *Mécanique céleste*	1872 Dedekind, *Stetigkeit und Irrationalzahlen*
1821 Cauchy, *Cours d'analyse*	1883 Cantor, *Grundlagen* of set theory
1822 Fourier on heat diffusion	1888 Dedekind, *Was sind Zahlen?*
1829 Jacobi, *Functionum ellipticarum*	1889 Peano on axioms for arithmetic
1905 Lebesgue on trigonometric series, and Baire on discontinuous functions	1910–1913 Whitehead and Russell, *Principia mathematica*
1932 Bochner on Fourier integrals	1931 Gödel's incompletability theorem
General mechanics	1934, 1939 Hilbert and Bernays, *Grundlagen der Mathematik*
1687 Newton, *Principia*	**History, general**
1743 d'Alembert, *Dynamique*	1799–1802 Montucla, *Histoire des mathématiques*
1788 Lagrange, *Méchanique analitique*	1892 Rouse Ball, *Mathematical Recreations*
1867 Thomson and Tait, *Treatise on Natural Philosophy*	1901 Hilbert, paper on mathematical problems
1894 Hertz, *Prinzipien der Mechanik*	

(*continued*)

TABLE 5.1 continued

Astronomy	**Dynamics**
1687 Newton, *Principia*	1673 Huygens, *Horologium*
1788 Lagrange, *Méchanique analitique*	1738 Daniel Bernoulli, *Hydrodynamica*
1799–1805 Laplace, *Mécanique céleste*	1889 Poincaré on the three-body problem
1809 Gauss, *Theoria motus*	1893 Lyapunov, *Stability Theory*
1889 Poincaré on the three-body problem	1927 Birkhoff, *Dynamical Systems*
Probability and statistics	**Mathematical physics**
1715 James Bernoulli, *Ars conjectandi*	1822 Fourier on heat diffusion
1718 De Moivre, *Doctrine of Chances*	1828 Green, *Electricity and Magnetism*
1763 Bayes on probability theory	1844 Grassmann, *Ausdehnungslehre*
1809 Gauss, *Theoria motus*	1873 Maxwell, *Electricity and Magnetism*
1812–1814 Laplace, *Probabilités*	1877–1878 Rayleigh, *Theory of Sound*
1854 Boole, *Laws of Thought*	1892 Heaviside, *Electrical Theory*
1900 Pearson on the chi-squared test	1904 Thomson, *Baltimore Lectures*
1925 Fisher, *Statistical Methods*	1909 Lorentz on electrons
1931 Shewhart, *Economic Quality Control*	1916 Einstein on general relativity theory
1933 Kolmogorov on the foundations of probability theory	1930 Dirac, *Quantum Mechanics*
	1932 von Neumann, *Quantenmechanik*
Topology	**Social and life sciences**
1889 Poincaré on the three-body problem	1871 Jevons, *Theory of Political Economy*
1923–1926 Urysohn and Brouwer on dimensions	1917 Wentworth Thompson, *On Growth And Form*
1934 Seifert and Threlfall, *Topologie*	1931 Volterra on mathematical biology
1935 Alexandroff and Hopf, *Topologie*	

CHAPTER SIX

Too Mathematical for Historians, Too Historical for Mathematicians

To the fond memory of Morris Kline (1908–1992)

1. Not for the Mathematicians

In 1992 the British Society for the History of Mathematics (BSHM) marked its twenty-first year of existence with a special issue (21, indeed) of its newsletter. It contained an introduction by me (president from 1985 to 1988), followed by a list of the 90-odd meetings held and 450 talks given during those years. The continued occurrence of such activity seems to speak well for interest in the subject, but the history of mathematics is a classic example of a ghetto subject: too mathematical for historians and too historical for mathematicians.[1] For example, in Britain there is virtually no professional basis for the subject outside the society itself and the recently constituted research group at the Open University. I explore this regrettable situation, using Britain as the main (but not the only) example.

"Historians are only amateurs," a senior British mathematician confided to foreign historians of mathematics in 1991. "They cannot read a modern mathematics

This chapter is a merger of parts of "A Residual Category: Some Reflections on the History of Mathematics and Its Status," *Mathematical Intelligencer, 15* (1993), no. 4, 4–6, and "Does History of Science Treat of the History of Science? The Case of Mathematics," *History of Science, 28* (1990), 149–158. Reprinted by permission of Springer Science and Science History Publications respectively. Compare also "Cottage Industry or Ghetto? The British Society for the History of Mathematics," *Annals of Science, 50* (1993), 483–490.

1. In a further misfortune, history of mathematics is both too historical and too mathematical for modern professionalized philosophers, despite the considerable influence that mathematics has borne on philosophy (many of the great philosophers were mathematicians) and still brings to many of its modern concerns (especially in the [mis]use of logical systems and theories). {Philosophers tend to carry out their own version of heritage on past figures (including philosophers), highlighting features consistent with their own stance, and playing down or ignoring the rest.}

But there has been some increase in interest in the history of philosophy recently (a society for the subject has been formed in Britain), and in response to an evident gap in the literature, I launched in 1980 the journal *History and Philosophy of Logic,* a topic that was too logical and philosophical for historians, and too historical for logicians and philosophers.

book, whereas any mathematician can read a history." While about 80 current members of the BSHM are professional mathematicians, such a view has been and is, I suspect, pretty normal in the mathematical community worldwide. But the idea of a typical professional mathematician appraising, say, the various Greek and Arabic editions of Hero of Alexandria is perhaps a slightly optimistic vision of that mathematician's historical capacities.

In a popular lecture given by a leading mathematician recently, the public was "informed" that partial differential equations were introduced into mathematics because of electrodynamics; and that complex numbers, dating from the sixteenth century, became important in this one because of certain needs in quantum mechanics. One would not expect a leading musicologist to inform us that symphonies were introduced into music because of Johannes Brahms, or that operas became important because Richard Strauss wrote some: but in the history of mathematics "anything goes," especially standards (figure 6.1).

In Britain such attitudes must have underlain the *need* for the BSHM to be founded in the first place. Only a year or two before its founding, in 1969, Sir Edward Collingwood had been elected president of the London Mathematical Society and had hoped to introduce some historical lectures and meetings into its schedules; but the rebuff from his council was resounding.[2] As far as I know, the other mathematical societies in Britain have rarely mooted such an interest in the first place.

An important manifestation of this attitude concerns the neglect of manuscript sources connected with mathematics. Every sketch scrap of, say, van Gogh will be rightly cherished; but entire Nachlässe of mathematicians will be thrown away for lack of interest, for example, most of Cauchy's in 1937 when the Paris Academy of Sciences refused to accept it from the family.[3] When the mathematicians take such attitudes, they cannot be surprised (and perhaps not be bothered) when the academic community despises their efforts at archival conservation "as a peripheral item with peripheral interest," to quote the director of Humanities Research Center at Austin (Texas) in 1983 on the transfer there of files of mathematics departments of U.S. universities (Kolata 1983).[4]

2. I draw here on personal information and discussions with Collingwood (who died suddenly in 1970) as well as with others at the time. In recent years the London Mathematical Society has been extremely generous in its financial support of BSHM meetings {and has organized the occasional historical meeting}.

3. Information kindly supplied by Bruno Belhoste (Paris). The library of the Institute of France, located about 100 yards from the academy offices, holds the Nachlässe of many important French figures, including scientists and mathematicians.

4. {In recent years the situation for the Archives of American Mathematics has much improved thanks to the Center for American History at the University of Texas at Austin and to the Educational Advancement Foundation.}

Abstract. Analyzing Kepler's law in two dimensions, Newton discovered an astonishingly modern topological proof of the transcendence of Abelian integrals. Newton's theorem was not really understood by mathematicians at that time, since it was based on the topology of Riemann surfaces . . .

* * *

The series of Special Articles was created to provide a place for articles on mathematical subjects of interest to the general membership of the Society. The Editorial Committee of the *Notices* is especially interested in the quality of exposition and intends to maintain the highest standards in order to assure that the Special Articles will be accessible to mathematicians in all fields. The articles must be interesting and mathematically sound. They are first refereed for accuracy and (if approved) accepted or rejected on the basis of the breadth of their appeal to the general mathematical public.

Items for this series are solicited and, if accepted, will be paid for at the rate of $250 per page up to a maximum of $750 . . .

Figure 6.1. Money for jam. Opening of the abstract, and attached editorial note, of {Arnol'd and Vasil'ev 1989}

The following, if true, would make a fair analogy with many mathematicians: musicians have no idea whether Giuseppe Verdi lived before, during, or after the time of Richard Wagner; have never heard a note of either composer; and have no interest in finding out. Such a scenario may accurately reflect some of the more lunatic ends of postserialist musicians, who listen maybe as far back as the early tapes of Pierre Boulez; but in this analogy the characterization is *normal* (in Thomas Kuhn's sense) for mathematicians.[5] The extent of this ignorance can be ludicrous: for example, various definitions and theorems are named after mathematicians (often the wrong ones), and textbook writers sometimes spell the names incorrectly anyway ("Caley-Hamilton theorem," "Lebesque's theory of integration," and other sics).

Further, mathematicians usually view history as the record of a "royal road to me"—that is, an account of how a particular modern theory arose out of older

5. If this sounds to be an extreme construal, let me recall an occasion in 1970 when I was allowed to examine the (largely disordered) archives of the Institut Mittag-Leffler, a mathematics research institute in the suburbs of Sweden. In the course of finding materials relating to my own concerns, I came across mounds of completely unknown manuscripts for a major figure of the nineteenth century, and smaller collections for several other such figures (Grattan-Guinness 1971). "Look what I have found now," I said to the director after one exploration, arms literally loaded down with bundles. "Oh really," he replied, "I could not care less."

By coincidence, just at that time the discovery of some rather trivial materials of Jane Austen was greatly exciting the literary and historical world.

theories instead of an account of those older theories in their own right. In other words, they confound the question "How did we get here?" with the different question "What happened in the past?" (ch. 2).

2. Not for the Educators

There is a similar resistance among mathematics educators to draw on history as a basic source, even though it is rich beyond belief for their purposes. I advocate history to be used in education principally through history-satire, where the mathematics of the past is used as a bank of results and methods for some modern educational purpose: the general historical record is respected (for example, changes in practice over decades), but the nuances of historiography are usually set aside (compare ch. 9). Similar techniques could obtain for education in other sciences.

Failure to note history can lead to unfortunate consequences. For example, the disaster of the "new maths" of the 1960s exchanged one set of shibboleths for an even less efficacious set (to use a noun associable with one of its worst features). Even the chosen *name* of the movement manifested historical ignorance; for most of the mathematics involved was then already 80 years old! "New maths" to whom? Is Bernard Shaw new drama? This development is a perfect example of an educational policy undertaken without any serious attention paid to the history of the mathematics involved—not as a means of providing material for educational strategies but as a source for pondering the suitability of the curricula in the first place.

A prominent place is given by the membership of the BSHM to educational questions, thereby suggesting that history of mathematics is gaining a foothold in the community of mathematics education (compare §2.13). Further, the founding of the society helped to launch in 1972 a study group on the "relations between the history and pedagogy of mathematics" at the second International Congress on Mathematics Education, due to take place the following August in Exeter. It still flourishes, publishing a widely circulated newsletter.

But such interests are exceptional rather than normal; in general, specialists in mathematical education are little more inclined to take history seriously than are professional mathematicians. As a result, a source of extraordinary educational richness goes a-begging: problems, motives, exercises, rigor at different levels, applications, interactions—everything that goes into learning real mathematics instead of fulfilling educational mores. Of the various associations of mathematics teachers in Britain, contacts for the BSHM with the Mathematical Association have been the strongest.

3. Elsewhere

Some comparisons with other countries would be instructive. Countries showing stronger representation, such as the United States, Italy, France, and Germany, do not have national societies for the subject, probably because it is integrated in a more solid way among the mathematicians and mathematics educators (and the historians of science to some extent). For example, a considerable number of people and some *équipes* work on the subject in Paris, where it is published in the *Cahiers du séminaire d'histoire des mathématiques;* meetings on it take place from time to time at the mathematics conference center at Luminy (Marseille). The corresponding venue in the Federal Republic of Germany, at Oberwolfach, reserves a week for it about once every two years, interspersed recently with some small-group workshops;[6] and the German Mathematicians' Union now has a history section, partly because of the absorption of the corresponding society of the former German Democratic Republic, where the subject was very well developed (but is now much destroyed). {The two Gruppen are now merged.} The change of the Soviet Union into Russia does not seem to have had comparable effects on the substantial work done especially in Moscow and Leningrad; for example, the journal *Istoriko-matematicheskogo issledovaniya*, started in 1948, is to continue.

Italy is very active, with many meetings taking place. The Italian Mathematical Society has sponsored since 1980 a research journal, the *Bollettino di storia delle scienze mathematiche*. Publications now include an excellent new bimonthly magazine on mathematics and its history, *Lettera Pristem* (the latter word is the acronym of a study group), partly modeled on the *Intelligencer*. The history of mathematics also appears occasionally in the similar publication *Informazione filosofiche*.

In the United States special sessions take place regularly at meetings of the American Mathematical Society and the Mathematical Association of America, and are well attended; and some years ago the former body launched a book series (now co-sponsored by the London Mathematical Society). Indeed, a century ago the country pioneered the use of history in mathematics education. In Mexico a group publishes the journal *Mathesis*, which is especially concerned with the history of foundational questions.

6. {Sadly, this practice has now ceased.}

4. Diagnosis

Nevertheless, among the world community of mathematicians, tens of thousands strong,[7] history is usually surplus to requirements. There seem to be several reasons for this normal distaste, which apply to educators also.

One reason is the mistaken inference that, since mathematics shows a continuity of many major concerns over the centuries (for example, with integers, lines, curves, equations), then it is a cumulative subject, and so only the modern top layer need be learned and the mistakes set aside. However, understanding of mathematical notions has enjoyed great changes—"revolutions," even (ch. 8)—in addition to increases and innovations in content; indeed, quite profound unfamiliarity of history is required to produce ignorance of such features.

A second reason follows from this one; that since continuity appears to obtain, the purpose of history to mathematicians (apart from anecdotes and so on) is to inject the modern versions onto the old texts in order to find out what the historical figures were *really trying* to do (see ch. 2 on history and heritage). For example, some things in Lagrange in the late eighteenth century look like group theory; hence Lagrange had to be trying to create group theory. Simple-minded determinism is unavoidable (it happened, therefore it had to happen); the illusion of accumulation is reinforced; the intentions of our predecessors are travestied; and the efforts of historians to reconstruct them are patronized. Of course modernizing old ideas is a perfectly legitimate process, even a good way to do research, but identification with the history of those ideas is a profound mistake.

Another reason concerns quality of problem; as several mathematicians have told me, deep reading of great predecessors has revealed the humdrum nature of their own contributions. Thus *history becomes subversive, at one remove.* However, a misunderstanding of the past arises; composers are not discouraged by exposure to Wolfgang Amadeus Mozart, and mathematicians would do better to recognize the past as an *active* source of inspiration, in various practical and even technical respects as well as by example.

The lack of awareness of history among mathematicians is what leads to the interesting mistake of calling it "mathematical history," which is a quite different subject (§2.1); in fact, it is part of applied mathematics, that is, history within which (maybe

7. A recent *Combined Membership List* of the American Mathematical Society, the Mathematical Association of America, and the Society for Industrial and Applied Mathematics (Providence, 1987), covers not only the United States but also (through joint memberships of other societies) a number of mathematicians for other countries. The book lists around 43,000 persons, with the majority of them based in institutions of higher education.

modern) mathematics is used as a research tool. Its topics include statistical methods in social history, cliometrics and prosopography, and combinatorial techniques of compensating for fragmentary information (Fogel 1982). Thus, for example, if some (ancient or modern) mathematical theories are applied to a piece of the history of mathematics, then one will have mathematical History of Mathematics.

All these factors apply in other sciences also as well as other countries. For example, scientists frequently call the history of science "scientific history," which again is quite different, being a part of applied science (and of history also, of course). Let us turn to historians of science.

5. Not for the Historians

One factor unique to the history of mathematics is the degree of abhorrence of it evident among historians. By "History of Science" I shall refer in the discussion below to the (partly) professionalized activity of today; the "history of science" means the record of what happened in the past in science. The thesis is that mathematics has played a major role, at times a dominating one, in the history of science; but by contrast, it is normally ignored in History of Science.[8]

Regarding relations with historians of science, I think that the BSHM's first contact with the British Society for the History of Science was attempted in 1982, when I was a member of the BSHM council. I brought along a proposal for a joint meeting, which was received with such shock that it was passed over at once (it does not appear in the minutes of that council meeting). A joint gathering did take place eventually, in 1987, on the history of British mathematical education, but the cooperation was approved in an "informal" manner.

Here are three other examples of the ignoring of mathematics in History of Science, of rather different kinds: they are chosen as being typical, not extreme. The first example is the joint meeting in 1988 of the British Society for the History of Science and the History of Science Society (USA). There were about 90 lectures, of which only two were mathematical in content.

The second example also relates to the American society: the volume on the origins of American sciences published to inaugurate the second series of the journal *Osiris* (Kohlstedt and Rossiter 1985). All the main sciences were covered—except,

8. As an example of the relative development of history of mathematics relative to histories of accepted sciences, comparison of the section of abstracts in *Historia mathematica* with the quarterly *Current Work in the History of Medicine* suggests that work in the latter area is the larger by between 15 and 20 times.

of course, mathematics, despite its presence in the history of American sciences and its interesting features for historians of science (a rather late development, from the 1890s on, various strong and different influences from German mathematics, the role of a company [Bell Telephones] in the rise of applied mathematics, what more do you want, etc.).

The last example is an article on marital collaborations in science (Ogilvie 1987). This very nice piece takes examples out of the histories of physics and the life sciences, but not even a passing mention is made of W. H. and G. C. Young, the first substantial collaboration in the history of mathematics, even though they form an unusual case (she drew him into research, whereupon he revealed the greater creative talent); there is a mass of manuscript material available in the Archives of Liverpool University concerning the manner in which the collaboration was effected (even the collaboration between man and woman was explicitly discussed), and secondary literature is available on their lives (Grattan-Guinness 1972).

As for professional historians in general, contact seems possible at all mainly for ancient and (a bit of) medieval and Renaissance mathematics, where the mathematics as such may be elementary (although the history is immensely difficult). For the rest, a historian would recognize from (ironically, mathematical) words like "1687" and "1821" that history is at hand; but the degree of his resistance to becoming acquainted with its content has to be experienced to be believed.

Of course skews and imbalances can be found all over History of Science: some topics have been well remembered, others are patchily handled, and many very neglected, so that the overall picture of the history of science furnished by History of Science is rather distorted. But the case of mathematics is substantially different in degree from those obtaining in (parts of) the other sciences, to a measure that deserves especial attention. Indeed, a difference of *kind* will be suggested, with educational origins.

I find the ignoring of mathematics especially strange in History of Physics: it is one of the strongest areas of History of Science, and yet most of its many talented practitioners will consider the pertaining mathematics only as far as, say, quoting a formula *without* regarding it as an object of historical attention as worthy as a physical law or as an experimental design. Only slightly less puzzling is the situation in History of Technology, where the pertinent mathematics is barely studied at all: in fact, this is the most neglected part of History of Mathematics.

When concern is taken, terms like "mathematicization" are thrown around occasionally by some historians of science, but at least this reader is usually quite uncertain which of dozens *markedly different* intellectual processes are under consideration, and how the historians will relate the process applying in their given case to the *variety* of mathematical issues with which they are concerned. Algebraizing mechanics, interpreting the passage of light in terms of non-Euclidean geometry, analyzing the flow of water in a canal, marking the daily passages of the Moon on an ox horn with a sharp instrument, calculating the path of a comet from observations, bringing statistical tests of significance into medicine, developing a decent theory of capillary flow for use in aortic regorgitation, analyzing the flow of traffic in a large city, finding properties of special functions in both real and complex variables for use in astronomy, studying different formulations of recursion for computing, seeking predator/prey models for biomathematics . . . These and numerous other matters fall quite legitimately under the rubric of "mathematicization," and between them they range, and have ranged, *across* much of the spectrum of scientific thought. Calling them all "analytical formulas," to quote favored *bavardage* that I have seen several times, does not help at all.

It is self-delusive, therefore, to treat mathematicization as if it were a single intellectual point, or to regard the assignment of mathematical (or non- or anti-mathematical) attitudes in historical figures as any more than the most *preliminary* appraisal. In the *history* of science mathematicization can straddle the rainbow of methods *used* in science and indeed has helped to form many of them in the first place.

6. Diagnosis

Why, then, should an obviously unsatisfactory state of affairs last for so long, and why should historians of science find it a matter of so little concern? A significant factor is mathsphobia among historians; and an interesting convolution of processes needs to be explained here. To work backwards:

1. Historians of science, like most of the population, do not like mathematics, or at least find nothing particularly interesting or appealing in it. Note as evidence the different social attitudes towards illiteracy and innumeracy: the former causes deep embarrassment, but the latter is merely a minor nuisance concerned with not being able to add up properly.
2. This distaste for mathematics comes principally from experience of the subject at school, where it appeared to be boring and difficult (and may well have been taught by a non-too-competent person).
3. This situation has lasted for a very long time, and so is itself subject to historical study. Unfortunately the history of school mathematics is not a well-pursued branch of History of Mathematics, but it seems clear that

misguided philosophies about rigor, certainty, and the primacy of foundations led to corrigible decisions and policies about the choice of material to be taught and the stages at which it should be introduced to students.[9]

4. These decisions on curriculum are guided significantly by ignorance among the educators and policy makers of the history of mathematics. More precisely, a certain distorted inheritance of knowledge is handed down—nonhistorically—and adopted without fundamental question.
5. Thus a vicious self-generating circle is initiated, with generations and communities of mathematicians and educators practicing their subject as if they had created it all themselves, and remaining completely ignorant of its historical and cultural background—or even that it has such a background in the first place.

7. Concluding Remarks

This last clause can be generalized without much exception. Every historian knows that there are imbalances and lacunae in almost all areas and regions of his discipline, but the case of mathematics is different in degree and, I suggest, even in kind. Within history, the history of ideas has been a very significant component. Within the history of ideas, science has often played a prominent and even dominating role, influencing (for better or worse) the development of other parts of human activity. Within science, the physical sciences have often been given preference (especially after the rise of physics among the sciences, which incidentally is another aspect of French science of the early nineteenth century). Within the physical sciences, mathematics has often been a leading realm—and a vast and varied realm it has been, with a wide spectrum of aims, methods, epistemologies, and techniques, frequently in competition with each other. So mathematics is somewhere near the summit of a summit of a summit of knowledge in its historical development.

By contrast, History of Mathematics is one of the least recognized or discussed branches of History of Science, which is itself still rather separated from the profession of history in general: thus it lies at the very outermost margins of the community of historians. Mathematics is as ancient, and as varied in its content and methods, as any other branch of knowledge and more so than most of them; it has

9. In a remarkable number of cases, research work in mathematics has been stimulated by a perceived educational need: the new theory appeared more in textbooks and lecture notes than in ordinary papers. But even here the educational policy could be highly questionable—the poor students were being served research-level mathematics as if it were appropriate for teaching at their level.

been extraordinarily pervasive in its influence, especially on science but also on most areas of human activity. As time advances, and recent periods become more historical, then the increasing presence of mathematics will render its neglect still more absurd. Historians (of science) should take it very seriously in many areas of their subject if they aspire to study the history (of science).

ACKNOWLEDGMENTS

This chapter draws on information and memories from colleague officers and other friends who have nurtured the BSHM through its 21 years. It makes use of my introduction to newsletter no. 21 (October 1992); I am grateful to the committee of the BSHM for permission to reproduce some passages here.

BIBLIOGRAPHY

Arnol'd, V. I., and Vasil'ev, V. A. 1989. "Newton's Principia read 300 years later," *Notices of the American Mathematical Society, 36*, 1148–1154.

Fogel, R. W. 1982. "'Scientific' history and traditional history," in *Logic, methodology and philosophy of science VI (Hannover 1979)*, Amsterdam: North Holland, 15–61.

Grattan-Guinness, I. 1971. "Materials for the history of mathematics in the Institut Mittag-Leffler," *Isis, 62*, 363–374.

Grattan-Guinness, I. 1972. "A mathematical union: William Henry and Grace Chisholm Young," *Annals of science, 29*, 105–186.

Kohlstedt, S. G., and Rossiter, M. W. 1985. (Eds.), *Historical writing on American science*, Philadelphia: University of Pennsylvania, Department of History and Sociology of Science.

Kolata, G. 1983. "Math archive in disarray," *Science, 210*, 940.

Ogilvie, M. B. 1987. "Marital collaboration: An approach to science," in P. G. Abir-Am and D. Outram (eds.), *Uneasy careers and intimate lives*, New Brunswick and London: Rutgers University Press, 104–125.

CHAPTER SEVEN

History of Science Journals

"To Be Useful, and to the Living"?

To the memory of Benjamin Britten (1913–1976)

As editor of the history of science journal Annals of Science, *I reflect here on the social utility of journals for a subject that attracts the interest of such a small proportion of humanity, in contrast with the popularity of (classical) music.*

1. Prelude

The death of Benjamin Britten on December 4, 1976, must have sent many people to think over the range of his marvelous achievements. It led me unexpectedly to notice a variety of points of comparison and contrast with the history of science. So I allowed myself the indulgence of discussing my views on the subject and my editorial work (though I occasionally refer to my research interests to illustrate more general points). In ¶2 I have also answered a number of questions that I have been asked about this journal.

The stimulus to write this essay came not from Britten's music but from the beautiful phrase that I have quoted in my title. It comes from an open letter to his fellow composer Michael Tippett, and here is the context: "What matters to us now is that people want to use our music. For that, as I see it, is our job—to be useful, and to the living" (Kemp 1965, 29).

The original version of this essay was first published in *Annals of science, 34* (1977), 193–202. Reprinted by permission of Taylor & Francis (www.informaworld.com).

This chapter looks at the history of science from a perspective based on my editorship of this journal, an improbable appointment explained in ¶2; I have omitted a few passages that give details of the functioning journal at that time.

Britten has long been a favorite composer. His death was lingering, so that on the day that it occurred, the BBC Radio broadcast a special memorial program that it had prepared. It included a broadcast talk by Britten, which included the phrase that I used in the title and quoted in the second paragraph. The instant that I heard it, this *entire* article came into mind, even the order and number of sections; the writing took two days. I have never had an experience on that scale before or since.

Britten's mission is abundantly expressed in his music. He had a facility for beautiful simple tunes in which perhaps only Richard Strauss is a rival in that century. I am amazed anew at the amount of Britten's music that I memorized on first hearing and have never forgotten. He probably surpassed any other composer of our time in his impact on the musical interests of ordinary people—especially the young, for whom he wrote several works. He did not shirk even from political pieces, especially in various vocal and choral works written shortly before the Second World War. But he also lost nothing to the intellectualists of the composing community in the structural interest of his works. For example, the *Prelude and Fugue for 18-Part String Orchestra* (1943) contains one of the most remarkable of all fugues and is very exciting to hear.

Naturally, there were also limitations. The concerns with cruelty and with pacifism are to me rather overdominating in his output, and I regret finding only rather pastel-colored love music in his operas (with the exception of a kind for *Death in Venice* [1973]). He also suffered, as do many other artists and not a few scientists, from his successes. The most conspicuous example is the *War Requiem* (1962), a setting of the requiem text for soprano, choir, and orchestra together with some of the antiwar poems of Wilfred Owen for baritone, tenor, and chamber group. The work was written for a great event, the opening of the new Coventry Cathedral, and for that reason may have been oversold before its first performance. I took part in the choir for some of the early London performances and well remember the extraordinary sense of occasion that attended them. Today some unevenness may be more apparent.[1] On the whole I prefer the *Cantata misericordium* (1963), far too little performed, a setting of the Good Samaritan parable that expresses the same kinds of sentiment in very similar musical style and with great concision.

But Britten's impact came not only from his compositions. His Aldeburgh Music Festival rapidly became a major annual event in music after its inception in 1948, and his English Opera Group was an important stimulus to the country's musical profession. He was a superlative executant musician-pianist, accompanist, and conductor, and not only of his own music. He was also a superb person to meet (as I had the good fortune to do a few times at Aldeburgh). A vibrancy of purposeful

1. With "Kleenex at the ready," Stravinsky thought that the *War Requiem* was "rather a soft bomb," using "an idiom derived in part from Boulanger-period Stravinsky" (Stravinsky and Craft 1966, 14). A comparison between Stravinsky and Britten is an excellent example of the contrast between innovative and traditional composers (a contrast that can be made between scientists, too: see also ch. 8). Stravinsky is still a towering master of twentieth-century music and has been much more influential than Britten on composers and professional musicians. But Britten may be closer to the public heart, and his death felt more deeply than was Stravinsky's, which occurred in 1971.

life came from him, together with a ready desire to receive ideas as well as dispense them, a remarkable combination of confidence and humility. Let me end this section by relating a story that I heard him tell at a sherry party in 1963.

As part of his commitment to the living, Britten disparaged the abstract intellectualism of his avant-garde contemporaries. In illustration of his views, he recalled that a few years earlier he had attended a festival of extremely modern music in Warsaw. A piece for soprano and orchestra was performed, in which the soloist was allowed to sing more or less any noises and tones that she liked. The audience sat silently through the performance, except at one point, where many laughed out loud. Noticing no particular humor in that passage, Britten inquired afterward of his Polish hosts about the cause of the laughter. They told him that, in the course of inventing the required random noises, the American soprano had accidentally but clearly sung the Polish word for "shit."

2. Some Editorial Practices on *Annals of Science*

Britten wished primarily to serve the living. Can the same be argued for the history of science, when so few of the living are interested in it? In particular, what are its journals for: for the living, or the ages, or both? Questions of this kind have often been on my mind during the time that I have been editing *Annals of Science*.

Naturally people were very surprised when the editorship of a senior journal in the history of science passed to someone who is not professionally engaged in the field. As I have been asked many times of the circumstances, I shall briefly set them down here.

The pressure of papers on the journal became so great that by 1973 it was clear that N. H. de V. Heathcote, who was bearing the brunt of the editorial work,[2] would need some assistance. In August he discussed with me the possibilities of my assisting him, and further meetings with other board members and with the publishers were planned. Unfortunately, various illnesses prevented their taking place, and late in the following May I was contacted directly by the publishers and asked if I was willing to join the board: the year's first issue was about to appear (the country's three-day working week had contributed to the delay), and they wished to know whether or not to print my name and address on the cover. I accepted the offer, and a few days later the publishers and I met to discuss my role in the future development of the journal. It became clear during our conversation that an early

2. However, Heathcote was not the titular editor; there was no such office until my appointment in 1974.

meeting with the editorial board was needed. The publishers telephoned Heathcote to arrange a suitable time and literally interrupted him typing out a letter in which he announced his retirement from the journal and proposed me to succeed him. The publishers accepted his suggestion; so, with one minute's warning, I became editor of the journal—indeed the first.

I instituted a new editorial system on the journal. The terms of reference, "the history of science and technology since the Renaissance" (as it then was—I changed it to "from the thirteenth century" with the 1976 volume), were so wide that they clearly needed dividing up into main areas, each one covered by an expert on the board. But how many areas should be created? I recalled the claim of group psychology that the sizes of the most effective groups are around 4, 12, and 40 members (and certain higher numbers). Twelve seemed to be the most suitable choice and has in fact been precisely the number that we have used since the 1976 volume, when a specialist in medieval science was added to the team.

Every paper submitted goes to the members of the board within whose areas it falls, and each member is free to call in referees to assist him if he wishes. (A referee might be another member of the board who has an interest in the topic of the paper, although it does not fall within his areas of responsibility.) The bulk of the discussion of revisions of a paper normally takes place directly between board member and author, without my involvement; for I wanted a system that would avoid the following possibility. I once submitted a paper to a history of science journal and received from the editor some excellent referees' reports. I rewrote the paper, but I had to tell the editor that to me one criticism was wrong both historically and mathematically. "Oh dear, I don't know any maths!" was the editor's response, and caused the kind of situation that I wanted to avoid as editor of *Annals of Science*: adjudicating on technical matters about which I know nothing.

I have also been asked why an "international" review has most of its board members based in the United Kingdom. I realized that the type of editorial scrutiny described above would involve a lot of contact between board members, so that it would be a great advantage to have most of them within relatively easy reach and personally familiar with each other. If board members were scattered over many different countries, then consultation would either be impossibly slow if manuscripts were sent by sea, or impossibly expensive if sent by air. However, we use referees from all parts of the world.

Mention of expense raises another matter now prominent in each editor's mind: the financial state of his journal. It is curious that so many of the new history of science journals started in the United Kingdom, where the organization of the subject is fragmentary (and was even more so in the 1960s, when they began). Readers of books and journals are skeptical about the prices asked nowadays, and while their doubts may in some cases be justified, the amount of work involved in producing publications in this field is much greater than is realized.

On this journal, for example, three sizes of type are used for the main articles: 11 point on 12 point for the main text, 10 point solid for tables, captions to figures and indented quotations, and 8 point on 9 point for footnotes. (For articles like this one in special departments, and for book reviews, only the latter two sizes are used.) For titles, 10 and 12 point boldface are used. The first proof, which is seen only by the printers, has the text printed in these various type sizes consecutively. The compositor then makes up pages, inserting the 10-point quotations into the 11-point narrative, placing the footnotes in their (correct) position at the feet of the pages, and keying in tables and figures in the places indicated by me on the manuscript. With the kind of material that we are publishing, this can be an extremely complicated process—and therefore time consuming and expensive, especially if hand setting of special symbols is also required. Authors receive these paged galleys for correction and return to me, and they appear in due course on the sheet proofs of the issue. After I have checked these, the pages are made up into signatures and the issue run off and bound with the cover.[3] All that work, and for only several hundred copies.

3. Orthodoxies and Fringes in the History of Science

The increasing work in recent years has brought with it a degree of professionalization to the subject, with departments or divisions being set up in some universities (especially in the United States). Along with the many undoubted benefits to scholarship and education have come some less welcome features. The progress has been very uneven among the different sciences: physics, chemistry, astronomy up to Newton, and parts of the earth and life sciences have become orthodox fields, while the other parts are either relatively neglected fringes or somewhat separate activities (or both). Especially conspicuous to me is the almost entire absence of the history of mathematics from the history of science as it is practiced (ch. 6). The histories of medicine and technology are also rather separate, and the latter is particularly underdeveloped (despite the efforts led especially by the excellent *Technology and Culture*) with regard to the modern developments that affected all our lives so fun-

3. I have been asked if the new cover design has any "significance." It is based on the golden section (also known as "the medial ratio"), which has manifested itself in both the arts and the sciences since antiquity. The point of intersection of the four quadrants on the cover divides the cross-diagonal medially, giving both horizontal and vertical medial division. The type-page size is 8 inches by 5 inches, so that medial ratio appears there also, via consecutive Fibonacci numbers.

damentally. For example, most of the files and records of commercial and industrial organizations, where so many of the technical inventions were produced, are still unexamined by historians of science.

Another imbalance that disturbs me is the excess emphasis on the achievements of the undoubtedly Very Great Men. Sometimes the impression is almost that only Euclid, Aristotle, Ptolemy, da Vinci, Harvey, Copernicus (the number of celebratory volumes around 1973 for him was ridiculous), Kepler, Galileo, Newton, Leibniz, Darwin, Einstein, and a few others had any brains. There are dozens of significant figures in my own tiny areas of interest alone on whom very little work has been done. In this respect it will be good if journals are not useful to the living, if it means accentuating these imbalances and continuing to neglect the very many other sides of the subject.

4. What Is the Place of the Social Aspects in the History of Science?

One of the most noticeable features of recent developments is the interest now afforded to the social history of science, covering everything from the social conditioning of Great Men through the history of scientific education and institutions to the diffusion and popularization of science to the general public. The rise of such an interest was long overdue and has brought freedom from the myopic concentration on the history of the structure and testing of theories. Indeed, one might argue that the social aspects of a theory are of major significance even for the intellectual history of science: for at its first appearance a theory may well be associated with all kinds of rivalry, intrigue, rejection of contemporary wisdom, and so on, all social components that fade away only when the protagonists are dead or have settled on some consensus the view about the theory, and its intellectual structure can be given greater attention. However, I have some reservations about certain of the current tendencies in sociological studies.

First, the social history of science is handicapped by the fact that our sociological models are still exceedingly crude: for example, it is not even clear what sort of mathematics, if any, should be used (compare §6.4 on mathematical history).[4]

4. To summarize very crudely a complex development, the tradition was to draw on the mathematics and physics, then the queens of the sciences. Hence, it was suggested, there were social "forces" pressing on the "inertia" of the "masses" in society, and so on. Now that the biological and medical sciences are taking over the throne, sentiment will surely move toward imitating "biomathematics." This can already be seen in catastrophe theory, although its inventor, René Thom, has expressed in lectures (and in conversation with me) the need for extreme caution in evaluating its possibilities. {See now also the next footnote.}

I agree entirely with Smith (1977, 52) that science affects society much more than society affects science and regret the absence of technical details of the science(s) involved from so many sociological analyses. I doubt that even the most doctrinaire social reductionist or behaviorist would assert that different areas of science are different only because they are associated with the different societies, journals, and lecture courses; but often I cannot tell where the social aspects are held to stop and the intellectual differences between the sciences are allowed to begin. For undoubtedly many kinds of social aspects are *strictly dependent* on some initial steps of an intellectual, nonsocial, character, although they may well affect further intellectual progress very substantially.

Second, and related to the above criticism, some social analyses are inadequate in logic. I refer especially to psychosociological studies that try to show how social influences on an historical figure determined his scientific interests. If proposition p expresses these influences and proposition q the interests, then these studies claim the logical structure

$$p; \text{ if } p, \text{ then } q; \text{ hence } q.$$

But I often find that only "p and q" is argued, from which "p implies q" cannot be inferred. It is shown that the historical figure had a certain background and developed his scientific interests, but it is claimed that the background led to the interests. Such a thesis may be true, but this is no proof of it.

Finally, the impression sometimes comes over that the social history of science is a Unique Field, an amalgam of social history and history of science not to be found anywhere else in the spectrum of historical work. But such kinds of history have been developed for a long time in the history of the creative arts, to take only one example. One can study, say, Britten's *Peter Grimes* (1945) as a work of art with a certain operatic structure, and as a major factor in the renaissance of British music after the last war, just as one might study the "internal" and "external" aspects of a scientific theory. (In fact, I think that analyzing *Peter Grimes* in this way is even more difficult than examining a scientific theory, for it involves aesthetic evaluation, about which so little is still understood.) It was very regrettable that all those dreary papers on the internalist/externalist issue were written, for they masked the completely obvious fact that in almost any historical study both components are necessary, and the "debate" thus suppressed the real historiographical question of how to use the two components simultaneously. To this question I now turn.

5. The Historiographies of Decision, Influence, and Originality

As an addition and perhaps alternative to some types of sociological study, I would advocate a focus of attention on three factors from which the intellectual and social aspects of science can be studied together, as they should be. These factors are decision, influence, and originality, and I shall treat them in turn.

5.1. Decision and Discovery

One of the philosophical prejudices of scientists, and one that philosophers and historians of science usually share, is that scientists discover things—substances, properties, laws, and so on. Now the word "discover" must be used carefully, for all that one can discover is what actually is, namely, the ontologically true. But scientists can never show that they achieve such success, even if they do; and of course they can never discover the false. The bulk of their theory-creating activity is concerned with decisions about how things are, and how these decisions can be backed up or overthrown by logical or experimental analysis. For example, Copernicus decided that the Earth moves around the Sun, just as Ptolemy had earlier decided to the contrary. Copernicus may also have discovered that the Earth moves around the Sun (though a rather sophisticated analysis of referential frames is needed to clarify the meaning of "moves around"); but the bulk of the ensuing discussion and controversy centered on his decision, even though the discovery aspect was often prominent (its bearing on Christian doctrine, for example). Now the idea of decision contains the intellectual and social aspects of science at once; the decision itself is an intellectual affair, but its taking and the later discussion are social matters, very possibly conditioned by all kinds of social factor bearing on the participants. There is also the decision of a scientist to study a particular topic in the first place, which is subject to the same kind of analysis.

A further feature of the historiography of decision needs emphasis. Although the historian may see science as primarily a decision-making activity, scientists themselves usually think of it in terms of discoveries, a view that historians must bear carefully in mind even if philosophically they do not support it. The same feature applies in philosophy and its history, for philosophers have also often laid great emphasis on scientific discovery. Even Popper's *The Logic of Scientific Discovery* (1959), which contains many criticisms of established philosophies of science, contains the word "discovery" in its title, although much of its contents are more intelligible (whether they are accepted is not at issue here) if read as a treatise on the logic of scientific decision.

5.2. *Influence*

Influence is another factor where intellectual and social aspects are intermingled. It is an immensely important branch of historiography and is of great complexity. Consider first an individual historical figure. Influences may come to him from his personal discoveries or his previous work, directly from other individuals or indirectly from them via intermediaries, from his educational training, from philosophical and technical traditions in his environment concerning the topic in which he is interested, and from the importation of practices from other environments; and any of these influences may be influences of acceptance or influences by reaction. Note that I have not used any specifically scientific terms in this paragraph, so that historians of science need not feel alone in analyzing influences; one could equally well be discussing influences on Britten, for example (a particularly complicated case, incidentally). Influences on a group contain all the above types for each of its members, together with inter-personal relationships within the group.

It might be possible to mathematicize the study of influence. Some time ago I tried to develop a mixed field and particle model of influence on a community of individuals, in which each member is represented as a particle in historical space-time, and the person's coordinates express a variety of geographical, philosophical, and talental factors; forces of attraction and repulsion between the individual and colleagues represent interpersonal influences; and field strengths of various kinds describe the social influences on the community. The model would be stated as a sequence of differential equations, whose solutions would describe the careers of each member over the period involved (compare Kendall 1975). Exact analytical solutions would probably be too hard to obtain, but for some classes of relatively simple differential equations there is now a well-developed knowledge of computer-programmable numerical approximations to solutions by eigenfunction expansion. The pitfalls of this model are partly that its differential equations would not be suitably simple, and especially that its variable and constant terms cannot be measured with sufficient accuracy. But the basic approach has some attractiveness.[5]

5. {Apart from recommending Kendall (1975), I now disown this proposed mathematical history, an attempt to emulate physical sciences in solving social problems such as construing influence; for example, in Grattan-Guinness (2007) I record the paltry success of emulating mechanics in neoclassical economics in the second half of the nineteenth century. To some significant extent the nonphysical sciences may need their own kinds of mathematics.}

5.3. Originality

Out of influence on the individual comes originality, another historiographical issue of great complexity. In both arts and sciences originality has been and is rated very highly, but as late as the Elizabethan period, *learning* was the human quality esteemed the most. Learning was associated with wisdom, the somewhat passive acquiring of past doctrines. One of the great changes during the Renaissance period was to throw still greater emphasis on *knowledge,* and the acquiring of new knowledge by original work: *ecce homo,* the standing human, thinking for himself (and about as many things as possible) rather than amassing wisdom while on his intellectual knees. So we find at this time the struggles for coexistence of astronomy with astrology, of chemistry with alchemy, of algebra with geometry, and their eventual autonomy of, and even triumph over, their elder fellow disciplines.

But during our century the move toward originality has become a dogma of its own, and humans are now often back on their knees again—certainly the extremely modern composer whose soloist accidentally amused the Warsaw audience, and the paper-scribbling scientist or scholar who measures his achievements by his page count. It is very noticeable that many original people have worked successfully within traditional frameworks; Britten is a fine example. Indeed, even the great innovators have often started out from a thorough knowledge and mastery of their inherited traditions (compare ch. 8 on convolutions). Stravinsky is a striking case, as was further confirmed by his recently published early *Piano Sonata* (1902), some of the best Tchaikovsky that Tchaikovsky never wrote. Similarly, the early Schönberg wrote in a musical style that he, the other great innovative composer of our century, was to replace by serialism.

I think that the history of the arts can provide historians of science with a wide variety of forms in which originality emerges out of influence-bearing traditions—both innovative originality and originality within those—and even the occasional example where influence seems to be very slight. Conversely, science has things to tell the arts about how learning and wisdom are and have been related to originality and knowledge. In both cases, as with everything else in this section, the distinction between internal and external aspects is itself of little importance (and that alleged to exist between the "two cultures" of arts and science is counterproductive).

6. Educational Aspects of the History of Science

In recent years the educational implications of the history of science have been of steadily greater interest to me, and I was glad to accept the suggestion of my board member Harold Sharlin that we have a teaching department in *Annals of Science*. The great benefit that history of science can bring to science education is in increasing the understanding of science instead of the current excessive emphasis on knowledge. It is not necessary to pack ordinary (or even extraordinary) students full of "facts"; if the fundamental motivations to, and structural features of, a theory are understood, then many of these details can be known simply by looking them up. (Hence, incidentally, students should be taught how to use indexes and libraries.)

Naturally, there are grave dangers inherent in such an approach. For one thing, if the students become bogged down in historical details, then they will never escape from old historical situations. The idea of history-satire is to secure the benefits of a historical approach while minimizing the dangers (§6.2, ch. 9). Here one would imitate in teaching the prime features of a historical development (both the theory and the experiments), without worrying overmuch about the particular minutiae of detail that happened to apply at the time. A cultural shock may indeed come to the students;[6] but better that it hits them then rather than when they are in research or in a position of responsibility—or, worst of all, never. I am sure that history of science journals can be very useful to both the living and the ages in encouraging articles on the educational implications of the subject.

7. Book Reviews

The purpose of the book review section of a history of science journal is of course to review books in the history of science. However, the diffuse nature of the subject means that it is not always clear what a history of science book is, or whether a (professional) historian of science should review it. I am convinced that one way of spreading interest in the subject is to use people outside it to review books, especially if the books belong to one of the many fringes or deal with relatively modern developments. Some reviews of this type that I have obtained from *Annals of Science* have been excellent; the (scientist-) reviewer has made technical criticisms at a level that a historian of science could not reasonably be expected to match.

6. This is an allusion to Brush (1974).

But poor reviews, whether by scientists or professional historians, may deter the potentially interested from the subject. When I began to run the book review section of *Annals of Science* in October 1974, I read the reviews published in recent years in several scientific weeklies and monthlies. I was disturbed to find that some of them were inane or even incompetent. I suspected as reasons that the reviews editors involved did not know whom to ask, and that they allowed the reviewers too few words and too little time to do justice to themselves or to the books. The importance of the latter reason was confirmed when I turned to the reviews in the history of science journals. As far as I could judge, the standard of reviews there was quite reasonable: the reviewer usually seemed to have tried conscientiously to cope with the book, even if the comments were not always very penetrating. And more than once I found a case where a reviewer had published a poor piece in a scientific magazine but a good one *on the same book* in a history of science journal. In between writing the two reviews, the reviewer had obviously had time to read the book properly. The magazines seem overanxious to serve the living in this respect, and the consequences are very unfortunate: for scientists are much more likely to read the reviews in large-circulation magazines than those in small-circulation journals.

This criticism naturally leads to the question of the value of having book reviews at all in history of science journals. Such reviews are only going to be of use to the living; and if few of the living see them, and they become outdated anyway by later publications in the field,[7] then why publish them? A fellow editor of a history of science journal once developed such a theme to me, concluding that while his journal would continue to publish book reviews, he personally was against them.

The case in favor of book reviews would point to the vastness of the subject and the numerous holes still left gaping in it and claim as a useful service that the living be told how well, in the reviewer's hopefully competent opinion, a particular book has defined and dealt with the task to which it was addressed. Further, the book review sections are probably the most widely read parts of history of science journals. The dilemma concerning usefulness to the living and the ages appears in a sharp form here, and I shall conclude by returning to it.

7. This last point is not necessarily applicable to reviews of reprints of old books (either scientific or historical works). It is common practice in history of science journals to have only brief notices of reprints, but I would not reject in principle a lengthy review of an old book, for the new generation of reviewers may have fresh and interesting things to say about it. There also ought to be film, play, and opera reviews when the subject matter is relevant.

8. Coda

The analogy between music and the history of science that I have been making throughout this article breaks down on an essential point: the extent of public interest. Britten could argue strongly for the usefulness of music to the living because music is an experience that many share; thus he would expect to receive a lot of reaction to his work.[8] By comparison, the history of science is an obscure activity (though big enough to make it hard to keep up with!), of significant interest only to a small percentage of the minority who are capable of understanding it. Thus it is impossible to obtain "feedback" in substantial quantity. Why bother with the living when so few of them are involved? Or is it those few to whom the editorial effort should primarily be directed? Perhaps compromise is the optimum solution: journals should attempt to achieve some balance between their duties to the living and to the ages. But the knowing attainment of such a balance is rendered difficult by the fact that one is oneself among the living, trying to judge the quality of one's own decisions. Britten's *Songs and Proverbs of William Blake* (1965) includes a setting of this proverb:

> The folies of the hour are measur'd by the clock;
> But of wisdom, no clock can measure.

BIBLIOGRAPHY

Britten, E. B. 1964. *On receiving the first Aspen Award*, London: Faber and Faber.

Brush, S. G. 1974. "Should the history of science be rated X?" *Science, 183*, 1164–1172.

Grattan-Guinness, I. 2007. "Equilibrium in mechanics and then in economics, 1860–1920: a good source for analogies?" in V. Mosini (ed.), *Equilibrium in economics: Scope and limits*, London: Routledge, 17–44.

Kemp, I. 1965. (Ed.), *Michael Tippett: A symposium on his sixtieth birthday*, London: Faber.

Kendall, D. G. 1975. "Review lecture: The recovery of structure from fragmentary information," *Philosophical transactions of the Royal Society of London, 279A*, 547–582.

Smith, C. S. 1977. "A highly personal view of science and its history," *Annals of science, 34*, 49–56.

Stravinsky, I., and Craft, R. 1966. *Themes and episodes*, New York: Alfred A. Knopf.

8. {See Britten (1964).} A very interesting category in public reaction to the arts is that of the critics, who as titular spokespersons for the audience play a role somewhat like that of book reviewers. Britten shared the disdain for critics common to creative artists; the passage from the letter to Tippett that I quoted in ¶1 is framed by these rather waspish comments: "Evaluations, comparisons; the whole apparatus; does it mean anything to you? It doesn't to me, much. Slaps or bouquets, they come too late to help, long after the work is over . . . Criticism likes to separate, to dislodge, to imply rivalries, to provoke jealousies."

CHAPTER EIGHT

Scientific Revolutions as Convolutions?

A Skeptical Inquiry

To Hans Wussing, 65 years young

A major factor in historical method is understanding changes of theories, especially major ones. The word "revolution" is often employed; proposals are made here to both temper and clarify its use.

1. Convolution versus Revolution

Thanks to the encouragement of Hans Wussing and the commercial courage of the houses of Birkhaüser (Basel) and the Deutscher Verlag der Wissenschaften (Berlin), I was able to publish in 1990 a large work of 1,600 pages entitled *Convolutions in French Mathematics, 1800–1840* (Grattan-Guinness 1990). It treats of a golden but little-studied epoch in the history of science, when France was by far the leading mathematical country in the world, and an extraordinary galaxy of major figures converted "calculus to mathematical analysis" and "mechanics to mathematical physics," to quote from the subtitle of my book. Major changes were wrought in the period that I studied, and the temptation to begin the title of my work "French revolutions in" was very great. Yet it had to be resisted: although priority rows and plagiarism charges occurred (following a distinguished Parisian tradition!), no *truly* revolutionary events happened, involving (for example) some sort of *illegal* activity or procedure.

Search for an alternative word brought me to "convolution," which seems to convey much more accurately than "revolution" the processes involved in "revolutionary" scientific activity. "Coiling, twisting, or winding," the dictionaries tell us, "rolled or coiled up," "enclosed in folds": this is surely a more perceptive image than

First published in S. S. Demidov and others (eds.), *Amphora: Festschrift für Hans Wussing zu seinem 65. Geburtstag,* Basel, Switzerland: Birkhäuser, 1992, 279–287. Reprinted by permission of Birkhäuser Verlag.

"revolutionary" changes, whether that word is taken in its traditional sense of rotation back to the original ideas or the more recent one of changes of ideas.

In addition, "convolution" has a dynamic connotation, which also seems suitable: it even recalls the word "convulsion," although the two words have different roots. In addition, it covers both the concepts of *innovation* (when something essentially new is introduced) and *replacement* (where some previously popular notion is set aside). One major way in which these processes work is in the deployment of analogies and homologues from a normal science to the new one; for example (in fact, a whole suite of them), the (mis?)use of mechanics and physics to articulate theories in mathematical economics (see ¶5).

During such "transitional" periods in science new ideas entwine and grapple with old ones until some new normal science emerges as the dominant approach; however, the previous science usually leaves some residue. The new theories, which began as outsiders—and were often proposed by (young) outsider scientists—become the new orthodoxy. Then, after a time, radically fresh ideas challenge this orthodoxy, and the processes begin again when the fresh crop of outsider ideas begin to work.

I deliberately used Kuhn's happy phrase "normal science" above to characterize the periods when the main lines of a science seem to be settled. There has been much discussion in the history of science in recent years about the concept of a revolution: how the meaning of the word has changed over the centuries, and how it is (over) used by historians today (Cohen 1985; Shea 1989). Kuhn's own book *The Structure of Scientific Revolutions* (1970) has justly been praised as an important contribution to the debate. However, it seems to me to be better understood as a study of "The structure of normal science," in that the processes and causes of revolutionary change are *not* convincingly conveyed there—especially for want of attention to processes of convolution.

2. The Normal and the Revolutionary

One feature of these processes seems to make the word "convolution" particularly apt: the outsider ideas often begin life *within* the framework of the current normalcy, and even in connection with some quite specific detail of it, *not* as a fully frontal assault of its basic principles. The processes of change are then much concerned with pursuing the consequences of the particular new insights as they wind their way around the science(s) involved.

Wussing's own major work *Die Genesis des abstrakten Gruppenbegriffes* (1969;

English translation 1984) shows similar features. A major change in algebra—namely, to think of working out from an uninterpreted axiomatic structure, deriving its theorems, and then seeking its interpretations—was actually achieved by a *reverse* sequence of convoluted processes: the interpretations arose first, as pieces of normal mathematics such as the roots of equations or the rotations of geometrical figures; and versions of their "abstract" features were stressed in a context when other aspects constituted normal algebra; then the new properties were systematized in each mathematical context, and structural similarities (compare ¶4) noticed across them; then that common factor was individuated as a group; then the group was "abstractified," so that the factor would be rendered independent of all its occurrences.

My French story is full of such examples: Cauchy starting out in mathematical analysis very much concerned with singular solutions to differential equations; Fresnel's (first) wave theory of light being tried out for diffraction; Fourier developing a theory of heat diffusion on discrete bodies, then straight rods, then curved rods, then solid bodies, then . . . before the radical innovation was clearly articulated (see also ¶3); Ampère working out the essentials of electrodynamics and electromagnetism in a similar style, from straight wires, then curved wires, then . . . ; Lazare Carnot and his engineer successors working out the consequences for mechanics in (very!) general from the consequences of impact and percussion effects. These and many other cases exhibit a philosophical process I call "desimplification" (Grattan-Guinness 1986): a very simple but not stupid case is worked out with some success, and then the complications of nature are gradually put back into the theory so that its "revolutionary" potential can be appraised. I introduced the notion of "desimplification" of a theory intended as an alternative to philosophers' excessively nervous discussion of ad hoc hypotheses; the two concepts are *not* coextensive.

Among many other cases, probability theory is rich in convolutions, for in an astonishing range of contexts in physical, natural, and social sciences, the novelty of probabilistic and statistical thinking wound its way out from among more traditional approaches. It is disappointing but instructive, then, from our point of view, that the single most important study of this history should be called *The Probabilistic Revolution* (Krüger and others 1988); for in this fine collection of essays some refute the title, even on occasion explicitly.

Mathematics is not the only science to show convolution; for example, the Darwin industry has revealed that his great idea of evolution was very much worked out from detailed work on specific contexts, such as barnacles and pigeon breeding.

The idea has seemingly wide range, even outside science itself; for example, much early Beethoven is like Haydn, and the young Stravinsky wrote some excellent neo-Tchaikovsky.

3. Biography and Education

One type of context in which the normal and the revolutionary can be examined together is biography. The (so-called) sociology of science became so rampant in the 1970s that Hankins (1980) had to come to the defense of biography. Luckily other historians adopted the same viewpoint; and meanwhile Wussing had already been producing heavily in this genre, especially a collective biographical volume (Wussing and Arnold 1975), which has seen several editions since and ought to be available in English. In addition, there were his own biographies; one on Copernicus (1973), followed by rather similar volumes in the Teubner (DDR) series, on Gauss (1974), Newton (1977), and Ries (1989). This quartet of historical figures divides into complementary pairs: Copernicus and Ries show convolution in particularly marked forms (Copernicus's planetary model is literally nearly as convoluted as Ptolemy's), while Newton and Gauss kept many of their most innovative ideas in manuscript and in this way showed convolutionary public faces.

In many cases revolution and conservatism are copresent, convoluting around each other; the major problem for the biographer of balancing synchrony against diachrony often has a basis in convolutionary issues. For example (from ¶2), when Fourier worked on heat diffusion, he was passé in his first, n-body model; conventional in deploying the Euler differential model in his second try; replacing when exposing previously unnoticed mathematical qualities in trigonometric series; and innovatory with the Fourier integral solution to the heat equation (and, modulo a scrap of Biot just before him, innovatory also in mathematicizing heat diffusion at all). All this is exhibited by one person within one problem area in six years (Grattan-Guinness and Ravetz 1972); luckily, in this case a pretty straightforward chronological presentation was possible. Many historical figures have shown such simultaneity: recognition of this fact would desimplify the shallowness of much discussion of the revolutions in science.

Another context important for biography is the educational situation of the historical figure: initially to (fail to) discuss his original training, then to (fail to) describe his teaching and research opportunities. The relationship between the educational learning experience and the later research career is a marvelous source for convolutionary processes, because both positive and negative influences may function. This is where Kuhn's notion of normal science seems to be particularly fruitful, since normally (sic) the teaching program will follow convention.

This scenario obtains most clearly from the late eighteenth century onward, when the institutional structure increased very greatly. For earlier times, especially the Middle Ages and the Renaissance, a variant situation obtained; the Church imposed normal positions against which some scientists (such as Copernicus) reacted and convoluted their ideas. In other words, the historiographical dogma of the simple warfare model between science and religion *will not do,* neither for that earlier period nor for any later one (see, for example, Moore 1979 for the particularly instructive case of post-Darwinism).

4. Theories and Concepts

Wussing has written explicitly on historiographical questions, largely from a Marxist standpoint (1970, 1975, 1976); the first article is the most pertinent here. In an attempt to analyze "the developmental history of scientific concepts" he considers the "factors and means" that generate "the 'feedback' between knowledge and societal needs." The burden of the analysis falls on the dynamics between three central historiographical categories: the "intension of a concept," namely, "the totality of all properties and relations that stand in an explicit conceptual specification (e.g., definition)"; its "extension," the "class of all those objects and phenomena of nature or else conceptual reflection upon them"; and "ostension," the "region of activity, as far as it is relevant for the conceptual history researched."

Wussing's stress on "developmental history" chimes in very nicely with the notion of convolution, and his diagrams of interaction between his three categories include some examples of desimplification. There is also a healthy acceptance of ontology, in a neutral Platonist sense; this for me is a central category of the philosophy of science, especially where abstract objects are concerned (Grattan-Guinness 1986). If there are differences between us, I would mention three.

First, I would be less inclined to emphasize concepts and give pride of place to the theories within which concepts are found; for it is from theories that concepts draw their significance. The distinction here is not just jargon: the role of essentialism (which I wish to diminish) in epistemology is quite crucial. In particular, for me the *early stages of theory building* are crucial for understanding the (non)growth of scientific knowledge; but then one has proto-concepts more than "proper" ones, notions like intensional definition are inapplicable (and indeed beg the question of whether such a stage is reached in/by a theory to start with), and the range of ostensional relevance is itself fuzzy. It is doubtful to me that one can use *any* model of Wussing's kind for such early stages at all, for it may not address the historiographical question of whether *or not* a scientific theory gets beyond them (an example is mentioned in ¶5).[1]

Second, Wussing quite correctly gives examples where conceptual broadenings have taken place (including from his work on algebra mentioned in ¶2); but I would be more inclined to highlight the situation of (dis)analogy between one not-that-general scientific theory and another one, for such processes seem to occur much more often. In the special case of mathematics, it has led me to emphasize the place of "structure-similarity" between theories, that is, the extent to which the structure of one theory "means" in another one (Grattan-Guinness 1992; and 1987 for a range of examples taken from French science).

Third, I give a central place in epistemology to *ignorance*: that is, to things not known (§2.5.4). This applies to the analysis of the problem situation of a historical figure: knowing that, for example, Euler had no matrix theory is a major point in not misunderstanding many parts of his applied mathematics. But it also bears on ahistorical philosophy: knowledge and ignorance go together; scientific advance often involves knowledge *of* ignorance, for then the problem is posed; and then (maybe) knowledge is produced in response to it.

5. Convolution and Feedback

The emphasis laid by Wussing on feedback is particularly welcome. The history of mathematics exhibits some particularly difficult (and interesting) examples of feedback desimplifying the dynamics of convolutions. Two principal reasons are that mathematics itself has influenced the developments in technology and economics from which those factors that encourage the growth of (mathematical) knowledge have been formed in the first place. A few remarks on each are in order.

The interaction between mathematics and technology is especially fascinating (but very little studied, unfortunately). I surveyed a wide range of examples in (post-) Revolutionary France (Grattan-Guinness 1990, esp. chs. 8 and 16), where the institutional structure gave a uniquely central place to technology, allowing its influence on science (*not* vice versa) to be noticeable and marked. Indeed, the "ingénieur savant" even became a special kind of professional scientist (Grattan-Guinness 1993).

1. {I have returned to this topic recently in Grattan-Guinness (2008), with a refutation of Eugene Wigner's highly influential thesis about the "unreasonable effectiveness of mathematics in the natural sciences."}

The (related) case of mathematical economics is more marked than might be expected; for almost all its classical and neoclassical theories were developed by strong analogies from mechanics, both for optimization (Mirowski 1989) and equilibrium (Ingrao and Israel 1990, chs. 2–3). The strength of the analogies was of course a major question (§7.5.2), and convolution was well evident; and these processes was exacerbated by significant missed opportunities, when real innovation was not perceived even by its creators and delayed near-revolutionary changes by a century (Grattan-Guinness 1994). These are cases of early stage science mentioned above: for all that time the more developed forms were never attained.

These are examples of profitable feedback. A railway enthusiast would enjoy the following analogy: iron (and then steel) is produced to be made into wagon wheels and railway lines and to be used in the manufacture of railway engines, so that trains can run on the lines between locations of coal to locations of iron ore, so that more iron can be processed to produce more carriage wheels and railway lines and engines. . . . The example is not a fancy; it expresses a major factor in the history of American railway technology in the late nineteenth and early twentieth centuries {(Schallenberg 1975)}.

6. Convolution and Metaconvolution

In general parlance, including in the history of science, the word "convolution" has a common but pejorative usage: to describe a line of thought, administrative procedure, or whatever, being complicated and even re-entrant in structure. But its original sense, as recalled in ¶1, which even encouraged its introduction into mathematics as a technical term, seems worth the serious attention of historians of science seeking language to express the substantial changes that they see take place in the sciences that they study.

Will such a change take place in the immediate future history of science? One cannot expect so, especially if it is uttered from the ghetto of the history of mathematics (ch. 6); much better, it seems nowadays, to expostulate on the hermeneutics of intellectual discourse or the sociomethodology of scientific achievement. This is much of normal history of science today, where the lunatic fringe is becoming the lunatic core; so the idea of convolution is itself in a state of metaconvolution, trying to wind its way out of current normal historical discourse and bring clarity where "difficulty" seems to be preferred.

Maybe educational needs can help; as experts in educational "theory" produce ever more well-subsidized and sophisticated models of pedagogy science and pupils become progressively more nonscientific and innumerate, then maybe history and its historiography will be seen as essential elements in real science education. *If* so, then Wussing (1979) is one of the best teaching aids of this kind available for the needs of mathematics teaching.

BIBLIOGRAPHY

Cohen, I. B. 1985. *Revolution in science*, Cambridge, Mass.: Harvard University Press.

Grattan-Guinness, I. 1986. "What do theories talk about? A critique of Popperian fallibilism, with especial reference to ontology," *Fundamenta scientiae, 7*, 177–221.

Grattan-Guinness, I. 1987. "How it means: Mathematical theories in physical theories. With examples from French mathematical physics of the early nineteenth century," *Rendiconti dell'Accademia del XL, (5)9*, pt. 2 (1985), 89–119.

Grattan-Guinness, I. 1990. *Convolutions in French mathematics, 1800–1840: From the calculus and mechanics to mathematical analysis and mathematical physics*, 3 vols., Basel, Switzerland: Birkhäuser; Berlin: Deutscher Verlag der Wissenschaften.

Grattan-Guinness, I. 1992. "Structure-similarity as a cornerstone of the philosophy of mathematics," in J. Echeverria, A. Ibarra, and T. Mormann (eds.), *The space of mathematics: Philosophical, epistemological, and historical explorations*, Berlin and New York: De Gruyter, 91–111.

Grattan-Guinness, I. 1993. "The *ingénieur savant,* 1800–1830: A neglected figure in the history of French mathematics and science," *Science in context, 6*, 405–433.

Grattan-Guinness, I. 1994. "From virtual velocities to economic action: The very slow arrivals of linear programming and locational equilibrium," in P. Mirowski (ed.), *Markets read in tooth and gear: Nature, economy, and science*, New York: Cambridge University Press, 91–108.

Grattan-Guinness, I. 2008. "Solving Wigner's mystery: The reasonable (though perhaps limited) effectiveness of mathematics in the natural sciences," *Mathematical intelligencer, 30,* no. 3, 7–17.

Grattan-Guinness, I., in collaboration with Ravetz, J. R. 1972. *Joseph Fourier 1768–1830: A survey of his life and work, based on a critical edition of his monograph on the propagation of heat, presented to the Institut de France in 1807*, Cambridge, Mass.: MIT Press.

Hankins, T. L. 1980. "In defence of biography: The use of biography in the history of science," *History of science, 17*, 1–16.

Ingrao, B., and Israel, G. 1990. *The invisible hand: Economic equilibrium in the history of science*, Cambridge, Mass.: MIT Press.

Krüger, L., and others. 1988. *The probabilistic revolution*, 2 vols., Cambridge, Mass.: MIT Press.

Kuhn, T. S. 1970. *The structure of scientific revolutions*, 2nd ed., Chicago: University of Chicago Press.

Mirowski, P. 1989. *More heat than light: Economics as social physics: physics as nature's economics*, Cambridge: Cambridge University Press.

Moore, J. R. 1979. *The post-Darwinian controversies*, Cambridge: Cambridge University Press.

Schallenberg, R. H. 1975. "Evolution, adaptation and survival: the very slow death of the American charcoal iron industry," *Annals of science, 32*, 341–358.

Shea, W. R. 1989. (Ed.), *Revolutions in science*, Canton, Mass.: Science History Publications.

Wussing, H. 1969. *Die Genesis des abstrakten Gruppenbegriffes*, Berlin: Deutscher Verlag der Wissenschaften. [English trans. 1984.]

Wussing, H. 1970. "Zur Entwicklungsgeschichte naturwissenschaftlicher Begriffe," *Schriftenreihe NTM, 7*, no. 2, 15–29.

Wussing, H. 1973. *Nicolaus Kopernicus*, Leipzig, Germany: Orania.

Wussing, H. 1974. *C.F. Gauss*, 1st ed., Leipzig, Germany: Teubner.

Wussing, H. 1975. "Zur Diskussion. Versuch zur Klassification des historischen Wechselverhältnisses zwischen Naturwissenschaften und materieller Produktion," *Schriftenreihe NTM, 12*, no. 1, 98–104.

Wussing, H. 1976. "Historiographie der Mathematik: Ziele, Methoden, Aufgaben," *Mitteilungen der Mathematischen Gesellschaft der DDR*, nos. 3–4, 120–132.

Wussing, H. 1977. *Isaac Newton*, 1st ed., Leipzig, Germany: Teubner.

Wussing, H. 1979. (Ed.), *Vorlesungen über die Geschichte der Mathematik*, Berlin: Deutscher Verlag der Wissenschaften.

Wussing, H. 1984. *The genesis of the abstract group concept*, Cambridge, Mass.: MIT Press.

Wussing, H. 1989. *Adam Ries*, Leipzig, Germany: Teubner.

Wussing, H. and Arnold, W. 1975. (Eds.), *Biographen bedeutender Mathematiker*, 1st ed., Berlin: Volk und Wissen.

PART 2

PATHWAYS IN MATHEMATICS EDUCATION

CHAPTER NINE

On the Relevance of the History of Mathematics to Mathematical Education

This chapter discusses the possible uses that can be made of the history of mathematics in mathematical education. The problem is discussed in general terms in ¶2; it is prefaced by a description of a lecture course for teachers in the history of mathematics, and followed by a discussion of some specific matters.

1. A History of Mathematics Course for Teachers

I ran such a course as a special topic in the part-time master of science (education) degree at the University of Western Australia in Perth during the winter term of 1976. I shall describe the course first, for it will be a fertile source of examples of matters to be discussed later. It ran as two two-hour sessions for each of the 10 weeks of the term. Each session began with an hour-long lecture by me, and then one session continued with a reading class using some classical mathematical literature (in English translation where necessary), while the other session was completed by a discussion period. The students were assessed by four essays of around 2,000 words each, which they prepared during, or shortly after, the end of term. For more details of the course, see Grattan-Guinness (1977a). Rather than fill the chapter with many references to the literature whose history I am about to describe, I cite the extensive bibliography in Grattan-Guinness (1980), the book on which the course was based.[1]

I decided that in the course we should concentrate in some detail on a particular

The original version of this essay was first published in *International Journal of Mathematical Education in Science and Technology, 9* (1978), 278–285. Reprinted by permission of Taylor & Francis (www.informaworld.com).

1. I edited this book; my five coauthors were the historians of mathematics H. J. M. Bos, R. Bunn, J. W. Dauben, T. W. Hawkins, and Kirsti Møller Pedersen {now Andersen}.

branch of mathematics and over a limited time, rather than flit about over the whole range of the history of mathematics (as is unfortunately done so often in courses of this type). I chose the development of the calculus from the early seventeenth century through its establishment as a central discipline of mathematics during the next century, and its absorption into mathematical analysis in the early years of the nineteenth century, to the introduction of mathematical logic and set theory at the beginning of the twentieth century. This story is relatively clear in its historical structure, and its content relates closely to mathematical education in schools and colleges (calculus and set theory) and also to degree-level training (mathematical analysis and mathematical logic).

We began with the calculus methods developed in the period 1630–1660 by Descartes, Cavalieri, and Fermat. We noticed that while each man devised techniques that were similar to the differential or integral calculus, none of them (nor any of their contemporaries) actually "invented" the calculus. The reasons were due chiefly to the fact that they did not realize that, given a function $f(x)$ of a variable x, the purpose is to calculate another function: the derivative $f(x)$ (or some equivalent) for differentiation, and the indefinite integral $\int^x f(u)\, du$ for integration. In the context of differentiation their methods hinged on locating particular points on the diagram (for example, the intersection of the required tangent with some given line), so they treated x as an unknown constant (usually the root of some equation) rather than as a free variable. (The distinction between an unknown constant and a free variable is discussed in more detail in ¶3.1.) In integration, their methods were confined to calculating definite integrals. We recognized that this failure was more fundamental than not realizing the inverse relationship between differentiation and integration, for without it the relationship cannot even be conceived, never mind used.

We then passed on to the surmounting of these obstacles by Leibniz in the 1680s, and the further development of his system during the next 100 years, especially by John Bernoulli, Euler, and Lagrange. Although the calculus was algebraic in its mode of operation, geometrical considerations were to remain prominent for a long time, especially with regard to dimensional homogeneity. In this style of calculus, if x is a line, then so is dx (an infinitesimally short one); if y is also a line, then ydx (= $y \times dx$) is an area, and so is $\int y\, dx$ (= sum of the infinitesimally narrow rectangles $y \times dx$). Limits are being avoided here, for in being a very short line dx is not a point, which is the limiting *but dimension-changing* case of a sequence of ever shortening lines. But there were conceptual difficulties involved in these infinitesimally small-but-not-zero quantities, and the even smaller higher-order differentials, and the problem somewhat mitigated against the great success achieved in applications and technicalities.

We then backtracked to Newton's creation of his calculus in the 1660s. Like Leibniz (and earlier than him), Newton surmounted the difficulties of his predecessors, but the form of his theory was very different for, albeit in a naïve form, it used the notion of the limit. From a function $x(t)$ of "time" t, he calculated the "fluxion" $(\dot{x}t)$, which is basically the same in definition as our derivative. Conversely, $x(t)$ was the "fluent" of $\dot{x}(t)$, and so on. In foundational terms, his system is no more rigorous than is Leibniz's (although the difficulties are rather different in kind), and we looked at Bishop Berkeley's famous criticisms of both systems during the 1730s.

As Newton's system was much less influential than Leibniz's, we did not follow the fortunes of his successors but returned to the Leibnizian tradition to look at two important applications. The analysis of the vibrating string in the mid-eighteenth century by forming and solving the "wave equation" was an important early case study in the applications of the calculus, especially in the development of partial differential equations. Functional solutions were suggested, although the extent of their generality was disputed. However, Daniel Bernoulli suggested (on physical grounds only) a solution by trigonometric series; but there was difficulty for everyone in interpreting their periodicity, and although the coefficients were calculated later by Euler, the "Fourier series" did not achieve prominence until Fourier's analysis (1805–1822) of heat diffusion, in which he solved the "diffusion equation" by these means. We looked both at his derivation of the equation and at aspects of his series expansion, especially those which differ from the modern textbook treatment.

At this time the calculus was absorbed into a more general subject, which became known as "mathematical analysis." Here a *theory* of limits—not merely the definition, which had been given several times in the eighteenth century—was the central feature, and it was used to embrace the calculus, the theory of functions, and the convergence of infinite series. Cauchy was the first mathematician to give (in the 1820s) a comprehensive presentation of mathematical analysis, and we examined his definition of limit, continuity, convergence, derivative, and differential, and some of his principal theorems. We noticed that, among the obvious similarities with the mathematical analysis with which we are now familiar, there were also some surprising differences. The chief of these arose from Cauchy's failure to realize that multiple limits require much more sophisticated techniques than the single-limit theory which he had successfully espoused. This was particularly clear in the general convergence problem of Fourier series, as we saw when studying the "proofs" offered in the 1820s by Fourier, Poisson, and Cauchy. Fortunately, Dirichlet was to do the problem justice, in 1829.

From this time on, the problems of chronology became more complex, and we tended to take themes. The first was the integral, which we traced from Cauchy's limit-style definition as the limiting value (if it exists) of a sequence of partition sums, through Riemann's 1854 necessary and sufficient conditions for integrability, to some of the early set-topological difficulties that were encountered in clarifying and then extending Riemann's definition. Next we returned to the problem of multiple limits to see Weierstrass introduce modes of uniform and nonuniform convergence, and we examined the need to make clear and careful statements of the relevant definitions. Then we took up the introduction of definitions of irrational numbers in the 1870s, learning why such definitions were desirable in the first place and analyzing in some detail Dedekind's "cut" definition, which is the easiest to understand in both motivation and form.

Our next topic was the installation of set theory into mathematics. Georg Cantor was largely (though far from wholly) responsible, and we traced his career from his 1870 theorem on the uniqueness of representability of a function by a trigonometric series (in which he extended some work of Riemann and Heine and came to naïve forms of both set topology and transfinite ordinals), through the analysis of the real line and problems with the definition of dimensions, to a more general conception of set. In this latter connection, which became steadily more prominent an aspect of his thought, he produced his formal theory of transfinite ordinals and the associated definitions of number-classes and alephs; but in the 1890s he also came across two paradoxes of set theory, and his hopes to lay a foundation of mathematics in set theory were badly shaken.

The first librarian of these paradoxes was Bertrand Russell, who invented an important one of his own and hoped to find a philosophical solution to all of them by diagnosing a common cause of malaise. Eventually he accepted the vicious-circle principle as the cause of the paradoxes, and regarded as built on it his "type theory," within which he developed a structuralization of mathematics justifying for him his "logicist" thesis, that pure mathematics is a branch of logic. We looked at some basic ideas of mathematical logic, especially the propositional function and quantification, and examined a few parts of his logicist structure.

The paradoxes were not the only source of interest in foundational studies at this time, and to conclude the lecture course we looked briefly at two other approaches. Each of them showed an influence from Cantor. Hilbert's program of metamathematics and proof theory, initiated in the 1900s, was based on a refinement of Cantor's formalist belief that the consistency of construction of a mathematical object guarantees its existence within the study of consistency (and other properties) of a mathematical system in metamathematics. Zermelo gave the first *full-scale* axiomatization of set theory, but in some ways it was related to Cantor's views, especially in sharing Cantor's belief that the paradoxes arose from assuming the existence of "too big" sets.

The last lecture of the course briefly summarized the content of the previous 19 and pointed out a number of educational matters that had arisen on various occasions. To such questions I now turn.

2. Theses on the Teaching and Learning of Mathematics

The course outlined above describes how (a part of) the history of mathematics might be presented to school and college teachers. Why should they be required to learn such material? How should it come down to undergraduates, trainee teachers, and school pupils? I shall argue toward a standpoint on such questions via a sequence of theses.

2.1. Knowledge, Understanding, and Motivation

The tradition in mathematical education has always seemed to concentrate on mathematical knowledge: definitions, theorems, proofs, and so on. To me, such knowledge is necessary but not sufficient for mathematical education to be successful. For sufficiency we need to take on the broader concept of mathematical understanding: not merely the knowledge itself, but also its motivations (historical and heuristic), the ways in which it may be created, its underlying logic and proof methods, and so on. In these respects the orthodox textbook (or lecture course) presentation is often very inadequate. Such books are fine as long as the reader has learned the mathematics already. They are excellent for those who are eventually going to be professional teachers or mathematicians but often seem to be irrelevant to the other 99 percent of the learning community. No wonder that it is common to find revulsion of mathematical education among ordinary people, and at least a feigned pride in their innumeracy.

Without deep, penetrating motivations, education is lost, and it is in providing these that I am sure that the history of mathematics can be of great service in mathematical education. Mathematics is human-made work, at least as a human activity (I am not picking a quarrel with the Platonists here), and it always takes place against the backcloth of a cultural heritage. Therefore it is an integral part of one's understanding to be aware of the source of the mathematics; but unfortunately it is not necessary for one's mere knowledge of it, and therein lies the methodological trap into which mathematical education often falls.

But of course the use of historical material is not straightforward. For example, the history of mathematics is full of blind alleys, laborious maneuvers, and unnecessary delays, and there is little point in living through all those agonies again. Furthermore, it is not necessary or even desirable for pupils and students to become bogged down by the complex historiographical issues that inevitably attend a serious historical study. (However, I think that their teachers should have some awareness of such matters, and point out clear-cut examples of them whenever possible.) Useful here is the idea of history-satire (Grattan-Guinness 1973), in which one uses the historical record as a bank from which to draw a sequence of developments as a means of teaching the mathematics involved. Above, ¶1 may be regarded as a sketch for a textbook on the calculus and set theory devised along these lines.[2] The order of material there provides a good chapter order: the principal events in the story supply the structure of the chapters, other historical details would generate good exercises, and some general feeling for the progress of thought over time could be conveyed, especially if the tale were spiced with some details of the personalities and careers of the main characters and of the history of mathematical education itself. {Grattan-Guinness (1980) is therefore a more detailed sketch; it is intended to complement textbooks and lectures.}

I conclude this subsection on the difference between knowledge and understanding with a historical remark. The philosophy of mathematical knowledge received much impetus from the foundational developments between 1870 and 1910, with whose history I concluded my lecture course at Perth. But these studies, in being knowledge oriented, were rather narrow in scope and purpose. Now, earlier in the last century, and especially during the period 1830–1870, the more general philosophy of mathematical understanding was being developed, especially in the contexts of non-Euclidean geometries, the calculus of differential operators, and vectorial and matricial mathematics. These studies, still far too little known historically, did not achieve the foundational detail or depth of their immediate successors; but they have not deserved the later virtual annihilation of their philosophical purposes. Their authors were true philosophers of mathematics, in my opinion.

2.2. *Foundations, Heurism, and History*

It is clear from ¶1 that in the course at Perth we tended to concentrate on the foundational aspects of the mathematics, which we handled. This was done for reasons of presenting a coherent story in the limited time available; but I chose to concentrate

2. {Bressoud (1994) is now such a book, and a very good one. It was followed by Bressoud (2008) on measure theory.}

on foundational rather than technical aspects in order to maximize the involvement with educational factors, for mathematical education is often much concerned with the presentation of the foundations of the mathematics being taught. From this point of view, my criticisms in ¶2.1 take the form of saying that foundations come from nowhere in such presentations, so they lead to results that are of no interest beyond their status as consequences of these foundations.

The historical dimension provides the following insight. In a formal (not necessarily axiomatic) presentation of a piece of mathematics, we begin with certain basic components, and proceed through deductions (and, possibly, extra definitions) to a sequence of results. Now it often turns out that to a striking degree the historical order of discovery of these results was the *reverse* of a formal presentation, with the basic components being the most recent things to have been obtained during the period concerned. (The history related in ¶1 is quite a good example.) This shows, incidentally, that foundations are dug down to rather than built up from.

From this state of affairs we may pose the following question. In mathematical education we have to choose some heuristic presentation of the mathematics. If formal and historical presentations are at two extremes of a spectrum, where between them should a heuristic presentation lie? My criticism in ¶2.1 is essentially that education normally aspires much too closely to the formal end of the spectrum, and that the historical end should be explored much more seriously. The schema shows this point diagrammatically:

- Historical presentations
- History-satire–oriented presentations
- Orthodox heuristic presentations
- Some forms of "new" mathematics presentation
- Formal presentations

It also contains a comment on the "new" mathematics on which I shall now expand.

2.3 Some Thoughts on the "New" Mathematics

I greatly admire the ingenuity of the games and other devices that have been introduced in the recent reforms to enliven mathematics for the very young; they are much better than the stuff that I received at that age. But the further up the age scale that we go, and especially when we come to the teenage level, where some degree of systematization is essential, I become more and more suspicious of the reforms, and to the extent of the criticism embodied in the figure. For some of the "new" mathematics schemes seem, in places, to be even more subservient to a formal presentation than do the old presentations that they replaced. Let me explain this point briefly with regard to set theory.

One of the main points that I emphasized in the latter parts of my lecture course at Perth was that the motivations to set theory were most erudite and were tied to certain philosophical and mathematical purposes that require considerable mathematical acquaintance to understand in the first place. Now, of course, later developments have provided many additional motivations to set theory on which we can draw for educational purposes; but it is still true that, far from being "simple" things to handle, sets are among the most intricate intellectual objects ever devised, and most inappropriate for presentation to the very young—or to anyone whose involvement with mathematics is not fairly considerable.

One result of the attempt to convey this material in mathematical education is that it is watered down to the extent of inanity. We have all seen examples of the silly nonsense—two quadratic equations can no longer have a common root, their solution sets have to have a nonempty intersection set, and so on—and we do not need to dwell on them any more here. A more substantial criticism is that the relationship of set theory even to the logic that it is supposed to help elucidate is unclear. In particular, the teaching of Venn diagrams is most ill-advised: their scope is very limited; the sets can only be used extensionally, so that their value for logical exposition is incoherent; and very often they are confused with Euler diagrams (Grattan-Guinness 1977b).

The advocates of new mathematics have made the wrong choices from set theory and logic. They should teach the following:

1. The distinction between theory and metatheory, between doing a piece of mathematics and talking about it (informally, of course; formal metamathematics should not be touched without a considerable mathematical background).
2. The notion of proof, including the ways that it has changed and developed over the years.
3. The notion of the propositional function as a means of formally stating properties (for an example, see ¶3.1).
4. The nature and purpose of definitions, and their various forms.

Another significant component of new mathematics teaching is abstract algebra, and my comments earlier in this subsection apply to it also.[3] In particular, the

3. {See now Cooke (2008) on common and abstract algebras, a textbook written in the same history-satire spirit as Bressoud's.}

inverse relationship between formal and historical presentations is very clearly exhibited, as is the excess imitation in teaching of a formal style of approach. It also exhibits a point that set theory shows, that the main steps in the development took place over seventy years ago. "New" mathematics—new to whom?

2.4. An "In-the-Middle" Approach: Whatever Happened to Geometry?

In the final paragraphs of this section I shall be more explicit about the kinds of alternative approach to teaching that I want to see explored. Quite obviously, far too little use is made of historical material, but its overuse will lead to erudition, which is just as bad (though different in form) as the current situation. So where is the compromise?

The schema in ¶2.2 provides the clue. The history-satire–oriented approach lies roughly centrally between the formal and historical extremes and should draw from both. The resulting treatment would not only mix these two together, but also show a blend of foundational and technical aspects of the mathematics being taught; each interacts with the other, and foundational studies are motivated by consequences of technical investigations as well as by lacunae in hitherto supported foundational systems.

An alternative approach would be taken to *rigor*. It would be seen as a process, designed to fulfill certain intellectual aims and changing with changes in those aims, rather than the arid and unmotivated excellence that it so often seems to be. Thus plenty of *bad and wrong mathematics* would be taught—and taught precisely for the purpose of showing the ways in which it is deficient, thus motivating the need for the superior replacement. Furthermore, whenever possible, the strictly methodological aspects of the learning process should be quite explicitly emphasized whenever a suitable mathematical example of them is taught.

Now, for an in-the-middle presentation such as this, geometry is by far the most fertile branch of mathematics to use. The decline of geometry has a longer history than might be imagined; during my lecture course I mentioned how the nonreliance on, or even avoidance of, geometry can be traced back to the textbooks of Euler and Lagrange of the mid- and late eighteenth century, and can be found prominently also in the teaching of Cauchy and Weierstrass in the nineteenth century. Indeed, one can criticize these authors to some extent for their failure to use geometrical intuition for heuristic purposes. Similarly, we must deplore the (near-) elimination of geometry in some forms of new mathematics education (Thom 1971). The main consequence has been to replace the experientially immediate with the intellectually abstract, the exciting with the erudite, the obvious with the obscure.

Geometry offers everything that a teacher could ask for. It interprets at once for the young in terms of shapes and things around them and reinforces their learning of spatial concepts. For the older pupils it provides an excellent vehicle for presenting other branches of mathematics, especially trigonometry, linear algebra, and the calculus, and is very useful for help in numerical analysis, probability, and statistics (which ought to be taught far more in schools than they are). At degree level geometry is often at least the link subject between "applied" mathematics and the physical applications themselves. And at the same time it yields a host of important foundational questions. There is the inadequacy of Euclid to provide sufficient axioms to justify many of the diagrams that one needs to draw, and the consequent need to add new axioms and demonstrate the consistency and completeness of the expanded systems. Then there is the relationship between Euclidean and non-Euclidean geometries (including group-theoretic differences), their own relationship to space and to physics, and the criterion of distinguishing between pure and applied mathematics (for geometry can be interpreted either way in a very clear manner). Incidentally, all the features mentioned in this paragraph can be traced in the history of geometry, especially during the nineteenth century—an exciting complex of developments.

3. Some Complementary Matters

The above discussion suggests that history can be substantially employed in the training of teachers but has to be diluted or "saturated" for pupil consumption. Furthermore, the accidents of historical development render some branches of mathematics more responsive to historical treatment than others. In this section, I complement these views with discussion of some specific matters.

3.1. Free Variables, Unknown Constants, and Propositional Functions

In ¶2.3 I advocated the early introduction of some ideas in mathematical logic. Here is an important example. Assume that in

$$y = f(x) \tag{9.1}$$

x and y are free variables ranging over some set S of values (which I take for convenience to be the real numbers). Set $y = 0$ (say), and we obtain

$$0 = f(x), \tag{9.2}$$

where x is no longer a free variable but an unknown constant standing for any of the zeroes of $f(x)$ (or, equivalently, the roots of [9.2], if there are any in S. In particular, mathematical identities (an unhelpful term) such as

$$x^2 - 1 = (x - 1)(x + 1) \tag{9.3}$$

are of the form of (9.2); the roots happen to be *every* member of S. Similarly,

$$x^2 - 3 = (x - 1)(x + 1) \tag{9.4}$$

is also like (9.2); it has no roots in S, which is one less than is possessed by

$$x^2 - 3 = (x - 2)(x + 1). \tag{9.5}$$

Now, if we modify (9.3) to

$$y = x^2 - 1 = (x - 1)(x + 1), \tag{9.6}$$

then we are back again to free variables for x and y, in a functional relationship stated in two different forms in (9.6). (A similar modification to [9.5] or [9.6] would be meaningless.) Again, if we quantify (9.3) universally with respect to x over S ("for all x in S," symbolized $\forall x$), then x becomes a bound variable, and we have the true sentence

$$(\forall x)\{x^2 - 1 = (x - 1)(x + 1)\}, \tag{9.7}$$

we can also quantify (9.4) and (9.5); (9.4) would be false for both universal and existential quantification, while (9.5) is false for universal and true for existential quantification.

There is an important class of cases where x seems to be both an unknown constant and a free variable at once, and great uncertainty will be caused to pupils if the point is not explained carefully (which it never is). There is an interpretation of (9.2) in which x is a free variable, namely, relative to the property that x is a zero of $f(x)$. To express this property we need to introduce a propositional function ϕx:

$$\phi x := f(x) = 0. \tag{9.8}$$

In (9.8), x is a free variable because it is the argument for $øx$; but relative to f alone, x is still an unknown constant, as it was in (9.2). (I choose not to add to the distinctions under discussion here by considering the difference between a function and a function letter.)

To conclude this brief summary, let me point out some ambiguities with functions of several variables. In

$$z = x^2 + y^2 \tag{9.9}$$

x, y, and z are free variables. Set $z = 8$, and we obtain

$$8 = x^2 + y^2. \tag{9.10}$$

We may regard (9.10) as an equation like (9.2), with (x, y) as the solution pairs, or else as free variables of the type in (9.1). Geometrically, this corresponds to interpreting (9.10) as a two-dimensional section of the paraboloid in (9.9), or as a circle in the only two dimensions under consideration. Analytically, it corresponds to taking both x and y to be independent variables in (9.9), and then only taking certain pairs of values in (9.10), or regarding only one of them as independent.

I have used this explanation of unknown constants and free variables to exemplify the usefulness of a notational means ($\emptyset x$) of expressing a property. There are many other examples, especially in algebra and calculus. It is also worth telling pupils that "mathematics" books often talk more about mathematics than do it. From simple points such as these, essential ideas about the difference between theory and metatheory can be illustrated.

3.2. Vector and Linear Algebra in School Education

I am much happier with the introduction of these topics than with the more prominently advocated novelties, especially if vector and linear algebra are motivated via geometry as directed line segments and their transformation in space. I also think that we should go beyond vectors and matrices of very low orders, for otherwise we commit the sin of excess oversimplification, which I mentioned in ¶2.3. In fact, this is a very good context in which to illustrate the difficult but vital fact that in mathematics we very often need more than three dimensions.

Incidentally, the history of vector and linear algebra will not be of much use for teaching purposes. They should have developed from about 1760 onward in the course of obtaining eigenfunction expansion solutions to systems of differential (and other contexts using similar mathematical techniques). Unfortunately, for reasons which I have not yet divined, nobody thought of it, and the main impetus came only from about 1830 onward, and then from rather general and even philosophical considerations. (A few parts of the theory of determinants came up earlier.) Here is another large area awaiting its historian—who had better have a good sense of humor.

3.3. Arithmetic in School Education

Here are a few remarks (and more in ch. 12) on the teaching of arithmetic.

1. Teaching pupils anything like Peano axioms is daft, as is emphasizing laws of distributivity, associativity, and commutativity. I have been told several times by teachers that pupils know that $3 \times 5 = 5 \times 3$, but not that $5 \times 3 = 15$. This is just awful.
2. Geometry probably does not have a role to play of the prominence visible elsewhere in mathematical education, but it is a useful site for objects to be counted and ordered. (For example, if you want to stress that $5 \times 3 = 3 \times 5$, then lay out a 3 by 5 grid of marbles, say, and then turn it at right angles.) Furthermore, the spatial ordering of objects in space, and the presentation of the concept of betweenness, is a good example of arithmetic and geometry working together. I prefer such teaching to the study of the differences between cardinals and ordinals, which smack too closely of the Cantorian worries of long ago, and do not matter much in the range of finite numbers.
3. Negative numbers cause trouble for pupils. We could take the lesson of history, and follow late medieval Continental businessmen in interpreting negative numbers as debts. Zero can be quite nicely handled here, too, for it means not having any money (§12.3, §12.9).
4. As for the manipulation of real numbers, the normal algorithms for multiplication and division are not the simplest; alternatives should be taught (§12.5–6).

3.4. The Question of the Role of Examinations

To err slightly on the side of complacency, I think that examinations are criticized far too much and are used as the scapegoat for deeper defects. *It is basically the syllabuses, and the attitudes behind them, that are at fault.* Change these, and many other difficulties, including examinations, will ease substantially. I also regard examinations as very useful in providing temporal signposts and targets along the line of a young life, which may otherwise be a meaningless meander. Of course, I know and share several of the criticisms of examinations. I am very interested in the many variants being tried and do not think that the changes I propose would bring any special difficulties with regard to assessment of pupils and students.

Examination papers could contain questions with a historical element, but I think that projects and essays that are historically oriented enough to penetrate beyond the level of anecdote should not be attempted until the last years at school, if at all there. But at the degree level, such assessment would seem most desirable, not only for the reasons that I have discussed earlier, but also as an opportunity for the student to learn to write literate prose, to dip in and out of books, and to use their indexes (significant parts of university learning on which students usually receive no help at all).

4. Concluding Remarks

It seems to be quite feasible for groups of teachers, mathematicians, and historians to collaborate on preparing new kinds of textbooks, source books and other kinds of teaching material along the lines that I have discussed. If more than one collaboration deal with the same area, then fine; a revised edition merging them both will probably be better than either.

But the practice depends on money and manpower, which in turn depend on *attitudes* among the decision makers and influence bearers, and thence on to the mathematical community in general; and in this respect the situation is very different. I doubt very much if the influential mathematicians in the world will do much, principally because they see no need for changes of the kind which I advocate, for these changes are grounded in philosophical and methodological issues, not in narrow mathematical concerns. Perhaps some pressure will come from engineers, scientists, and economists, who have to study mathematics and often loathe the experience.

BIBLIOGRAPHY

Bressoud, D. 1994. *A radical approach to real analysis*, Washington, D.C.: Mathematical Association of America.

Bressoud, D. 2008. *A radical approach to Lebesgue's theory of integration*, Cambridge: Cambridge University Press.

Cooke, R. 2008. *Classical algebra: Its nature, origins, and uses*, Hoboken, N.J.: Wiley.

Grattan-Guinness, I. 1973. "Not from nowhere: History and philosophy behind mathematical education," *International journal of mathematical education in science and technology, 4*, 421–453.

Grattan-Guinness, I. 1977a. "A history of mathematics course for teachers," *Historia mathematica, 4*, 341–343.

Grattan-Guinness, I. 1977b. "The Gergonne relations and the intuitive use of Euler and

Venn diagrams," *International journal of mathematical education in science and technology, 8*, 23–30.

Grattan-Guinness, I. 1980. *From the calculus to set theory, 1630–1910: An introductory history*, London: Duckworth. [Repr. Princeton, N.J.: Princeton University Press, 2000.]

Thom, R. 1971. "'Modern' mathematics: An educational and philosophic error," *American scientist, 59*, 695–699.

CHAPTER TEN

Achilles Is Still Running

In this chapter I discuss the Achilles-tortoise argument, with particular reference to assumptions that it does not contain. Classes of models are displayed for which the argument is valid, thus disproving the widely held belief that it is bound to be wrong. Although the paper is not specifically historical, some intriguing remarks of C. S. Peirce and Lewis Carroll are discussed.

1. Before the Race: The Distinction between Paradoxes and Contradictions

The literature on the Achilles-tortoise paradox is vast. This chapter is adding to it only because nevertheless certain essential features of the problem are still misunderstood or distorted.

Let us begin with a logical clarification. The terms "paradox" and "contradiction" are used in an extremely loose fashion, not least by logicians. Symbolically, a contradiction arises when opposing conclusions are derived from the same premises, that is,

$$A \to P \text{ and } A \to \neg P. \tag{10.1}$$

Often contradictions appear in the condensed form given by $A = P$, namely,

$$A \to \neg A. \tag{10.2}$$

Reductio ad absurdum proofs in mathematics have this logical structure (in the form of either [10.1] or [10.2]), and from them one deduces $\neg A$ as a theorem.

But a paradox arises when the assumption of $\neg A$ also leads to a contradiction:

$$\neg A \to Q \text{ and } \neg A \to \neg Q, \tag{10.3}$$

The original version of this essay was first published in *Transactions of the C. S. Peirce Society, 10* (1974), 8–16, after rejection by four philosophy journals and three history journals. Reprinted by permission of the editor of the *Transactions*.

or the corresponding condensed form given by $\neg A = Q$:

$$\neg A \rightarrow A. \tag{10.4}$$

Thus a paradox is a *double contradiction*; it arises from the conjunction of (10.1) or (10.2) with (10.3) or (10.4) and takes an especially sharp form when (10.2) and (10.4) apply together:

$$\neg \mathrm{A} \leftrightarrow \mathrm{A}. \tag{10.5}$$

2. During the Race: Estimates of Zeno's Paradox

2.1. The "Paradox" Is, if Anything, a Contradiction

If Zeno's argument is faulty at all, then it is a contradiction. I am not concerned here with the difficult historical question of what Zeno might have said, but simply with the type of argument that he proposed. The following formulation will suffice: "The slower will never be overtaken by the swifter, for the pursuer must first reach the point whence the fugitive is departed, so that the slower must always necessarily remain ahead."

The "paradox" is deduced as follows. The premises A are a collection of statements formulating the race, while the conclusion S is the statement "the slower will never be overtaken by the swifter." Thus we have

$$A \rightarrow S. \tag{10.6}$$

But we know that a swifter object will overtake the slower in the situation described; that is,

$$A \rightarrow \neg S, \tag{10.7}$$

and the contradiction arises from the conjunction of (10.6) and something wrong with A? Or is it (10.6) or (10.7) that is faulty?

2.2. A Suggestion of C. S. Peirce

Some of the commentators on the paradox have thought that perhaps Zeno was right, although they have not been sure why. C. S. Peirce was not one of them, and yet he had within his grasp the means of showing that the paradox was only apparent. In a manuscript of 1902, he asked the following question: "What is the natural mode of measuring time? Has it absolute beginning and end, does it reach or traverse infinity? Take time in the abstract, and the question is merely mathematical.

But we are considering a department of philosophy that wants to know how it is, not with pure mathematical time, but with the real time of history's evolution."[1]

He then considered three spirals in polar coordinates (r, θ), and interpreted r as the "measure of evolution of the universe" (that is, "real time") and θ as its measurement (in some form of "mathematical time"), where a revolution of θ would denote a year. Of interest here is the second spiral, given by the equation

$$\theta = 2\pi \tan (\pi r/2). \tag{10.8}$$

It starts from $\theta = -\infty$ with $r = 1$ (using the right-hand value $\tan (\pi/2 + \theta)$ of the tangent function) and passes to $\theta = +\infty$ and $r = 3$; then it starts afresh with a new infinite progression on θ as r advances from 3 to 5: "and so on endlessly . . . We must not allow ourselves to be drawn by the word 'endless' into the fallacy of Achilles and the tortoise. Although, so long as r has not yet reached the value 3, another year will still leave it less than 3, yet if years do not *constitute* the flow of time, but only *measure* that flow, this in no wise prevents r from increasing in the flow of time beyond 3."[2] In other words, our mathematical theory of time (θ) interprets things in such a way that a certain situation (r not advancing beyond 3) arises; but in "real" (r) time things may be different.

Peirce's philosophy was greatly influenced by Immanuel Kant, and his analysis shows a Kantian approach in conceiving of time as a continuous flow constructed by us to organize our sense experience. Thus he spoke of "the real time of history's evolution" and contrasted it with "years," which only mathematically "measure" that flow and so do not "constitute" it. Hence Zeno was wrong in constructing a measure of the flow of time, which implied that Achilles would not overtake the tortoise. For Zeno's theory was only θ-theory; in real t-time Achilles *would* overtake.

This is not the solution of Zeno's argument. But it teaches us something impor-

1. See Peirce's manuscript book "Minute Logic" written around 1902 in his *Papers* (para. 1.276). Cf. paras. 7.369–7.372, 7.481–7.483, and 8.274 for other discussions and applications of spirals.
 Another of the spirals of para. 1.276 may be written

$$2/(2r - 1) = 3 \log (1 + \text{antilog} (\pi/(2\theta - \pi))).$$

According to Peirce, θ starts negatively when $r = 1/2$ and passes to $\theta = +\infty$ with $r = 2$; then it begins another progression as r moves from 2 to 3½, and so on. He did not draw his spirals, and this analysis seems to be inaccurate. The spiral starts at $r = (1/2 + 1/(3 \log 2))$ when $\theta = -\infty$ and passes to $r = \infty$ when $\theta = \pi/2$; then it starts again at $r = 1/2$ when $\theta = \pi/2$ and passes to $r = (1/2 + 1/(3 \log 2))$ when $\theta = +\infty$.

2. I was alerted to this passage of Peirce by the excellent Murphey (1961), but he sides with Peirce in writing that "hence Zeno's paradoxes arise from the confusion of measurement with thing measured" (p. 285). This is not a valid point in connection with time, for the "thing measured" in this case is only one particular theory of time, not a "thing" in the Kantian sense of an already given category.

tant; that *we may construct theories of time of our own,* with their own properties. In particular, the Kantian flow of time is in fact only one of these theories, based on regular events such as pendulum swings. What happens in the world is what happens; it is *our theories about what happens* that can vary and be different, and the "flow of time" is one such theory. We can construct various time theories and relate them to each other, if we wish; or we may simply construct one theory and draw conclusions from it—as Zeno had done.

2.3. A Suggestion of Sir Karl Popper

During a discussion at the London School of Economics in 1965, Sir Karl Popper suggested that Zeno's argument was in fact simply the construction of a theory of time in terms of the occasions when Achilles reaches the places that the tortoise has recently left. *Relative to this theory, Zeno's conclusion is valid.* "Zenoic time," as we may call it, is one way of interpreting the race; it is of course not the only way, but other ways do not refute it. It does not prove that Achilles cannot overtake the tortoise, but it does prove that it is impossible to show that he must do so. The paradox arises only if we add extra premises that are not in Zeno's formulation, such as the orthodox theory of time, or the imposition on the competitors of uniform speed.[3]

Popper gave an example in which Achilles never catches up with the tortoise not only with respect to Zenoic time but even with respect to orthodox time. Let us suppose that the competitors become tired while racing under the hot sun and slow down in such a way that the moments when Achilles passes the tortoise's previous location occur at regular intervals. Then the race will indeed continue "forever" in the orthodox sense.

In terms of the logical structure of ¶1, we may say that the statement that opposes Zeno's (10.6), $A \rightarrow S$, is not (10.7), $A \rightarrow \neg S$, but the statement

$$B \rightarrow \neg S, \tag{10.9}$$

where B includes the additional assumptions. So something was wrong with (10.6) and (10.7) after all.[4]

3. This point is made in Whitrow (1961, 150–151), although the paradox is not properly resolved. The philosophical consequences of this important book bear closely on the question of the different theories of time, for his detailed discussions of "relational time," "biological time," "psychological time," and so on show that in many of our problems, we do not in fact work with orthodox time, or at least we use other theories in conjunction with it.

4. Quine (1962) classifies paradoxes into three types: (1) "Veridical paradoxes," which lead in fact to true conclusions (Reductio ad absurdum proofs are an example, and contradictions, in our sense, seem to be included also); (2) "Falsidical paradoxes," which are faulty deductions (such

2.4. *Zeno's Argument, Limit-Taking, and Continua*

Most commentators believe with Peirce that Zeno was wrong, and often they assert that Zeno did not "understand the concept of a limit," or some such phrase. But it seems obvious that he had at least an intuitive understanding of limiting processes.

However, Zeno did not possess the blessings of late nineteenth-century analysis, especially the Weierstrassian quantificationistic approach to limits in terms of epsilon-delta formulations, or the discoveries by Cantor about the topology of the real line. And it is these that allegedly reveal the true source of Zeno's error.[5] But Popper's counterexample shows that it is not so; epsilontics may enrich the classes of models for which the falsity of Zeno's argument may be demonstrated, but they all operate only with the aid of additional assumptions such as uniform velocity.

I hope on a later occasion to discuss the exaggerated claims made for Weierstrass-Cantor mathematics in the context of problems such as Zeno's. But the literature involved is extensive and complicated, and here I shall only mention another of Peirce's arguments, which had at least the merit of eschewing such mathematical techniques. In a manuscript of 1911 he relied mainly on physical considerations and employed as his only mathematical aid the summation of a geometric progression.[6] But he again presupposed uniform velocity, and so once more failed to characterize the problem properly. He even failed to judge Zeno's own possible situation properly, for he commented of the summation of the series that Zeno had a "want of arithmetical skill that was inevitable at his time." But even if you do believe that Achilles will catch the tortoise within a finite period, then you only need to establish the convergence of your series; it is not necessary to calculate its sum.

Peirce is a good example of someone affected in his assessment of the problem by making additional assumptions in it, for he could have applied other of his interests to it. One of the chief features of the Weierstrassian school was its elimination of infinitesimals from the processes of limit taking. The implication of this feature is that limit taking and infinitesimalism are *exclusive* choices, *separate* techniques. But

---

as dividing by zero); and (3) "Antinomies," which are paradoxes in our sense. He says of Zeno's "paradox" that "one man's antinomy is another man's falsidical paradox, give or take a couple of thousand years." But I would suggest that the argument is a veridical paradox.

5. A *locus classicus* for such a view is Russell (1903, esp. chs. 39–42). For a survey of the history of the examination of Zeno's paradoxes in the context of theories of limits, see Cajori (1915).

6. See Peirce's "A Sketch of Logical Critic" in *Papers* (paras. 6.117–6.183); and compare paras. 2.666–2.667, 5.202, and 8.333 for other manifestations of the assumption of uniform motion. More interesting are the dots argument of para. 4.202 and the maps-within-maps argument of para. 8.122.

this is not the case; one can take limits and use infinitesimals (in a dynamic variable form) *simultaneously,* as was very often done in the analysis of the early nineteenth century by Cauchy and those whom he influenced.[7]

I use the term "limit-avoidance" (as an abbreviation of the more accurate but clumsier "limit-value-avoiding") to emphasize that basically limiting values are defined solely by the property of avoiding by arbitrarily small amounts the members of the sequences for which they are the limits. An important consequence arises from the discussion above; that limit-avoidance is independent of the continuum of values over which the limit is being taken.[8]

Now Zenoic time is constructed in a limit-avoiding manner, since Achilles, on reaching the position that the tortoise has recently left, is avoiding the tortoise's present location by a nonzero distance. These distances do not necessarily become infinitely small; even if they do, the corresponding intervals of Zenoic time could still be large; and even if those intervals become infinitely small also, then the limit has still not been achieved. For then we have a case of infinitesimal limit-avoidance, in which the separation is susceptible of a further infinity of subdivisions, "and so on endlessly," as Peirce had written when he was so near to the point.

In other words, Zeno's paradox has infinitesimal as well as noninfinitesimal models, and this is why Peirce might have brought other of his interests to bear on it. For he was one of the few mathematicians of the Weierstrassian period who took infinitesimals seriously. But he never applied them to the paradox, for he always seems to have seen it as involving uniform velocity, when the infinitesimals do not affect the solution.

2.5. The Status of Zeno's Paradox

Some of the literature on the paradox has been concerned with the possibility of performing an infinitude of operations in a finite time, including nonuniform cases of the argument.[9] Now these are important and difficult questions, but in my view they are *quite irrelevant* to the paradox itself. For Zeno offers us here a thought

7. See Grattan-Guinness (1970), esp. chs. 2–4 on the Cauchy period and ch. 6 on the emergence of the Weierstrassian techniques.

8. See Grattan-Guinness (1970), esp. ch. 3. Another useful feature of limit-avoidance for noninfinitesimalist analysis is that the various arrangements and rearrangements of quantifiers take place within, and in fact are subclasses of, the class of limit-avoiding expressions.

9. A significant recent example is Grunbaum (1969). But in Grunbaum (1967, 100–109), he assumes uniform speed in considering this paradox.

experiment, independent of the physical features of this planet. Even if as a matter of fact we do live in a world of discrete quanta, then Zeno's argument is not refuted (and neither are mathematical theories using infinitesimals). The confrontation of the paradox with physical properties of the world may illuminate our understanding of the world but does not affect the paradox itself. In fact, relative to the paradox it again amounts to making additional assumptions.

Incidentally, and independently of the paradox, the question of the possibility of performing an infinitude of operations in a finite time is in my view not even a well-formed question *until the method of interpreting the phenomena as the operations involved, and the definition of the theory of time used, are explicitly stated.* For example, Zeno's argument states the occurrence of an infinitude of events in an infinitude of Zenoic time, while the usual misinterpretation of it concerns the alleged occurrence of the same events in a finite measure of orthodox time.

3. After the Race: A Note on a Contradiction of Lewis Carroll

In April 1895 Lewis Carroll published a famous article in *Mind* in which Achilles and the tortoise discussed the principles of logical inference (Carroll 1895). The race had finished; the tortoise was somewhat surprised and retaliated with an argument the reverse of Zeno's in that, instead of containing an infinity of steps which (allegedly!) could be completed in a finite time, it contained three steps which seemed to require an infinite time. He asked Achilles to note down the premises in his notebook:

A. Things that are equal to the same are equal to each other;
and
B. The two sides of this triangle are things that are equal to the same.

But in attempting to deduce the conclusion that

Z. The two sides of this triangle are equal to each other,

the tortoise showed that there were unexpected difficulties. For, as Achilles admitted, the logical principle

C. If *A* and *B* are true, then *Z* must be true

was undeniably relevant and therefore had to be entered in the notebook. But if *C* was needed to deduce *Z* from *A* and *B*, then this fact had to be written down also:

D. If *A* and *B* and *C* are true, then *Z* must be true.

Thus an infinite intermediate sequence of statements *C*, *D*, . . . was set up, implying that *Z* could never be deduced from A and B. But this contradicted the fact that we make deductions like this constantly.

Lewis Carroll was one of the profoundest logical thinkers of his time; his famous *Alice* stories, for example, show the quality of his logical thinking (as well as an interest in time). This argument stated with greater clarity than any other contemporary work the problem of logical inference. Like Zeno's paradox, it is a contradiction of the form of (10.6) and (10.7) and is solved in the manner of (10.7) and (10.9), in this case by a distinction between the metalinguistic rules of inference (*C*, *D*, . . .) and the object-linguistic premises and conclusion (*A*, *B*, and *Z*). This distinction, and the consequent theory of hierarchies of languages and formal systems, has been one of the most fertile sources of progress in logic during this and the last centuries.

But the relevant point here is the relationship between Carroll's argument and Zeno's paradox itself. Not only is the logical structure the same; the infinite conversation that passes between Achilles and the tortoise, if allowed to take place *during* rather than after the race, can be interpreted as a model of Zeno's argument. In the discussion mentioned in ¶2.3, Popper suggested a model in which the tortoise puts down pebbles and Achilles picks them up (and of course Achilles is "always" picking them up). Now in a variant of this model, the tortoise puts down a pencil every time he introduces one of his intermediate statements *C*, *D*, . . . and Achilles picks it up in order to write down that statement in his notebook. Then Achilles will "always" be writing down these statements; the conclusion *Z* will "never" be deduced; Achilles will "never" overtake the tortoise.

Thus I end with a conjecture concerning Carroll's paper. When he pointed out the problem of logical inference, did he also hide behind it a message concerning Zeno's paradox? Although he set the conversation after the race, did he wonder if Achilles was still running? I found some confirmation of this conjecture in a manuscript by Carroll in the Library of Christ Church, Oxford, dated November 22, 1874, and entitled "An Inconceivable Conversation between S. and D. on Indivisibility of Time and Space." S. (presumably Socrates) claims that no mathematical argument can prove that Achilles will not overtake the tortoise. But D. (Dodgson?) attacks the reliance of this conclusion on indivisible quanta of space and time, for if these quanta are unequal, then a larger one is in fact divisible into some smaller one and its complement; but if they are equal, then while Achilles traverses a space-quantum in a time-quantum, the slower tortoise can cover only a part of another in the same time.

BIBLIOGRAPHY

Cajori, F. 1915. "The history of Zeno's arguments on motion: Phases in the development of the theory of limits," *American mathematical monthly, 22*, 1–6, 39–47, 77–82, 109–115, 143–149, 179–186, 215–220, 253–258, 292–297.

Carroll, L. 1895. "What the tortoise said to Achilles," *Mind, new ser. 4*, 278–280. [Various reprintings.]

Grattan-Guinness, I. 1970. *The development of the foundations of mathematical analysis from Euler to Riemann*, Cambridge, Mass.: MIT Press.

Grunbaum, A. 1967. *Modern science and Zeno's paradoxes*, Middletown, Conn.: Wesleyan University Press.

Grunbaum, A. 1969. "Can an infinitude of operations be performed in a finite time?" *British journal for the philosophy of science, 20*, 203–218.

Murphey, M. 1961. *The development of Peirce's philosophy*, Cambridge, Mass.: Harvard University Press.

Peirce, C. S. 1931–1958. *Papers: Collected papers*, 8 vols., Cambridge, Mass.: Harvard University Press.

Quine, W. V. O. 1962. "Paradox," *Scientific American, 206*, no. 4, 84–96. [Also in *Mathematics in the modern world*, San Francisco: Freeman, 1968, 200–208.]

Russell, B. A. W. 1903. *The principles of mathematics*, 1st ed., Cambridge: Cambridge University Press.

Whitrow, G. J. 1961. *The natural philosophy of time*, London and Edinburgh: Nelson.

CHAPTER ELEVEN

Numbers, Magnitudes, Ratios, and Proportions in Euclid's Elements

How Did He Handle Them?

For a century or so much Greek mathematics has been interpreted as algebra in geometrical and arithmetical disguise. But especially over the last 25 years some historians of mathematics have raised objections to this interpretation, finding it to be misleading and anachronistic, and even wrong. Accepting these criticisms, I consider Euclid's Elements *in this context: if it cannot be read in this algebraic manner, how did he conceive and handle his various types of quantity? The question is not merely of historical interest, for it raises issues about basic relationships between algebra, arithmetic and geometry.*

1. Introduction: Euclid as an Algebraist

Although Euclid's *Elements* was not the first work of its kind when it was written around 300 B.C., it seems to have soon become a focal point for later writers, with innumerable commentaries, editions, and translations over the centuries. For a long time it was an active source for new results; then it became a classic, upheld by some as a fine tool for mathematical education.

In the course of this change of status, a change in interpretation gradually emerged. The *Elements* was usually taken at face value to be an account of geometry and arithmetic; but when algebra began to develop among some Arabic mathematicians from the

The original version of this essay was first published in *Historia mathematica, 23* (1996), 355–375, with a printing correction at *24* (1997), 213. Reprinted by permission of Elsevier Rightslink. In this printing I have used the Greek alphabet without diacriticals.

This paper was motivated by my writing the general history of mathematics (§5.2). No Greek scholar I, I was ready to run through the usual story about geometric algebra when handling Euclid. But when I came to read the *Elements*, I experienced one shock after another; for his mathematics had *nothing at all* to do with common algebra. I found out that specialists in Greek mathematics had already set aside this reading, but the news had not got through to all the laypersons; so this article was written as a guide for us. The issue also helped me much to understand how much was involved in the distinction between history and heritage (ch. 2).

eighth century and in Europe from the late Middle Ages, parts of the work were held to be algebraic in character and were even rewritten in algebraic terms and notations. An interesting example from the early nineteenth century is provided by François Peyrard, librarian of the Ecole Polytechnique, who produced editions and translations of various Greek mathematicians. Those of Euclid (partial in 1804, full in 1814–1818) did not contain much commentary;[1] but in the notes added to his translation (Peyrard 1807) of Archimedes he rewrote several results as algebra and was widely praised for so doing by various contemporaries, especially J. L. Lagrange, who advocated the conversion of mathematical theories into algebraic forms as a general principle.

Such praise, and the status of France as by far the leading mathematical country of that time, must have given the algebraic reading of Greek mathematics much greater status. In any event, from the mid-nineteenth century a line of thought evolved in which much Greek mathematics was seen as *algebraic in disguise*. Expounded by Georg Nesselmann in a volume on "the algebra of the Greeks" (Nesselmann 1847), this thesis took a still more specific form in the 1880s especially by Paul Tannery in (1882) and Hieronymus Zeuthen four years later (Zeuthen 1886). They interpreted much of the *Elements*, and some other Greek mathematics, as "geometric(al) algebra" (their phrase): that is, common algebra with variables, roughly after the manner of Descartes though without necessarily anticipating his exact concerns, and limited to three geometrical dimensions. The earlier books of the *Elements* were held to deal with simple algebraic identities and forms, while many of the later constructions correspond to the extraction of roots from equations, normally quadratic but sometimes quartic. For some commentators the pertaining operations constituted a "geometric arithmetic"; for convenience I shall subsume this aspect under the term "geometric algebra."

This interpretation of much Greek mathematics, especially the *Elements*, soon became popular. An important example is the commentaries given by Sir Thomas Heath in his English translation of 1908, based on a text established by J. L. Heiberg in the 1880s. I shall use it here, in the second edition (Heath 1926), citing an item by its book and proposition (or definition or postulate) number; thus (2.prop.1) is Book 2, proposition 1, while (*2.prop.1) refers to that proposition as an *example* of a point also evident elsewhere in the work.

Around 1930 the interpretation was adopted by figures such as Otto Neugebauer, who then sought the algebraic origins in Babylonian mathematics. It was picked up

1. Peyrard's full edition (3 volumes, 1814–1818, including books 14 and 15 and the *Data*) was based on a Vatican manuscript; a quotation from his introduction appears in ¶11. Only one of the *Data* problems was algebraized in his commentary. A distinguished French example just before Peyrard is J. E. Montucla (1799, 278); he multiplied magnitudes, unfortunately, but at least he used "::" in proportions (see ¶8).

by his follower Bartel L. van der Waerden, whose articles and especially the book *Science Awakening* (van der Waerden 1954), first published in Dutch in 1950, have been influential sources. Heath's follower Ivor Thomas has been a similar influence, especially with his edition (Thomas 1939–1941) of Greek texts. It came to be normal historiography of mathematics during this century and remains influential; for example, most general histories of mathematics adopt some form of it in their account of the *Elements*, usually without much discussion.

A typical simple example of the interpretation is (2.prop.4): "If a straight line be cut at random, the square on the whole is equal to the squares on the segments and twice the rectangle contained by the segments." This is held to be, at base,

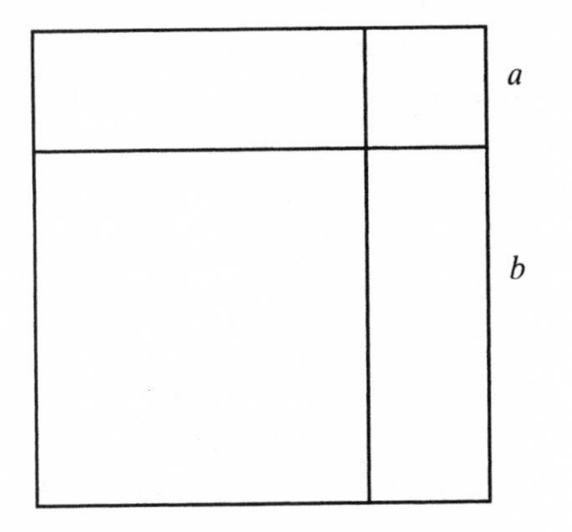

Figure 11.1

$$(a + b)^2 = a^2 + 2ab + b^2, \tag{11.1}$$

when the sides are the lines a and b as shown. A more sophisticated example is (6.prop.28), which uses an important construction called "application of areas," where the task is to set a parallelogram P of given size upon a given line as base, and similar to but smaller than another given parallelogram. In geometric algebra it comes down to solving a quadratic equation in x, a length important in the construction of P. We return to this example in ¶11.

In this chapter I examine the credentials and verisimilitude of geometric algebra as an interpretation of the *Elements*. Especially during the last quarter century, critical voices against geometric algebra have been heard and answered by defenders. Here, I find in favor of the critics. First I shall summarize the main objections and introduce three categories of algebraization that refine considerably the clarity of the positions on each side of the dispute. Then I state three features of the *Elements* that seem to deserve a central place in the discussion.

The main purpose of this chapter is to present, in a manner convenient for fellow nonspecialists in Greek mathematics, a *general* answer to the questions posed by my title: the quantities and propositions with which Euclid works, and the manner in which he handles them. I develop my interpretation in ¶4–¶9, where I shall suggest reasons, drawing especially on the three features just mentioned, *why,* and not just *that,* the *Elements* cannot be read as geometric algebra. Finally, the general methodological questions are reviewed in ¶11, using four examples from the dispute literature. Throughout I use quotation marks not only for quoted texts and names but also to indicate the mention rather than the use of words or symbols.

Some limitations of the chapter need explanation. First, I am not a Greek scholar, but I have checked various technical terms with experts. Heath's reputation for producing a (welcomely) literal translation from the Greek appears to be quite justified; indeed, if points of my interpretation falter because of his translation, then we are all in trouble! Similarly, I shall not discuss any philological or etymological questions, or scribal practices in manuscript editions.

Second, while there have been editions of Euclid based on sources other than that established by Heiberg and translated by Heath, my examination of some others suggests that the differences between them are not significant for my interpretation, although of course, differences of detail arise. Hubertus Busard and Menso Folkerts have recently published an important comparative edition (Busard and Folkerts 1992) of an influential medieval Latin version where such minor variants are evident, but no more.

Third, I take the *Elements* as it stands (in its 13 books, the two numbered 14 and 15 being later interpolations). I am not concerned with the pre- or posthistory of the work, the consequences of its interpretation to other figures, or possible sources in Babylonian mathematics: David Fowler has recently questioned most standard wisdom on these issues (Fowler 1994). Further, I shall mention only one other of Euclid's own works.

Fourth, there are well known epistemological difficulties in the *Elements*: definitions that are not really definitions, constructions and theorems muddled together (6.prop.28 is an example), and so on. I shall pass over these matters as much as possible.

Fifth, being only concerned with the general principles of Euclid's handling of his quantities, I shall ignore most of the theories developed from them: perfect numbers, the "binomial" and "apotome" magnitudes of Book 10 (surd expressions to the algebraists), and so on. I shall also not discuss his various methods of proof.

Finally, while I use some algebraic letters to represent Euclid's procedures and results, I try to minimize the differences from his arithmetic and geometric concerns.

These limitations do *not* reflect any dismissal of my part of the merit of the issues raised; they require the skills of specialists. Several of them are addressed or noted in the literature cited below.

2. The Case against Geometric Algebra

Heath himself saw limitations to reading parts of Euclid as geometric algebra (1926, vol. 1, 373–374): "The algebraical method has been preferred to Euclid's by some English editors; but it should not find favour with those who wish to preserve the essential features of Greek geometry as presented by its greatest exponents, or to appreciate their point of view." At the time of his comments in the mid-1920s, some nonadherents to geometric algebra began to express themselves. For example, E. J. Dijksterhuis avoided it in his presentation of the *Elements* in Dutch in his edition (1929–1930). During the mid 1930s, in an important article on "Greek Logistic and the Origins of Algebra," Jacob Klein was guarded in its use, precisely because he compared and contrasted Greek mathematics with the genuine symbolic algebra of the European Renaissance. His essay was Englished as Klein (1968), appearing at a time when more modern critics began to appear. Arpad Szabo attacked geometric algebra in 1969 in an appendix to a largely philological examination of the origins of Greek mathematics (Szabo 1969).

The most forthright critic was Sabatei Unguru, who wrote a polemical paper (1975) against it. He received replies of similar tone from three mathematicians with historical interests: van der Waerden (1976), Hans Freudenthal (1977), and André Weil (1978). The culprit replied in another journal (Unguru 1979), and with David Rowe he developed his position in more detail soon afterward (Unguru and Rowe 1981–1982). Other recent critics of geometric algebra include Wilbur Knorr on the evolution of the *Elements*, with a special emphasis on incommensurability (Knorr 1975); his study of 1986, where much Greek mathematics is viewed as concerned with geometric problem solving (Knorr 1986); and Ian Mueller, who proposed in 1981 a different kind of algebraic interpretation. The year 1987 was a rich one for further discussions in various contexts: by Fowler (1987), who stressed analogies between Euclid's theory of ratios and continued fractions (without imposing the latter *theory* on the Greeks); by R. H. Schmidt (1987), on the issue of analytic and synthetic proof methods; and by Roger Herz-Fischler (1987), in a detailed study of the method of "division in extreme and mean ratio."

The quality of the dispute over geometric algebra has been somewhat dimmed by the frequent failure on each side to make clear the epistemological place that algebra is held to obtain (or not) in the *Elements*. A very useful distinction was made by

W. R. Hamilton in his paper (1837) on irrational numbers. He distinguished these three categories of algebraic mathematics:

- — "Practical," like an instrument; algebra just produces a convenient set of abbreviations by letters or simple signs for quantities and operations, and rules for the subject at hand (for example, arithmetic);
- — "Philological," like a formula; algebra furnishes in some essential way the language of the pertinent theory; and
- — "Theoretical," like a theorem; algebra provides the epistemological basis for the theory.

Weak forms of the claim of geometric algebra (and arithmetic) invoke the practical category, that we can use algebra simply as a means of representing or abbreviating (some of) Euclid's theorems and definitions. Strong claims, which have been the center of concern, state that the *Elements* uses algebra theoretically (although not philologically): that is, assert the advocates and deny the critics, the work is at root algebraic in its conception, even though arithmetical and geometric in its content.

The principal criticisms of the claim may be summarized as follows.

1. The algebra is simply the *wrong* style: there are no equations, or letters used in an algebraic way in the *Elements.* In other words, the absence of the philological category is quite crucial.
2. Had Euclid been thinking algebraically, he would have presented constructions corresponding to easy manipulations of (11.1), for example, which in fact are absent from the *Elements.*
3. Information is lost when the algebra is introduced, in particular concerning shapes of regions. Thus using "$p + q$" to denote adding, say, two rectangles does not distinguish between them being adjoined at the top, bottom, left, or right (figure 11.1 at (11.1) gives examples). Again, theorems about parallelograms are often (mis)written in terms of corresponding theorems about rectangles (*6.prop.28).
4. Common algebra is associated with analysis in the sense of reasoning from a given result to principles already accepted. Euclidean geometry goes in the reverse, synthetic, direction. Hence proofs may well be warped.
5. Euclid never *measures* a geometrical magnitude of any kind. For example, there is nothing in the *Elements* directly pertaining to π, in any of its four roles for circles and spheres; apparently such mathematics was not Element-ary for him. Hence the association with algebra leads to an emphasis on arithmetic, which cannot be justified.

6. If the Greeks really possessed this algebraic root, why did they not bring it to light in the later phases of their civilization? Why, one might add, did that philosophically sophisticated culture not introduce a word to denote, even if informally, this important notion? This point is strengthened by Klein's real history (1968) of algebra from later Greek figures (especially Diophantos) through the Arabs to the Renaissance and early modern Europeans, for a gradual process in three stages is revealed: (1) using and maybe abbreviating words to denote operations, and known and unknown quantities; (2) replacing these words by symbols or single letters; and (3) allowing letters also to denote variables as well as unknowns, and extending notational systems for powers. The interpretation of Euclid as a geometrical algebraist requires him to have passed all three stages; and while he might have skated through them with greater ease than did his successors, the total silence over his achievement among his compatriots is indeed surprising.

3. Three Features of the *Elements*

To me these criticisms are quite correct; to borrow a word and notation from Lagrange's algebraic version of the calculus, the geometric algebraist constructed from the *Elements* is not Euclid but Euclid′, a fictional figure derived from Euclid's text by means that he helped to inspire in successors but did *not* possess himself. I return in ¶11 to the general methodological point involved here.

The following three features of the geometry in the *Elements* seem to provide clear evidence of central differences between Euclid and Euclid′.

3.1. No Multiplication of Lines

In his geometry, Euclid *never* multiplies a magnitude by a magnitude; for example, the line of length b is never multiplied by itself to produce the square b^2. This is particularly clear in (1.prop.46), where he constructs a square (already defined in 1.def.22): not even there, and nor anywhere else in the *Elements*, is it stated, assumed, or proved that the area of the square is the square of a side. Thus, for example, Pythagoras's theorem, which follows at once with its converse (1.props.47–48), states that two squares are equal to a third one, and the well-known proof works by shuffling around regions of various shapes according to principles of congruence and composition; nowhere are area *formulae* involved. To make an analogy (and no more) with arithmetic, this theorem deals with, say, 9 + 16 = 25, but not with $3^2 + 4^2 = 5^2$. The same point applies to his extension (6.prop.31) of the theorem, concerning similar rectangles laid out on each side of the triangle.

In other words, in Euclid's geometry *the square on the side is not the square of the side, or the side squared;* it *is* a planar *region,* which has this size. In the same way, he never construes the side of a square as the square root of the square; a square can *have* an associated side (*10.prop.54). Heath always translates Euclid's description of such figures by phrases like "square on" [τετραγωνον από] *AB* (*1.prop.47), or the rectangle "contained" [περιεχόμενον] by its sides *AB*, *CB* (*2.prop.4); but unfortunately, in his own commentaries, he writes algebraic equivalents of AB^2 and $AB \times CB$. The standard German translation of the *Elements* by Clemens Thaer (1933–1937) is even more disappointing in this respect, for there terms such as "AB^2" are used in Euclid's text. Of course, the Greeks knew the area property as well as anyone (especially in the taxation office), so its avoidance by Euclid is deliberate. Indeed, he avoided multiplying all kinds of geometric magnitude, for reasons explained in ¶5.

To take another important example, the method of "application of areas" to a line means that a rectangle (or maybe parallelogram) is constructed with that line as a side and equal to some given region, perhaps under further conditions also (*6.prop.28 in ¶1). But no theorems about areas *as areas*—that is, quantities with arithmetically expressible properties—are involved. This eschewal of geometrical products seems to refute the reading as geometric algebra on its own, in all three of Hamilton's categories. For example, (11.1) is an algebraic travesty of Euclid's geometry, since none of its four magnitudes involved appears in his theory.

By contrast, in Euclid's arithmetic, numbers *can* be multiplied: for example, he even calculates triples of numbers, involving squares, that satisfy Pythagoras's theorem (10.prop.28, lemmas 1–2). Thus the algebraic version of his arithmetic is free of this objection (though not of others, as we see in ¶10).

3.2. No Equality of Ratios

While he speaks of the equality of numbers and of magnitudes, Euclid never says that ratios are "equal" to each other, only that they are "in the same ratio," [εν τω αυτω λόγω] or that one ratio "is as" the other in a proportion proposition. Thus, the use of "=" for ratios, normal in geometric algebra, is again a travesty (¶7).

3.3. Compounding, Not Multiplication

Euclid's method of compounding ratios is not at all the same as multiplication, although the two theories exhibit structural similarity (¶7). Similarly, his theory of a lesser integer 1 being "parts" of a greater one *g* is *not* one of rational numbers 1/*g* smaller than unity.

Following the critics of geometric algebra, I claim that Euclid studies numbers, geometrical magnitudes, and ratios in the *Elements*. I shall use "quantity" as a neutral umbrella term to cover these three "types," and when clear I shall just say "magnitudes" for the second one. They are different from each other as objects, and they have a different ensemble of means of combination and of comparison, although some structural similarity is evident between them. In the next six sections I shall amplify this claim, treating the quantities in the above order. Like most other commentators on Euclid, I find mysterious the order in which the books lie: numbers are formally treated in Books 7–9, but they appear in many of the other 10, which are basically concerned with magnitudes; ratios first appear in Book 5 and regularly thereafter, usually within proportions (which are described in ¶7).

4. Euclid's Numbers

Euclid presents a theory of positive integers [αριθμόσ] starting with 2; 1 is a unit, and there is no zero (7.defs.1–2). The basic combinations and comparisons are given in Book 7.[2]

Numbers can be combined under the operations of addition, subtraction "of the lesser from the greater" to ensure that the resultant number is positive, and multiplication (7.defs.5, 7, and 15, and props.18 and 3–4). Numbers may be "equal to" [ισος], "greater" [μειζων] or "less" [ελασσων] than each other; the basic properties of equality are covered by three basic "common notions" in Book 1.

Euclid does not divide integers to produce rational numbers (contrary, once again, to the geometric algebraists' reading discussed in ¶2.5). Instead, a lesser number 1 is "part" [μερος] or "parts" [μερη] of a greater one *g* whether 1 "measures" g or not (7.defs.3–5)—to us, whether 1 is a factor of *g* or not. For example (my own), 3 is part of 9, 3 is parts of 7, and 6 is the "same parts" (*7.prop.6) of 14; but the rational numbers 3/9, 3/7, and 6/14 are not constructed thereby. Similarly, Euclid's rule for finding the least common multiple of numbers, ratios of them, and properties of part (7.props.36–39) cannot be so read.

The only exception is the use of part numbers, mostly "a half" (*13.prop.13), and

2. {Does this construal of integers as multiples of the unit have any connection with one of the original senses of "paradigm," namely, making another (for example) column exactly like the one(s) already made?}

occasionally cases such as "a third part" (12.prop.10 on the volume of a cylinder) and a "fifth" (lemma to 13.prop.18, on measuring angles). They correspond to 1 as part of 2 (or 3, or 5, or . . .) rather than to unit fractions. Presumably their Egyptian parentage gave them a special status (Knorr 1982); an explanation from Euclid would have been helpful! In addition, he uses numbers *in connection with* geometrical magnitudes in a way described in the next section.

In (7.defs.16–18) and thereafter in the books on arithmetic, Euclid presents numbers on occasion as lines, their squares as geometrical squares, and curves as cubes; and he says that a square number has a "side" [πλευρα] (8.prop.11). This overloads the link, however, for he is treating arithmetic within geometry, which is the realm of his theory of magnitudes—and a quite different theory, as we shall now see.

5. Euclid's Geometrical Magnitudes: Glossary

Euclid's theory of numbers deals with discrete quantities; the continuous ones are handled in geometry. Here he works with ten kinds (another general term of mine) of magnitude [μεγεθος], in five pairs, again with no zero;[3] many of them are defined, or so he thought, in (1.defs.1–23). The pairs divide on the property of being straight or curved for each magnitude; in fact, the latter are confined to circles and spheres, although the ideas seem applicable somewhat more broadly, at least to simple concave or convex curves and surfaces.

The pairs of magnitudes are listed below, with an example of each given in brackets. As usual, I distinguish a magnitude from any possibly arithmetical value that it may take: a line has length, a region or surface has area, a solid has volume, and an angle has measure.

Kind	Straight	Curved
lines	straight (line)	planar curved (arc of circle)
regions	planar rectilinear (rectangle)	planar curvilinear (segment of a circle)
surfaces	spatial rectilinear (pyramid)	spatial curvilinear (sphere)
solids	rectilinear (cube)	curvilinear (hemisphere)
angles	planar	solid planar[4]

3. In book 3 Euclid deals with tangents to circles and touching circles; while he skirmishes with curvilinear angles (3.prop.16), he does not seem to use a zero as such.

4. "Solid angles" (Euclid's name) appear only in (11.def.11 and props.20–26), and in the proof that there are only five regular solids (addendum to 13.prop.18); but they are handled in the same way as other magnitudes. Similarly, the theory of solids is far less developed than that of lines and regions—to Plato's distraction in *Republic* 528a–d (Fowler 1987, 117–121).

In these examples I distinguish between, say, a circle as a closed curved line and as a convex region, or a cube as a closed surface and as a convex solid; Euclid does the same, without confusion. He does not consider curves in space, or surfaces or solids set upon some kind of rectilinear base; probably he had no examples in mind of the latter.

The distinction between straight and curved magnitudes is a natural one; for example, it plays a key role in the theory of limits. Its place in Euclid (and many other Greek mathematicians) is evident in Book 10, where proofs of theorems relating recti- and curvilinear solids work by double contradiction ($A = B$ because both $A < B$ and $A > B$ lead to absurdities) rather than a direct process of limit taking, which would cross this conceptual boundary.

In addition, Euclid has on occasion "multitude" [πληθος] of a quantity, an informal idea referring to an unspecified number of them (*7.prop.14 for numbers; *10.def.3 for magnitudes). He also speaks of "equal in multitude," which we would treat as a 1-1 correspondence between members of two such collections (*5.prop.1 for magnitudes).

This leaves the status of points. The famous (1.def.1), defining a point "as that which has no part," is well recognized as a failed definition. I take it to be a principle of *atomicity*, asserting the existence of a point as a "primitive" part of a magnitude; like some thing inside a sphere, for example, or the place where two lines intersect or where a line has an extremity. It is striking to note that Euclid actually wrote of a "sign" [σημειον] and in particular cases said "this *A*," not "the point *A*." Perhaps inspired by Plato, the change from "point" [στιγμη] was maintained by most of Euclid's contemporaries and successors, although Aristotle used both words (Mugler 1958, 376–379). But the old tradition was to prevail again from the fifth century, when in the spirit of Roman empiricism, translators and commentators Martianus Capella and Boethius rendered "σημειον" as "punctum" (Heath 1926, vol. 1, 155–156). However, the reformers had had a good case; "sign" makes clear that the "objects" denoted do not admit the means of combination or comparison to which magnitudes are subject, and which we now examine.

6. Euclid's Geometrical Magnitudes: Handling within Each Kind

The key feature of Euclid's treatment of magnitudes is that, with an important exception to be noted in ¶9, the means of combinations and comparisons are treated *between magnitudes of the same kind*—in my view, *consciously and intentionally.* Since his magnitudes vary considerably in character (from a line to a solid), he does not always use the same word for the same sort of combination, and in some cases no word at all; but they may be fairly characterized as follows.

Magnitudes of the same kind may be added or subtracted, the latter combination restricted to positive resultant magnitudes.[5] They may also be multiplied by numbers; when the number is an integer, a "multiple" of the magnitude is produced, but there may also arise, for example, the one-and-a-half of a square in (13.prop.13) of ¶4. The converse never happens; that is, numbers are never multiplied by magnitudes. This difference shows on its own that they are different types of quantity.

Their comparisons are "equal to," "greater than," or "less than" (1.common notions.4–5, post.5). Euclid uses the same words as for numbers, but only structurally similar notions are involved, not identical ones (¶11). He does not equate a rectilinear region with a curvilinear one; indeed, in connection with the famous problem of squaring the circle, his commentator Proclus (fifth century A.D.) explicitly mentioned this possibility as a worthwhile *research* topic (Proclus 1970, 334–335). Euclid may well have deemed this problem, and similar ones such as squaring lunes, as not Element-ary—and with good justice.

Once it is recognized that Euclid handles magnitudes of the same kind, the reasons for his avoidance of their multiplication become clear. First, *the kind would be changed*; for example, the product of a straight line and a straight line is a rectilinear region. Second, many of the multiplications cannot be defined anyway (angle with angle, line with angle, solid with line, and so on); so for uniformity he omits *all of* them.

This point must not be confused with the fact that many theorems *involve* magnitudes of different kinds at the same time; for example, there are many in Book 10 in which lines and rectilinear regions appear together. However, no means of combination or comparison *between* kinds occurs.

7. Euclid's Ratios and Proportions

The "ratio" [λογος] is Euclid's third type of quantity; it is specified only between two numbers or between two magnitudes of the same kind. Following the seventeenth-century English astronomer Vincent Wing, I shall write "a:b" for the ratio of two such quantities a and b.

Euclid's presentation is not clear. The ratio of numbers is not formally defined at

5. Euclid would not have regarded *every* possible combination as meaningful: for example, an arc to a circle (considered a rectilinear curve), or at the end of an infinitely long line (as permitted to exist [1.def.23] by the parallel postulate [1.post.5]).

all, but it creeps in first in (7.def.20) in a proportion about the sameness of two pairs of ratios, and is used in a theorem about subtraction (7.prop.11). Further, he does not stress that the ratio, say, 3:7 is *different in type* from 3 being parts of 7, which is a property *within* arithmetic (¶4).

Something like a definition of the ratio of magnitudes appears in (5.def.3): "A ratio is a sort of relation in respect of size between two magnitudes of the same kind." The word "size" [πηλικότης] denotes the measure of the ratio. This definition is best interpreted as creative, although, of course, he had no theory of the sort that has been developed in this century, that a definition is creative relative to a theory if there exist theorems in it which cannot be proved without using the definition.

Comparisons between a pair of ratios are expressed in the companion theory, that of "proportions" [αναλογια] between a pair of ratios. In addition to bringing some clarity to the role of these mysterious ratios, the main advance of proportion theory over magnitude theory is that ratios of magnitudes of *two different kinds* can be compared (for example, between lines and regions in [6.prop.1]), or compared with ratios of numbers, by comparison of their sizes. For magnitudes a ratio may be "in the same ratio as," "greater than" (5.defs.5–7) or "less than" (not formally defined) the other one; the comparison in question is established in Eudoxus's manner by examining (in)equalities between multiples of the magnitudes involved.[6] The two theories of sameness are quite different, showing again that numbers and magnitudes are distinct types of quantity.

Euclid's total avoidance of the word "equals" (ιδος) for ratios shows in the clearest manner possible that he saw them as a third type of quantity distinct from numbers and magnitudes (or perhaps as a relation between its components). Hence I follow the practice initiated by Wing's wise contemporary William Oughtred in the seven-

6. Euclid never presents proportions between trios or greater multitudes of numbers or of magnitudes of the same kind. Presumably the reasons were that the case of nonsameness cannot be defined (sameness presents no difficulties), and that anthyphairesis becomes very hard to execute. Similarly, his theorems in book 8 on numbers in "continued proportion" (geometric progressions to us, and also to Weil [1978] in his rather marginal objection to Unguru) work by taking only neighboring pairs of numbers together. A geometric version of such sequences would be easy to produce, as pairs of bases of these triangles, which have successive right angles at *A*, . . . *G* . . .

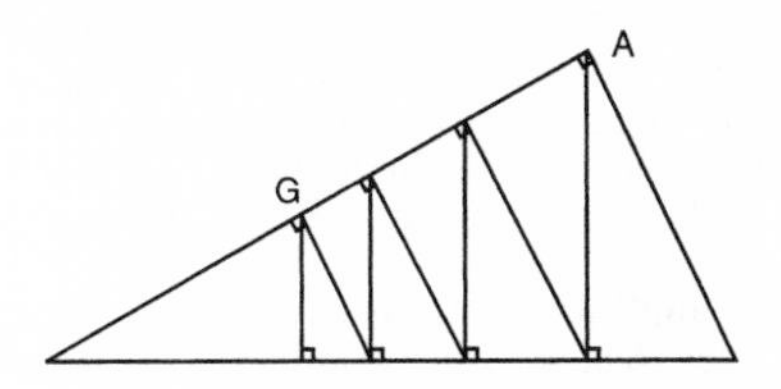

Figure 11.2

teenth century in symbolizing sameness by "::". Writing a, b, c, d for magnitudes, we never find equations such as, say,

$$a : 2b = c : d, \text{ but the proportion } a : 2b :: c : d. \tag{11.2}$$

The habit of the geometric algebraist to write "=" between ratios is inadmissible, as is the habit of drawing consequences involving the multiplication of magnitudes, such as

$$a \times d = 2b \times c. \tag{11.3}$$

One of the most extensive uses of magnitudes and ratios in the *Elements* is the theory of (in)commensurable and (ir)rational[7] magnitudes (mostly straight lines and rectilinear regions) in the virtuoso Book 10. These properties are based on "anthyphairesis," the successive subtraction of lesser from greater magnitudes of the same kind so as (not) to produce a ratio the same as that of two integers respectively (10. props.2 and 5–6). Note that it is not a theory of irrational *numbers.* Indeed, Euclid proves that two magnitudes are incommensurable if and only if their ratio "have not to one another the ratio which a number has to a number" (10.props.7–8). Therefore an error in type of quantity is committed when it is compared with Dedekind's theory of irrationals, unless one invokes from Dedekind's side the so-called Cantor-Dedekind axiom to set each irrational number in isomorphism with a geometric length. Indeed, Euclid may have taken up some existing theory of such numbers and translated it into geometrical terms, since only in geometry can the *continuum* of magnitudes upon which it depended be guaranteed.

However, numbers do have a role in anthyphairesis, as counters of repetitions. For example, the magnitude 7 subtracts from 46 six times, with remainder 4; 4 from 7 once with remainder 3; and so on, generating the sequence 6, 1, . . .[8] These ellipsis dots can encompass an unending sequence of remainders, sometimes periodic, with remarkable properties (Fowler 1992). They are still not well studied, although the (arithmetical) theory of continued fractions bears some structural similarity to them (Fowler 1987, chs. 5 and 9).

7. With little enthusiasm, I follow the tradition of translating "(α)λογος" as "(ir)rational"; something like "(in)expressible" is much better (as Heath himself notes in 1926, vol. 3, 525). Euclid defines irrationality from incommensurability by assigning a line as a basis relative to which such ratios can be defined his way (the rather messy [10.defs.3–4], which cover also [in]commensurability in square).

8. Maybe anthyphairesis inspired Euclid's use of the curious term "parts" for integers not in factorial form (¶5): 7 is parts of 46 because 46:7 leads to repetition numbers 4, 1, . . . whereas 7 is part of 42 because 42:7 yields 6 "straight off." The term may also have come from talk of the time on unit fractions and calculations.

8. Euclid's Ratios: A Musical Background?

Why did Euclid always avoid speaking of equal ratios? It is plausible that, like his contemporaries and predecessors, he understood ratio theory as *generalized music,* at least culturally (Barker 1989) and perhaps even philosophically. Whatever the Pythagoreans did or did not do (and doubtless more sources of information were available then than now), properties of strings and comparisons between tones and lengths were a major part of the mathematics of their time, and had remained so until Euclid's.[9] In such contexts, the notion of equality is not as natural as with numbers or magnitudes. *We* usually say that the intervals F♯–A and B–D are the same interval (here a minor third) rather than that they are equal—for the first is placed a major fifth above the second one. Further, the trio of Euclidean quantities, number-magnitude-ratio, surely bears an intentional cultural correlation with three subjects of the Aristotelian quadrivium, arithmetica-geometria-harmonia.

This association with music is evident also in the method of combination to which ratios are subject: they are to be "compounded" [συγκειμενον], tardily in (6.prop.23). In the simplest case, the ratio $a{:}b$ may be put together with $b{:}c$ to produce the ratio $a{:}c$ (5.def.9), the "duplicate ratio" of the original pair. This procedure is clearly similar in structure to taking the musical intervals, say, D♯–F♯ and F♯–C to produce D♯–C. Further, as with music, the process may be repeated, to produce the "triplicate ratio" $a{:}d$ after three stages (5.def.10), and the "ex equali" proportion that arises from the ratio $a{:}n$ achieved after N stages (5.def.17).

Clearly, one cannot follow (some) geometric algebraists and identify this procedure with multiplication, although there is obvious structural similarity between the two means of combinations (in the practical category but not the theoretical, Hamilton might say). But note that even if the quantities are all numbers, a compounded ratio results, not an arithmetical product. For example, with the symbol "•" denoting compounding, note the difference between the propositions

$$(3{:}7) \bullet (7{:}11) :: 3{:}11 \text{ and } (3/7)(7/11) = 3/11; \tag{11.4}$$

not only do they differ by type of quantity, but the latter proposition cannot even be stated in the *Elements* (recall ¶4).

Further, unlike a general definition of multiplication, Euclid does not offer

9. In an undeservedly neglected essay, McClain (1978) argues that ratios were a fundamental theory in many cultures, and he suggests novel interpretations for several aspects of Plato. Before Euclid, other applications had been made of ratios, for example, to astronomical motions and calendars.

the *general* definition of compounding *any* two ratios of magnitudes, for it would read

$$(a{:}b) \bullet (c{:}d) :=? ([a \times c] : [b \times d]), \tag{11.5}$$

which involves the forbidden product of magnitudes. The product could be replaced by regions in certain cases, such as the rectangles contained by *a* and *c*, and by *b* and *d* if all four magnitudes were lines—but *not* in full generality, for the reasons explained in ¶6. Nevertheless, it appears in (6.prop.23), the interesting but isolated proposition that substitutes for the multiplication (11.1) of sides: it states that the ratio of two equiangular parallelograms is the compound of the ratios of the respective pairs of sides.[10] In the proof, the given parallelograms *ADCB* and *CGFE* are aligned so as to admit the intermediate parallelogram DHGC in the (gnomon-like) figure 11.3, so that

$$\square ADCB : \square CGFE :: (\square ADCB : \square DHGC) \bullet (\square DHGC : \square CGFE) \tag{11.6}$$

$$:: (BC : CG) \bullet (CD : CE). \tag{11.7}$$

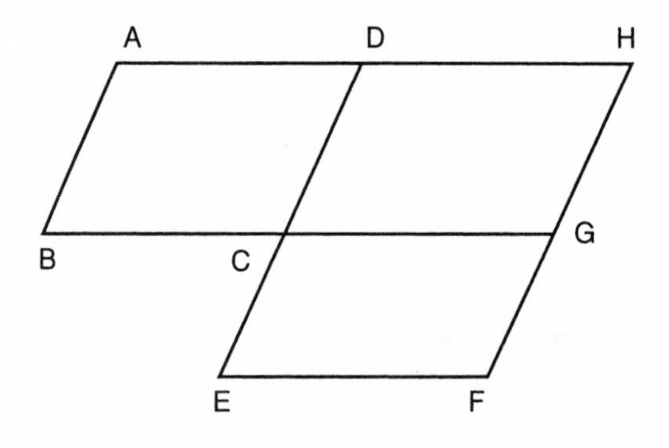

Figure 11.3

10. This is the proposition in which Euclid first uses the word "compounded"; (6.def.5) is an interpolation, not always included in editions. While the theory as such seems clear, his presentation is not good, either at (6.prop.23) or elsewhere. Compounding occurs in (6.props.19–20) on taking ratios of areas of triangles and polygons; in (8.prop.5) on plane numbers and (8.props. 11–12) on determining mean proportions between numbers; in (12.props.12–18 *passim*) on various propositions mentioned in ¶9 concerning the volumes of cones, cylinders, and spheres; and in (13. prop.11) in making a square 25 times bigger than another one. Contrary to Heath's editorial remarks, I do not see compounding necessarily present in (5.props.20–23), where various theorems on ratios of the same value are proved. Saito provides a good discussion of compounding (unfortunately notated "+") in Euclid and also in Apollonius; however, he considers possible theorems involving (11.5) (Saito 1985, 32, 38, 59) and uses "=" in proportions.

Compounding ratios is not to be confused with their composition (5.def.14): given *a*:*b*, form (*a* + *b*):*b*.

9. Mixed Kinds: The Alternation Theorem

One feature in Euclid's theory of ratios needs to be resolved: the alternation theorem (5.prop.16), according to which, in the notation of (7.1) for magnitudes,

$$\text{if } a : b :: c : d, \text{ then } a : c :: b : d. \tag{11.8}$$

There is a surprise: Euclid asks only that the pairs a and b, and c and d, be each of the same kind, not that the whole quartet be so. Thus, as it stands, the theorem asserts in its second proposition the sameness of two "mixed ratios" (my name), where a pair of ratios of the same different magnitudes are related. This feature has often been regarded as a slip (by Heath, for example, in his commentary on the theorem [1926, vol. 2, 165]). In the influential earlier edition by Robert Simson, the seemingly missing clause was even interpolated into the text.

But in my view Euclid has *not* made a mistake here (although, regrettably, he supplies no explanation either); for in one context he uses mixed ratios. This is in Book 12, where he presents his version of Eudoxus's theory of exhaustion of curvilinear regions by rectilinear ones. The first occasion arises when he proves that "circles are to one another as the squares on the diameters" (12.prop.2). There he constructs regular polygons inside each circle and then uses (11.8) to show the sameness of the ratios of each polygon to its parent circle.[11] The same procedure occurs in the companion theorem about spheres (12.prop.18), and in two theorems about cones and cylinders on circular bases (12.props.11–12). The proofs work by double contradiction, the type mentioned in ¶5.

Euclid's extension of proportion theory to mixed ratios is minimal, in that the magnitudes involved are at least of the same dimension each time. But it goes beyond the practice elsewhere in the *Elements*—and as an extension it escaped Heath's attention.

10. Euclid's Ontology

Euclid treats his quantities in a very direct way: he *has* numbers, angles, lines, regions, ratios, and so on. No constructed object depends on the means of construction in

11. There may be a slip in the diagram attached to this theorem. Two "areas" S and T play roles in the proof, and mainly in the comparison of equality to circles. Therefore, even granted that both recti- and curvilinear regions are involved in the proof, surely the latter shapes are more appropriate than the rectangles that have been drawn. This feature seems to be of long standing; for example, it is evident in the recent medieval Latin version of the *Elements* edited by Busard and Folkerts (1992, vol. 1, 294). Little seems to be known about the history of the diagrams in the *Elements*.

a reductionist sense, such as a square *as* the product of its sides. Thus, directness is also evident in his *Data*, where he presents a string of geometric exercises; for each one is expressed in the form that "if so-and-so properties of (say) a triangle are given, then such-and-such properties are given." The phrase "are (is) given" [δεδομενον] is the leitmotif of the entire work.[12]

In his work Euclid uses three different types of quantity, often together but with their theories distinguished. Much similarity of structure obtains between them; but there are also essential differences, so that identity of content *cannot* be asserted. This point is an important point in the philosophy of real mathematics when analogy plays a prominent role (Grattan-Guinness 1992; §7.5.2). Unfortunately the geometric algebraists miss it, for they assert that the *same* algebra obtains for each type of quantity; as Nesselman put it (1847, 154): "Allow us, however, to consider and to treat as arithmetical under our kind of thinking this theory that Euclid proposed geometrically . . . thus it would be hard to lay down a strong boundary between form and content as a ground for division."

While a perfectly legitimate reading of Euclid′, it is a distortion of Euclid and can lead to confusions—for instance, in identifying his theory of irrational magnitudes with modern ones of irrational numbers, or regarding numbers as special kinds of magnitude (see, for example, Heath [1926, vol. 2, 124, 113] respectively). Each type of quantity in Euclid has to be considered *separately* for its possible algebra. Converse to the phrase "geometric algebra," I shall use "algebraic" as an *adjective* in each case, to show that it qualifies the succeeding quantity noun.

First, there might be a case for algebraic arithmetic, though at most only for the practical category and *only* for integers and unit fractions. However, the fact that (unfortunately) Euclid gave numbers a geometric interpretation reduces the quality of the case even there—which seems to be the only place in the *Elements* where the phrase "geometric arithmetic" could be justified. Among his successors, a philological category for algebraic arithmetic begins to emerge only with Diophantos in the third century A.D., and then not fully.

Second, any theory of algebraic magnitudes must be radically different from geometric algebra. The avoidance of multiplication of magnitudes rules it out on its own, and the other criticisms rehearsed in ¶2 are very powerful. Mueller (1991) has given a cogent account of several of them; however, his alternative, a sort of logico-

12. Taisbak (1991) argues that "are given" refers to static states of affairs rather than dynamic ones of potential movement. This may also be a motive, though I prefer to give priority to a more simple explanation—that he means just what he says. Maybe "be the case that" would be a better translation. See also Schmidt (1987) on this matter.

functorial algebra, seems also to take us far from Euclid's numbers and magnitudes, although in a different direction from that of the geometric algebraists—to Euclid″, say.[13] Abstract algebra, such as groups and fields, would set us off down yet another irrelevant track.

Finally, a case for algebraic ratios might be argued, as long as compounding is not identified with arithmetical multiplication. However, the background in music provides a far more faithful orientation.

If one wishes to pursue algebraization at all, symbols corresponding to the shapes might be introduced, such as "□ *a*" for the square on side *a* and "○ *a*" for a circle with that diameter; Heath used such notations occasionally (*12.prop.7). However, even modern computers do not always readily supply the required symbols. Another good strategy is that of Dijksterhuis (1929–1930), who used functorial letters such as "*T(a)*" for the square on side a, and "O(*a*,*b*)" for the rectangle with sides *a* and *b*.

Should algebralike symbols be desired, different symbols *must* be used for the "same" means of combination and comparison when applied to different quantities. A possible choice could be these:

Quantity	Addition	Subtraction	Division[14]	Multiplication/ compounding	Equality/ sameness	Greater than	Less than
Numbers	+	−	:	×	=	>	<
Magnitudes		~	:		≈	»	«
Ratios				•	∷	⸬	⸬

Operations on magnitudes apply within each kind, of course, with the extension to cover exhaustion theory as described in ¶9. The same symbol ":" is proposed for the ratio of numbers and of magnitudes precisely because a ratio is created on each occasion. In the case of magnitudes and ratios, the means of combination are to be understood in a rather abstract, operation sense. While still far away from Euclid, a closer sense to him than geometric algebra is furnished by the connotation of the interior and exterior products that Hermann Grassmann proposed in his *Ausdehnungslehre* of 1844. It is a great irony that his work was gaining publicity in the 1880s and 1890s, *exactly* when Heiberg and Tannery were trying to convert Euclid into Descartes!

13. Mueller (1991) has, for example, SIMSOLID (*k*,1) for solid (that is, cubed) numbers *k* and 1, followed by a definition in terms of mathematical logic. But a significant anachronism is evident here, for his use of quantification deploys set-theoretic interpretations of the quantities involved, whereas Euclid's own treatment of collections follows the very different part-whole tradition of handling collections.

14. I use the word "division" in its modern sense, not in a way alluding to anthyphairesis.

11. The Tenacity of Algebraic Thinking

As the case of Euclid ≠ Euclid′ exemplifies, algebra is not among the ancient roots of mathematics, and to impose it on Euclid distorts and even falsifies his intentions. Moreover, the point is not restricted to the Greeks, for claims of similarity between their mathematics and that of the Babylonians and the Chinese have been grounded in the alleged common factor of geometric algebra. The principle seems to be a (mistaken) application to history of Euclid's own (1.ax.1), that "things which are equal to the same thing are equal to each other."

Although not an ancient source for mathematics, algebra and algebraic thinking and styles have long assumed a central role in much mathematics and mathematical education. (Recall how often children identify mathematics with equations and formulae, often in disgust.) The case, and even the *name* of "geometric algebra," shows the tenacity of the algebraic style (Grattan-Guinness 1996), for it characterizes an algebra, which is geometric, not even a geometry (or arithmetic) that might be algebraic. Note, by contrast, that nobody in the late nineteenth century suggested that Aristotle's syllogistic logic was a logical algebra, although it influenced substantially (though partly by reaction) the development of algebraic logic at that time (Grattan-Guinness 1988).

The underlying philosophy behind the interpretation of the *Elements* as geometric algebra, common to mathematicians' and sometimes even historians' understanding of the history of mathematics in general, is *historical confirmation theory* (ch. 2). Suppose that a theory T_1, created during epoch E_1, is followed by later theory T_2 in epoch E_2; then at epoch E_3 the (non)historical reading is proposed that T_1 was conceived by its creators as an *intended* draft of T_2, and thereby confirms it. Roll on history, deterministically (Popper 1957).

While this reading might be correct or at least well arguable in some cases, it should be treated with much caution, since it may well propose as actual developments those that were only potential. For example, Euclid's *Elements* undoubtedly influenced the real development of algebra among the Arabs and then (in a rather different form) during the Western Renaissance; but it does not at all follow that Euclid himself had been trying to be a geometric algebraist (Unguru 1979). The history of the various readings that Euclid has received over the centuries, especially regarding his types of quantity and their handling before the era of geometric algebra, would be a valuable contribution to historiography. He may have been better understood by (some!) later Greeks and in the Middle Ages than later, when algebraization began to develop in the stages outlined in ¶3. The final quartet of specific examples illustrates the point.

First, Heath's edition provides a frequent oscillation between Euclid in English and Euclid′ in algebra, sometimes to such an extent that he actually attributes a different procedure to Euclid. The example (6.prop.28) of ¶1 on constructing a certain parallelogram is a very good case. "To exhibit the exact correspondence between geometrical and the ordinary algebraical method of solving the equation," Heath sets up a Euclid′-style quadratic equation (in which Euclid's parallelograms are—consciously—replaced by rectangles), and after some working he calculates an expression for a certain line *GO* in the original proof. However, two pages earlier Euclid mentions *GO* only once (Heath 1926, vol. 2, 262–265).

Second, in his reply to Unguru, van der Waerden clearly explains that he uses the word "algebra" "for expressions like $(a + b)^2$, and how to solve linear and quadratic equations" (1954, 200). He then takes (2.prop.1), in Heath's translation: "If there be two straight lines, and one of them be cut into any number of segments, the rectangle contained by the two straight lines is equal to the rectangles contained by the uncut straight line and each of the segments," and comments (figure 11.4): "Geometrically, this theorem just means that every rectangle can be cut into rectangles by lines parallel to one of the sides. This is evident: everyone sees it by just looking at the diagram. Within the framework of geometry there is no need for such a theorem: Euclid never makes use of it in his first four books."

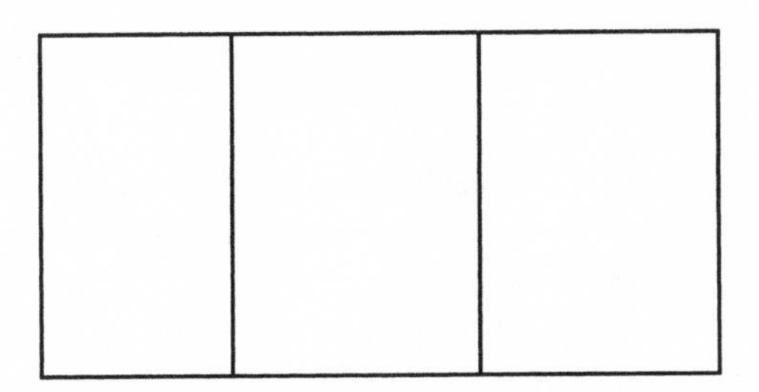

Figure 11.4

But Euclid's text and van der Waerden's version say different things: Euclid builds up the full rectangle *R* from its components to *form* rectangles and state a property of them (Saito 1986, 54–60); van der Waerden breaks it up into those components from *R*. Of course the theorem is very simple; hence its location at the head of Book 2. But the construction embodied here (recall from ¶2.4 that Euclid often conflated them with the theorems themselves) underlies or at least relates to many theorems from Book 2 onward, starting with (2.prop.2), long before Book 5 is reached.

Third, Freudenthal's reply (1977) to Unguru defending geometric algebra is unintentionally amusing, for he explicitly praises Dijksterhuis's edition as a source to read Euclid, and then makes all the mistakes that Dijksterhuis avoids—multiplication of magnitudes, equality between ratios, compounding as multiplication, and so on!

The final pair of examples are small but interesting ones, since they come respectively from a critic and an agnostic of geometric algebra. Both involve the unfortunate practice of setting "=" between ratios in a proportion. Szabo makes it in the very book that later on contains his attack on geometric algebra mentioned in ¶2 (1969, 131). So does Benno Artmann, in a nice recent short survey of the *Elements*, in which he regards the issue of geometric algebra as "moot" (1991, 7, 45–47).

Many other examples of this and other dubious procedures can be found, even in several other items in the bibliography here. Algebraic thinking holds on tight; as Peyrard declared already in (1814, ix), "Mr. Lagrange . . . often repeated to me that Geometry was a dead language."

ACKNOWLEDGMENTS

For advice on Euclid and the literature, I am grateful to various friends, especially David Fowler, Erwin Neuenschwander, Maria Panteki, and Imre Toth. Sharp refereeing helped me to tighten things up nicely.

BIBLIOGRAPHY

Artmann, B. 1991. "Euclid's *Elements* and its prehistory," in Mueller 1991, 1–47.

Barker, A. 1989. (Ed.), *Greek musical writings*, vol. 2, *Harmonic and acoustic theory*, Cambridge: Cambridge University Press.

Busard, H. L. L., and Folkerts, M. 1992. (Eds.), *Robert of Chester's (?) redaction of Euclid's Elements, the so-called Adelard II version*, 2 vols., Basel, Switzerland: Birkhäuser.

Dijksterhuis, E. J. 1929–1930. *De Elementen Van Euclides*, 2 vols., Groningen, Netherlands: Noordhoff.

Fowler, D. H. 1987. *The mathematics of Plato's Academy*, Oxford: Clarendon Press. {2nd ed. 1999.}

Fowler, D. H. 1992. "An invitation to read Book X of *The Elements*," *Historia mathematica, 19*, 233–264.

Fowler, D. H. 1994. "The story of the discovery of incommensurability, revisited," in K. Gavroglu and others (eds.), *Trends in the historiography of science*, Dordrecht, Netherlands: Reidel, 221–235.

Freudenthal, H. 1977. "What is algebra and what has it been in history?" *Archive for history of exact sciences, 16*, 189–200.

Grattan-Guinness, I. 1988. "Living together and living apart: On the interactions between mathematics and logics from the French Revolution to the First World War," *South African journal of philosophy, 7*, no. 2, 73–82.

Grattan-Guinness, I. 1992. "Structure-similarity as a cornerstone of the philosophy of mathematics," in J. Echeverria, A. Ibarra, and T. Mormann (eds.), *The space of mathematics: Philosophical, epistemological, and historical explorations*, Berlin and New York: De Gruyter, 91–111.

Grattan-Guinness, I. 1996. "Normal mathematics and its histor(iograph)y: Characteristics of algebraic styles," in E. Ausejo and M. Hormigon (eds.), *Paradigms in mathematics*, Madrid: Siglo XXI, 203–213.

Hamilton, W. R. 1837. "Theory of conjugate functions, or algebraic couples . . . ," in *Mathematical papers*, vol. 3, Cambridge: Cambridge University Press, 1967, 3–96. [Original publication 1837.]

Heath, T. L. 1926. (Ed. and trans.), *The thirteen books of Euclid's Elements*, 2nd ed., 3 vols., Cambridge: Cambridge University Press. [Repr. New York: Dover, 1956.]

Herz-Fischler, R. 1987. *A mathematical history of division in extreme and mean ratio*, Waterloo, Canada: Wilfred Laurier University Press.

Klein, J. 1968. *Greek mathematical thought and the origins of algebra*, Cambridge, Mass.: MIT Press. [German original 1934–1936.]

Knorr, W. R. 1975. *The evolution of the Euclidean Elements*, Dordrecht, Netherlands: Reidel.

Knorr, W. R. 1982. "Techniques of fractions in ancient Egypt," *Historia mathematica, 9*, 133–171.

Knorr, W. R. 1986. *The ancient tradition of geometric problems*, Basel, Switzerland: Birkhaüser.

McClain, E. G. 1978. *The Pythagorean Plato*, York Beach, Maine: Nicolas-Hays.

Montucla, J. E. 1799. *Histoire des mathématiques*, 2nd. ed., vol. 1, Paris: Agasse. [Repr. Paris: Blanchard, 1968.]

Mueller, I. 1981. *Philosophy of mathematics and deductive structure in Euclid's Elements*, Cambridge, Mass.: MIT Press.

Mueller, I. 1991. (Ed.), ΠΕΡΙ ΤΩΝ ΜΑΘΗΜΑΤΩΝ, Edmonton, Alberta (*Apeiron, 24*, no. 4).

Mugler, C. 1958. *Dictionnaire historique de la terminologie géométrique des grecs*, Paris: Klincksieck.

Nesselmann, G. H. L. 1847. *Die Algebra der Griechen*, Berlin: Reimer. [Repr. Frankfurt: Minerva, 1969.]

Peyrard, F. 1807. *Oeuvres d'Archimède, traduites littéralement, avec un commentaire*, Paris: Bouisson.

Peyrard, F. 1814. *Les Oeuvres d'Euclide, traduites en latin et en français, d'après un manuscrit très-ancien qui était inconnu jusqu'à nos jours*, vol. 1, Paris: Patris.

Popper, K. R. 1957. *The poverty of historicism*, London: Routledge and Kegan Paul.

Proclus. 1970. *A commentary on the first Book of Euclid's Elements* (trans. and ed. G. R. Morrow), Princeton, N.J.: Princeton University Press.

Saito, K. 1985. "Book II of Euclid's *Elements* in the light of the theory of conic sections," *Historia scientiarum*, no. 26, 31–60.

Saito, K. 1986. "Compounded ratio in Euclid and Apollonius," *Historia scientiarum*, no. 30, 25–59.

Schmidt, R. H. 1987. "The analysis of the ancients and the algebra of the moderns," in *Recipients, commonly called the Data*, Fairfield, Conn.: Golden Hind Press, 1–15.

Szabo, A. 1969. *Anfänge der griechischen Mathematik*, Muinch and Vienna: Oldenburg.

Taisbak, C. M. 1991. "Elements of Euclid's *Data*," in Mueller 1991, 135–171.

Tannery, P. 1882. "De la solution géométrique des problèmes du second degré avant Euclide," in *Mémoires scientifiques*, vol. 1, Paris: Gauthier-Villars, 254–280.

Thaer, C. 1933–1937. (Ed. and trans.), *Euklid. Die Elemente*, 4 vols., Leipzig, Germany: Engelsmann (*Ostwalds Klassiker*, nos. 236, 240, 241, 243). [Several reprs., usually in 1 vol.]

Thomas, I. Bulmer-. 1939–1941. (Ed.), *Selections illustrating the history of Greek mathematics*, 2 vols., London: Heinemann.

Unguru, S. 1975. "On the need to rewrite the history of Greek mathematics," *Archive for history of exact sciences, 15*, 67–114.

Unguru, S. 1979. "History of ancient mathematics: Some reflections on the state of the art," *Isis, 70*, 555–565.

Unguru, S., and Rowe, D. E. 1981–1982. "Does the quadratic equation have Greek roots? A study of 'geometric algebra,' 'application of areas,' and related problems," *Libertas mathematica, 1*, 1–49; *2*, 1–62.

van der Waerden, B. L. 1954. *Science awakening*, Groningen, Netherlands: Noordhoff. [Repr. 1963, New York: Wiley.]

van der Waerden, B. L. 1976. "Defence of a 'shocking' point of view," *Archive for history of exact sciences, 15*, 199–210.

Weil, A. 1978. "Who betrayed Euclid?" *Archive for history of exact sciences, 19*, 91–93.

Zeuthen, H. G. 1886. *Die Lehre von den Kegelschnitten im Altertum*, Copenhagen: Fischer-Benzon. [Repr. Hildesheim, Germany: Olms, 1966.]

CHAPTER TWELVE

Some Neglected Niches in the Understanding and Teaching of Numbers and Number Systems

Cultural questions have attended arithmetic since it began to develop in ancient times. They include possible differences between integers and non-integral numbers and in operating with them, religious and mystical uses and interpretations, the roles of zero, extensions to infinite numbers, and representing numbers by numerals in ways that aid calculation (including the use of algebra). The selection of historical examples given here concentrates on aspects of numbers that are not well known but that could be used in teaching, either at school or undergraduate level. Comments on educational utility are given, mostly at the end of each section; the word "student" refers to learners of all ages.

1. Ancient Number Things

The origins of numbers and their arithmetic cannot be known for certain, but it is likely that both arose in all cultures from counting and commercial exchange. In such contexts numbers can be regarded philosophically merely as marking stages in counting. Ancient texts, especially Greek and Chinese, give the impression that, say, 3 is about three *things* of some kind, even if no particular kind is mentioned in general explanations. The word "calculus" in Latin means "pebble," hinting at the role that tokens had played in the development of counting, and maybe even of writing itself (Schmandt-Besserat 1992).

The original version of this essay was first published in *Zentralblatt für Didaktik der Mathematik, 30* (1998), 12–19; ¶5 and ¶6 were first published as parts of "A Note on Some Elementary Arithmetic Algorithms," *International journal of mathematical education in science and technology, 3* (1972), 307–310. Reprinted by permission of the editor of *Zentralblatt für Didaktik der Mathematik* and of Taylor & Francis (www.informaworld.com).

Fractions and ratios raise further conceptual difficulties, for a rational number seems to be a different sort of thing from an integer, even though the rules for adding and subtracting are similar (Benoit and others 1992). And irrational numbers are even more "risky" ("lawless" would be a better translation for the Greek word "alogos" than "irrational").

In applications to abstract concepts (in science, for example), the perplexities can become great. Regarding multiplication and division, a long tradition stemmed from the Greeks of working with ratios and proportions (that is, sentences relating a pair of ratios) rather than equations. Let a, b, c, and d be numbers, and contrast the proportion

$$a{:}b :: c{:}d \text{ with } ad = bc \text{ and } a{:}c :: b{:}d. \qquad (12.1)$$

If a and b refer to the same kind of thing (force, say, or volume), and so do c and d (but *not* necessarily the same kind), then $a{:}b$ and $c{:}d$ will be *dimensionless* ratios, so that their (in)equality with each other can(not) be asserted in $(12.1)_1$ via "::" without qualms. This was the way Euclid worked with ratios and proportions, relating, say, a pair of lines to another pair of lines or to a pair of numbers (§11.3). However, $(12.1)_2$ poses a conceptual difficulty about dimensions; ad and bc may be equal, but in what units? Similarly, the brother proportion $(12.1)_3$ was maybe better avoided if the ratios were of two different kinds of thing.

Such emphases on proportions may have been transferred to metaphysical views about the construction of the heavens, the chronology of myths, and other cultural concerns in which rhythm or periodicity was held to apply (McClain 1978). The phrase "music of the spheres" may carry more overtones than is now realized.

In several cultures an early interest in proportion arose from musical harmony. The octave was set at 2:1, the perfect fifth at 3:2, and so on—but how can they fit together, since no power of 2 can equal any power of 3/2? Various systems of "temperament" have been devised to place the 8 tones or 12 semitones within the octave, especially in the late Middle Ages (Cohen 1984); for example, the devilish character of the midnote "tritone" (the augmented fourth) was associated with the irrational ratio $\sqrt{2}{:}1$.

During that period the theory of ratios became rather more arithmetical and less linked to geometry via dimensions. In the new approach it was permitted to state, say,

$$a : b = c \text{ and to deduce that } a = b \times c. \qquad (12.2)$$

The distinction between a fraction and a ratio tended to disappear, and proportions were replaced by equations (Sylla 1984). Indeed, for a long time the words "ratio" and "proportion" have been treated as synonyms, a sloppiness that was not practiced in earlier centuries.

The use of ratios was not confined to ancient or older times. For example, mechanics from the eighteenth century and mathematical physics in the nineteenth century sometimes used ratios to avoid exactly the same kinds of questions about force, say, or electrostatic charge.

Comment

Ratios deserve a much greater role in teaching than they normally receive, for their long history shows them to be a very natural way for human beings to compare. The well-known difficulties in teaching fractions can be alleviated by converting to ratios (so that 4/7 becomes 4:7) and pointing to examples in the classroom and ordinary life where they occur; music is a particularly good source for the latter context, or even both of them.

2. Integers with Properties

The autonomy of integers and their arithmetic from empirical factors became more evident when they began to be regarded as objects in some (usually unspecified) sense; that is, they possessed *properties* such as being prime, or factors of other integers. Diophantos (flourished around A.D. 250) used such properties to find solutions to linear equations, many of which involve rational as well as integral solutions.

But the importance of proportions just described is an example of cultural contexts in which integers were treated as objects. They involve forms of numerology, in which a community granted special status to certain integers by associating them with their metaphysical or religious beliefs (ch. 14). Probably they saw integers as invariants, which have always been a strong theme in mathematics; things that remain the same while other things change around them. A closely related doctrine was gematria, created when a culture had developed a written alphabet; each letter was associated with some integer, and a word or phrase of the language took the integer given by the sum of those of its constituent letters.

Several Greek thinkers, especially Plato and followers such as Iamblichos (flourished in the third century A.D.), advocated numerology. One important category was the "triangle numbers," integers starting with 3, 6, 10, . . . that can be written as 1 + 2 + 3 + . . . ; the name indicates the property that they can be represented spatially by triangles of tokens. Maybe this property was one source of the 3ness of orthodox Christianity, in which Jesus is held to have lived for 3 × 10 years in obscurity before prosecuting a 3-year ministry and advocating the Trinity prior to dying in a 3some on a cross but coming back to life 3 days later. After his resurrection Jesus appeared 3 times before the remaining apostles; the second time was before 7 (sic) of them on the Sea of Galilee, whereupon Peter catches fish to the total of 153 (John 21:11)—the triangle number of 17, which was the number of principal sects and societies in the Jewish Kingdom of the time.

The full account is given in the New Testament, which contains 27 books: 3^3, the Trinity propounded across the 3 dimensions of space. Such features embody profoundly meaningful metaphysics to cognizant Christians. There were many of them in the Middle Ages, deeply aware of both numerology and gematria (Hopper 1938); but now it is usually ignored or even derided. Thus believers do not know a significant factor in their belief system.

These examples shows connections between mathematics and (a) religion, a rich aspect of mathematics that is often ignored but that is not even confined to arithmetic (mechanics and probability theory are among other branches). The Jewish tradition is very strong here. For example, the 22 letters of its alphabet are understood as "Aleph" followed by 3 7s respectively of grace, mercy, and strict justice, while 10 features as the number of connected circles in the sefirot tree of the Kabbalah (Judaica 1971). For Hindus, 9 is especially important, as the number of parts of the religion itself and of the points in their special symbol, the swastika.

Comment

The arithmetic involved here is trivial; but its cultural weight is substantial and gives a route into multiethnic education. Chapters 14 and 16 contain more details on numerology.

3. Algebra within and beyond Arithmetic

The advocacy of algebra from the late Middle Ages onward was helped by its prowess in finding the roots of polynomial equations; but it also served to generalize arithmetic by expressing the basic properties of any numbers, known or unknown real ones, and even complex. This aspect was also encouraged by the neohumanist movement of the late sixteenth century, which extolled the merits of the inheritance from Greek culture (Klein 1968). An important example is Rafael Bombelli (1526–1572), who was shown a manuscript by Diophantos and rewrote part of his *l'Algebra* to include solutions to Diophantine equations before publishing it in 1572. As his bald one-word title shows starkly, he wished that now algebra would be regarded not merely as a means to aid arithmetical calculation but as a brother discipline in mathematics.

Another line emanating from Bombelli was continued fractions, an arithmetical analogue of the Euclidean algorithm. Let $b > a$, and divide and use the successive remainders c, d, . . . as follows:

$$b/a = C + c/a = C + 1/(D + d/a) = C + 1/(D + 1/[E + e/a]) = \ldots \qquad (12.3)$$

Some of the new integers C, D, . . . may be zero; if a remainder b, c, . . . is, then the algorithm stops, and b/a is a rational fraction. Properties of continued fractions have enriched both arithmetic and algebra, although the topic remains curiously fugitive; for example, it is rarely taught.

Bombelli's contemporary François Viète (1543–1603) was also influenced by neohumanism, for he viewed the new art of algebra as "analytic" in the sense of the proof method advocated by Greeks such as Pappos, which starts out from the theorem and ends up with axioms or with previously known results, in contrast with the "synthetic" proof method, which proceeds from assumptions to theorem and was associated with geometry. This pair of associations endured in the philosophy of mathematics, although often to little benefit as geometry turned more analytical and algebra used synthetic proof methods in its more advanced theories. Further, the philosophy of algebra was still tied down by geometrical considerations; for example, since space had only three dimensions, then for Viète $xxxx$ was the "plano-planum" power of x, not its "quartum." Liberation from such constraints came only in the qualm-free "x^4" used by René Descartes (1596–1650) in his *Géométrie* (1637); however, as his title shows, algebra served as handmaiden to his creation of analytic geometry. This was *not* coordinate geometry, by the way, where a system of axes is imposed on a diagram; G. W. Leibniz (1646–1716) was a pioneer here, in the development of his differential and integral calculus from the 1670s (§13.1).

Attached to the development of algebra was the status of negative numbers, which have suffered a nervous press over the centuries. For some mathematicians (including Descartes) their status was linked to that of complex numbers, since both kinds of number arose in connection with solving polynomial equations. Although one can naturally think of interpretations of negatives (as financial debts, for example, or as numbers marked off in the direction opposite to that of the positive numbers), their legitimacy as self-standing objects has frequently been questioned. Consider the two equations

$$5 - 3{\cdot}5 = 1{\cdot}5 \text{ and } 3{\cdot}5 - 5 = -1{\cdot}5. \qquad (12.4)$$

Philosophically speaking, can the second equation stand on its own, or is it only a way of really saying the first? The second position was preferred by many mathematicians from the days of early algebra until the nineteenth century, especially in England. Rather greater confidence for negative numbers was evident among Continentals. Immanuel Kant (1724–1804) argued in 1763 that negation should be construed as the dialectic opposite to positiveness (it is curious that one cannot say "position"!) rather than as expression of its absence. Later, the abbot Condillac (1714–1760) saw algebra as the language par excellence within his semiotics, and so treated negative quantities on a par with positive ones in his account of ordinary algebra, published posthumously in 1807 under another striking title: *La langue des calculs.*

Comment

This is a case where the historical lesson is negative; for the two natural interpretations are easy to teach (the latter as a staple in financial mathematics, which is becoming a trendy topic in education). The failure to draw upon such cases in the past reflects, I suspect, *the lack of model-theoretic thinking in mathematics in those times.* This view, which has flourished only from the twentieth century, allows negative numbers to be interpretable in such contexts while not in others. It is worth teaching explicitly, though the technicalities of model theory should be left until well into the undergraduate career.

4. Number Systems and Calculation

The Babylonians worked with a number system based on 60, which still leaves its traces on our system of time keeping. They also used 10 as their base, as have most cultures. A system of counting based on 12 is more convenient, since it can be done on one hand by counting the knuckles of the fingers with the thumb; but sadly this system never became popular, although it is richer in that 12 contains more factors than does 10. Note that both 10 and 12 are both factors of 60, which may have been a common source for both integers: against frequent statement, the linguistic evidence is *not* strong that 10 came from the numbers of fingers and thumbs or of toes; maybe its status as the triangle number of 4 also played some role.

A variety of other number systems has been known; for example, the Mayans took 1 and 5 as their basic units in a 20-place arithmetic (Lounsbury 1978). Not all nonstandard systems are ancient; for modern science is still developing a library of adjectives (mostly derived from Greek) for these numbers 10^n:

n =	. . . −18	−15	−12	−9	−6	−3	3	6	9	12	
name:	. . . atto-	femto-	pico-	nano-	micro-	milli-	kilo-	mega-	giga-	tera-.	(12.5)

Some systems of numerals reflect the operations carried out. For instance, if in our Hindu-Arabic system (hereafter, HA) I add 6 apples to 2 apples, then I obtain 8 apples; but in Roman numerals I go from ii and vi to viii, and the sign "viii" exhibits the process of adding ii and vi. This feature seems to have appealed to accountants and others in the sixteenth century as a means of controlling honesty in financial records; so they opposed the introduction of HA. However, HA has other advantages and became dominant. Simon Stevin (1548–1620?) pointed out one benefit: if HA were extended to the use of decimal expansion of numbers, then magnitudes could be easily compared. Indeed, how else can one readily tell whether, say, 83/35 is less or greater than 71/29?

Although virtuosi on the abacus could be found until recently in Soviet supermarkets, gift at calculation has been most frequently found among the exponents of HA. Many interesting and indeed little-understood questions in psychology arise; for example, "lightning calculators" often do not know themselves how they calculate at such speed, and most of them have no particular gift for mathematics itself (Smith 1983). At the other extreme come innumerate people, for whom arithmetic is a frightening subject. While social conditioning from bad educational experience is a major factor, mental incapacity also seems to be involved, especially when it occurs in some forms of calculation and not in others.

A major issue in the development of methods of calculation in the Middle Ages was *whether or not the details were retained, for purposes of checking.* The nonretentive methods, such as the use of tokens mentioned in ¶1, which advanced to moving pebbles to various positions on an abacus (Latin for "flat surface") to represent values, became known as "algorist"; the other methods were "abbacist." (Note the two *b*s now used to avoid the clash of names; the latter sense came from Fibonacci's "Liber Abbaci" [1202], an important source for Europe of Arabic and Hebrew arithmetic and algebra.) Gradually abbacist methods came to prevail, mainly because of the advantages of being able to check; figure 12.1 shows a classic illustration of the triumph of an abbacist over her plodding algorist companion.

Comments

The issue of checking calculations bears on modern education in a way that is often overlooked: namely, that *an electronic calculator is an algorist device,* in that no means are available to check the answers delivered. Abbacist principles should be empha-

Figure 12.1. Happy abbacist, sad algorist (as in Greek drama?): propaganda in G. Reisch, *Margarita philosophica* (Freiburg, Germany: Schott, 1503), unpaginated frontispiece to book 4 on "speculative arithmetic." The algorist has these numbers on the board: on our left, 2 + 30 + 50 = 82; on our right, 1 + 40 + 200 + 1,000 = 1,241.

sized in teaching, at least that the student should have some idea of the order of magnitude of the answer before pressing the buttons.

Errors in calculation can provide amusing but instructive cases. For example,

$$\not{6}4/1\not{6} = 4 \text{ can be generalized to } \not{a}b/1\not{a} = b, \tag{12.6}$$

which is an indeterminate problem and so requires the student to examine options rather than proceed to the answer(s) in the usual robotic manner. A little work on factors in $(12.6)_2$ will furnish also

$$\not{9}5/1\not{9} = 5; \text{ and there are variants, such as } \not{9}8/4\not{9} = 8/4. \tag{12.7}$$

5. An Overlooked Arithmetic Algorithm for Multiplication

This algorithm is based on the place-value representation of a number as the sum of its placed digits each multiplied by the corresponding power of the modulus of the arithmetic. For convenience we shall represent the modulus (usually 10, of course) by x and remove the decimals by an understood multiplication of the number by the appropriate power of x. This gives to each number A the form

$$A = \sum_{r=0}^{\infty} a_r x^r, \tag{12.8}$$

where we assume a potentially infinite sequence of (zero or nonzero) coefficients a_r. Then in the form of (12.8) the multiplication of two numbers A and B to produce C is

$$\sum_{r=0}^{\infty} a_r x^r \sum_{r=0}^{\infty} b_r x^r \sum_{r=0}^{\infty} c_r x^r = \text{ , where } c_n = \sum_{r=0}^{\infty} a_{n-i} b_i. \tag{12.9}$$

Thus the simplest way to multiply two numbers together is to calculate the expressions in $(12.9)_2$ successively from right to left.

Example:	6 2·9 3 8 7 × 3 9·6 2
First place (x^0):	2 × 7 = 14; write 4, carry 1;
Second place (x):	(2 × 8) + (6 × 7) + 1 = 59; write 9, carry 5;
Third place (x^2):	(2 × 3) + (6 × 8) + (9 × 7) + 5 = 122; write 7, carry 11;

and so on. The simplest layout is:

Multiplicans		0	0	0	6	2 ·	9	3	8	7
Multiplicandum		0	0	0	0	0	3	9 ·	6	2
Product	2	4	9	3 ·	6	3	1	2	9	4
Carry-over terms		6	9	12	12	14	12	5	1	

Thus 62·9387 × 39·62 = 2493·631294. (12.10)

The decimal point is inserted by inspection, or from the fact that the number of decimal figures in the product must be the sum of the number of decimal figures in the multiplied terms (here, 4 + 2, giving 6).

Comment

A long-standing and continuing difficulty in mathematical education and practice in all cultures is the learning of laborious methods of multiplication and division in arithmetic. {Therefore, after the acceptance of the place-value system in a culture, why did this algorithm not become recognized? For example, on Britain see Yeldham (1926, 79–84; 1936), where it is not mentioned.} Mathematicians often have had a strange disinclination to explore conceptual (as opposed to mechanical) aids to computation: read on!

6. An Overlooked Arithmetic Algorithm for Division

In division, given the dividend D and the divisor E, find the quotient Q; using the notation of (12.8), calculate its coefficients q_r. The most natural way is to reverse the above algorithm. The coefficients are calculated from left to right; thus we must reinterpret the representation (12.8) as a *descending* power series, with x usually taking the value 1/10 rather than 10. The analogue to (12.9) may be written

$$\sum_{r=0}^{\infty} = e_0q_0 + (e_1q_0 + e_0q_1)x + (e_2q_0 + e_1q_1 + e_0q_2)x^2 + \ldots \qquad (12.11)$$

Equating coefficients, we have

$$e_0q_0 = d_0,\ e_1q_0 + e_0q_1 = d_1,\ e_2q_0 + e_1q_1 + e_0q_2 = \mathrm{d}_2, \ldots \qquad (12.12)$$

This is a triangular linear system, easily solvable for the q_r. From $(12.12)_1$ we obtain e_0q_0; in $(12.12)_2$ we subtract e_1q_0 from d_1 and obtain e_0q_1; and so on.

But there is an important limitation in this algebraic representation; the forms of the coefficients do not take into account the carry-over terms. Thus in general the coefficient of x^N in the product term d_N in (12.11) will not be an integer between 0 and 9; it lacks the carry-over term from d_{N-1}, and includes the term to be carried over to d_{N+1}. Thus when we carry out this procedure we shall obtain remainders that need to be carried back to the next power below. Now the magnitude of these carry-over terms may, on occasion, be such that the process of subtraction yields a negative quantity. Hence the multiple of e_0 to be subtracted from it, which is the value of the corresponding q_r, will also be negative. But these negative q_r can be removed from

Q at the end of the analysis by carrying back one unit from the next higher power, so that, for example, . . . $8\,\bar{6}\,9\,\bar{8}\,3$ . . . becomes . . . 74823 . . .

Finally, the argument does not presume that the dividing factor e_0 is necessarily between 0 and 9. The same is also true of d_0 and, in principle, of any d_0.

Example, taken from (12.10): calculate 2493·631294 ÷ 62·9387.

We choose e_0 (arbitrarily) to be 62 and indicate below the coefficients of E and D over the appropriate integers, bearing in mind that the correspondence is inexact because of the carry-over terms. Each line of the argument is described to its left, and the calculated coefficients q_r are listed on the right.

$e_0 \quad e_1\, e_2\, e_3\, e_4$	$d_0 \quad d_1 \quad d_2\, d_3\, d_4 d_5\, d_6\, d_7$
$E = 62 \cdot 9\,3\,8\,7$	$D = 249\;3 \cdot 6\;\;3\;\;1\;\;2\;\;9\;\;4$
Divide e_0 into 249 four times	248 Thus $q_0 = 4$
Remainder and next digit of D	1 3
Calculate $e_1 q_0$	3 6
Subtract	2 3
Divide e_0 into 23 minus once	6 2 Thus $q_1 = \bar{1} = -1$
Remainder and next digit of D	3 9 6
Calculate $(e_2 q_0 + e_{11})$	3
Subtract	3 9 3 and so on.

In fact $q_3 = 2$ and $q_4 = q_5 = q_6 = 0$, and therefore

$$2\,4\,9\,3 \cdot 6\,3\,1\,2\,9\,4 \div 6\,2 \cdot 9\,3\,8\,7 = 4\,\bar{1} \cdot 6\,2 = 3\,9 \cdot 6\,2, \qquad (12.13)$$

which checks with (12.10).

This algorithm is easily modified for the extraction of square roots; put $E = Q$ in (12.11). In principle it may be extended to the calculation of cube and higher roots, although the terms to be subtracted become much more complicated.

Comment

The division algorithm seems to have been given first in Western mathematics by Joseph Fourier (1768–1830), in his posthumous book on equations, together with a rather unclear theoretical justification and a complicated criterion for detecting the imminent arrival of excessively large subtraction terms that the version above shows to be unnecessary after the allowance of negative coefficients (Fourier 1831, 189–197). The rule for extracting square roots was developed from Fourier's result by (Schaar 1851), and both methods were expounded in some detail in the intriguing but little-known lectures of 1883 on "mathematical methodology" by J. F. Dauge, apparently Schaar's successor at the University of Ghent (Dauge 1883, 56–64, 87–91).

These algorithms are known in Vedic mathematics, although their history there is not at all clear. An important exposition by Swami Bharati Krsna Tirtha, intended as introductory to a series of volumes but terminated by the author's death in 1960, reveals a plethora of delightful algorithms conveyed from the past in the form of sutras, or aphoristic instructions, and here expanded (without historical and at times technical detail, and often with idiosyncratic English and abominable printing) into a treatment of not only arithmetic but also simple linear systems and solutions of polynomials. All three algorithms are there (Tirthaji 1965, chs. 3, 27, and 34), without the algebraic backing.

Otherwise there seems to be only a scattering of passing and incomplete references in the literature to algorithms such as these. Thus, although they are by no means unknown, they have never received the attention they deserve, least of all in the context of mathematical education, where they could serve such a useful role.

7. The Logic and Set Theory of Arithmetic

The arithmetic of positive integers received two new levels of foundation in the 1880s. C. S. Peirce (1839–1914) and Giuseppe Peano (1858–1932) both put forward axiomatizations based on the notion of 1 as the initial integer and on the operation of successorship (2 as the successor of 1, and so on). One of the axioms was the principle of mathematical induction, in the form that if a property applies for $n = 1$, and if applicable for any value of n then is also for its successor, then it applies for all n. The formulation was simplified in 1907 by Mario Pieri (1860–1913). Richard Dedekind (1831–1916) gave a somewhat similar treatment in 1888, in terms of a notion of chain (a relation mapping a set of integers onto itself in such a way that all the successors of some initial integer were obtained); but he went further in proving a theorem that legitimated proofs by mathematical induction by providing a justification for inductive definitions (van Heijenoort [1967] contains many of these original texts).

In 1884, Gottlob Frege (1848–1925) had gone a layer below this one when he gave nominal definitions of integers, which Bertrand Russell (1872–1970) was to find independently in 1901. They defined integers as sets of sets, starting with 0 and the unit set of the empty set Ø, 1 as the set of sets isomorphic (that is, in one-one correspondence) to {Ø}, and so on upwards but not in a vicious circle. The ascent was infinite for Russell, since he accepted the theory of transfinite numbers (which is described in the next section).

Frege proceeded to further definitions of real numbers (whether integral or not) via expansions of their nonintegral part in the bicimal power series

$$\sum_r a_r 2^{-r}, \text{ where each } a_r = 0 \text{ or } 1; \tag{12.14}$$

Russell defined rational and irrational numbers as certain sets of integers and of rationals respectively. Arithmetical operations were defined in terms of the corresponding combinations of sets (for example, addition from the union of disjoint sets). Both men also had means of defining negative numbers.

Both men also defined sets in terms of propositional functions, for these procedures formed part of their logicist theses, that arithmetic (for Frege) or all "pure" mathematics (for Russell) could be derived solely from logical principles and procedures. A modern variant of their approach regards integers as quantifiers: the sentential form "there are . . . apples here" is bound into a (true) sentence by the insertion of the integer in question (Bostock 1974, 1979).

Comments

1. The vision behind these enterprises, especially Russell's, was the unification of arithmetic, mathematical analysis, and geometries under the umbrella of set theory and mathematical logic. The purpose was epistemological, concerned with restructuring and "justifying" known theories by locating them within such foundations; there was no educational or heuristic content, although some parts can be taught in late student years at university. Sadly, half-understood versions percolated down to the educational community, granting set theory a central place in the "new" mathematics of the 1960s (§9.2.3). One factor was played by Jean Piaget's misunderstanding of Russell's enterprise, especially in giving such a grossly exaggerated place to isomorphisms in "the child's conception of number" (Piaget 1952).
2. The adjective "new" shows already the absence of historical knowledge (after all, Sibelius was hardly new music at that time), never mind the original purpose and its limitations. The basic error in the approach was *premature rigor and generality,* solving problems that the student cannot have encountered in the first place (Grattan-Guinness 1973). Further, mistakes made in the texts testify to the subtlety of set theory, such as distinguishing properties of a set from those of its members. And the idea

that the theory is the most general way of handling collections *is doubly mistaken.* First, it assumes membership always to be well defined (that is, true or false): fuzzy set theory takes care of the many exceptions (Dubois and Prade 1980). Second, it allows members to belong only once to the set—an elementary limitation, as evidenced by multiple roots of a polynomial equation: multiset theory is needed instead (Rado 1975).

8. Transfinite Arithmetic

From 1883 Georg Cantor (1845–1918) also defined integers from sets but by a very different means: from a given set M, abstract (that is, mentally ignore) the nature of its elements to leave behind the "order-type" $\overline{M}$ in which its members lie; then abstract the order to leave behind the cardinal number $\overline{\overline{M}}$ of M (Dauben 1979). The role that he gave to mental acts was rejected by most of his contemporaries, although some residue lies in the phenomenological interpretation of integers proposed by Edmund Husserl (1859–1928) from 1891, partly under influence from Cantor (Willard 1984).

Further, Cantor realized that infinities came in different sizes, so he introduced the concept of inequalities between infinitely large integers. He defined the sequence of transfinite ordinals, starting with the smallest of them, ω assumed to exist after the finite ordinals, and with no predecessor ordinal:

$$1, 2, 3, \ldots \omega, \omega + 1, \ldots 2\omega, 2\omega + 1, \ldots 3\omega, \ldots \omega^2, \ldots \omega^3, \ldots \qquad (12.15)$$

He also found a way of dividing this literally infinite sequence into "number classes" whose sizes (that is, cardinalities independent of order) were shown to be different. The first one, comprising the finite integers, took the smallest such cardinal number, $\aleph_0$; the second class, starting at ω and defined by the property that the count up to any ordinal member could be reordered into a set of size $\aleph_0$, was itself of the next larger cardinality, $\aleph_1$; and so on, for infinitely ever.

Another of Cantor's innovations was to realize that the members of a set may be ordered in different ways. In order to preserve the generality of arithmetic, he asserted the "well-ordering principle," that any set could have its elements arrayed in the order exemplified by (12.15). This was to be one of his unsolved problems, of which the proof (in 1904) turned out to require the axiom of choice (Moore 1982). The main advocate of this axiom, which caused much controversy among mathematicians and philosophers for its various forms and especially for its nonconstructive character, was Ernst Zermelo (1871–1953); but Russell found a form slightly earlier, in the context of defining the infinite product of numbers.

Cantor developed an arithmetic for each kind of integer, different from each other and from that of finite arithmetic; for example,

$$\omega + 1 > \omega, \quad 1 + \omega = \omega; \quad \text{and} \quad \aleph_0 + \aleph_1 = \aleph_1 \tag{12.16}$$

However, his sequence (12.15) also led to paradoxes of the supposedly greatest ordinal, and the greatest cardinal, for either such number N, both $N = N$ and $N > N$. Avoiding these paradoxes, and those like Russell's concerning sets alone, led to modified definitions of numbers as sets of sets. In particular, the type theory in Russell's postparadox logicism required that integers be defined and their arithmetic be developed for each type.

Comments

This material forms a good undergraduate course. I find that the axioms of choice are rather more fruitful sources of appreciation than the details of transfinite arithmetic, partly because the range of consequences for mathematics far exceeds set theory and arithmetic, and partly because the wide range of equivalent axioms exposes fine examples of apparent (non)constructivity in mathematics.

9. Much Ado about Zero

The Babylonians usually indicated zero by an empty space; but then it is hard to distinguish, for example, 3 5 from 3   5 (in HA, 305 from 30,005). So a sign was needed. They had one, looking something like ⪡; but it seems to have been used only as place marker for the blanks. The great step of using it *also* as a number that could be combined arithmetically with other numbers seems to be of Indian origin.

Culturally, the status of zero, its misidentification with nothing, has been a widespread concern (Rotman 1987). An important example is the playwright who wrote as "William Shake-speare," especially in his *King Lear*, where nothing is everything. He was working at the turn of the sixteenth and seventeenth centuries, precisely when HA was coming into general use in Britain; doubtless this process heightened his awareness.

{A curious and little noticed use of zero occurs when we say on, for example, Tuesday that it will be Thursday in two days' time, for this counting takes Tuesday to be the zeroth day. In keeping with the Romans' number system, they said the equivalent of Thursday being the third day after Tuesday.}

As for signs, HA has "0"; its origins are not known for certain, for it may lie in

Greek, Indian, and Chinese mathematics (possibly independently), perhaps from the second century A.D. onward. In addition, "o" is the first letter of the word "ouden" in Greek for nothing. It may also have been proposed as a sign for the vagina, as the nothing out of which things are born, for it is well known that fertility and sexual symbols played a prominent role in cultural and religious life of ancient civilizations. Other symbols used were similar to "0"; a dot •, and in Greek sometimes either "φ" or "θ."

Even when the need for a sign was recognized, much philosophical perplexity surrounded zero, especially for its failure to satisfy the cancellation law ($a \times 0 = b \times 0$ without $a = b$). In addition, it was usually associated (too) closely with nothing. Even in their formulations of arithmetic reported in ¶7 and ¶8, Cantor and Dedekind both always began with 1: Cantor could not define 0, for he would have had to abstract from the empty set $\emptyset$ (which makes surprising his notation "$\aleph_0$"!) Only Russell and Frege clearly understood the difference between nothing, $\emptyset$, and 0. Since then the distinction has become firmly established, even in systems where a mathematician may *choose* to conflate them. For example, the system of ordinals proposed in 1923 by Johann von Neumann (1905–1958) goes as follows:

$$0 := \emptyset, 1 := \{\emptyset\}, 2 := \{0, 1\}, \ldots, \omega := \{0,1, \ldots\}, \ldots \tag{12.17}$$

Comments

1. The teaching of zero is often deplorable, especially when it is stated to be nothing. This is quite incorrect; for zero has properties, such as $7 + 0 = 7$ ($7 +$ nothing is not defined).
2. One can also use zero to emphasize the distinction between equality and identity; in addition to standard cases such as

$$2 + 2 = 4, \text{ examples such as } 0 + 0 = 0 \tag{12.18}$$

show the difference still more starkly (two zeros on the left hand side, only one on the right, hence not identical). Identity is a difficult philosophical concept with which to work; for educational purposes in mathematics, it is best presented in terms of identification (of 4 as the sum of 2 and 2, say).

10. Formalisms and Incompleteness

The word "formalism" occurs regularly in connection with the philosophy of arithmetic, but various types should be distinguished. There is a type of formalism in

Cantor in that he saw that the consistent construction of transfinite numbers guaranteed their existence. He had no metatheory in which consistency could be proved, and this lack may have been one of the stimuli for David Hilbert (1862–1943) to devise metamathematics as a theory in which it (and completeness) of an axiomatized theory could be studied.

Hilbert's formalist program,[1] which flourished mainly in the 1920s and 1930s after launch at the beginning of the century (Detlefsen 1986), took arithmetic in a form broadly similar to Peano's axiomatization mentioned in ¶7 and viewed its foundation as sufficient to provide the foundations of much (maybe "all") mathematics. But his hope of demonstrating the consistency of arithmetic was set back by a theorem proved in 1931 by Kurt Gödel (1906–1978), for its corollary shows that consistency can be proved only in a metatheory richer than arithmetic itself, contrary to Hilbert's vision of a more primitive metatheory, then still more primitive meta-metatheory, . . . Further, Gödel's main theorem refuted Russell's (intuitive) belief that his logicist axiomatization of arithmetic would be complete.

In addition, Gödel's proof method based on expressed metatheory in arithmetical terms was a principal stimulus for the study of recursive functions, which itself became a leading technique in the development of computers with Alan Turing (1912–1954) and has formed lasting links with both logic and mathematics (Davis 1965). Finally, the distinction between theory and metatheory required by Gödel's proof brought home to logicians the care with which they needed to observe this distinction. For mathematicians, however, his theorems were of marginal importance, since Gödel worked with a much stricter concept of proof than that with which they were (or still are) accustomed.

However, most recent philosophy of arithmetic has built on versions of Hilbert's or on a set-theoretic approach; some has been constructivistic or intuitionistic in character. Some other treatments are nominalistic, in trying to avoid giving numbers an abstract status (Field 1987). In addition, social interpretations of numbers and arithmetic have gained some favor (Livingstone 1986), in which numbers are experienced, arithmetical operations are performed, and proofs are actions; "2 + 2 = 4" is less than a traditional Platonic truth, but more than a matter of personal opinion. Numbers are ancient things (or symbols, or sets, or acts, or . . .); their statuses have always been obscure and are likely to remain so.

1. {Hilbert was only a formalist with respect to his metamathematics and *never* used that word to describe his general philosophical position. This latter reading was inferred in 1927 by L. E. J. Brouwer, as a criticism; and Hilbert took it as such (van Dalen 1990).}

Comments

1. Some aspects of formalism, especially recursion and computability, link nicely to under- and postgraduate courses in computing. But richer contexts arise in more elementary contexts, especially in stressing the difference between numbers and numerals, the brute sense of formalism. Relationships such as "7 > 3" should be studied carefully, and also the nonsense of wondering whether "3," "3" or "3" is the true number three; for then the more abstract character of numbers relative to numerals can be clearly indicated, whatever status one chooses to assign to them.
2. In addition to numbers and numerals, *digit strings* should be stressed. We have long been familiar with them as telephone numbers, but now they come also in barcodes, PINs, and so on. While a digit string can easily be associated with its corresponding number (numerologists often do this, such as the orthodox Christians taking 111 as the Trinity number), they are treated differently, with much less mathematical content; in particular, they have no associated arithmetic. We even say strings in a different way ("one one one" rather than "one hundred and eleven"), and write them differently; for example, there are various ways of stating telephone numbers in different countries.
3. There are a few overlaps; one is the manner of reading decimal expansions, where 27.27 is understood as "twenty-seven point two seven" since the latter part is read from left to right. There is a psychological point involved here also; because of the geographical origins of HA, we read integers from right to left, contrary to that of the words of our language. We do this too often to notice, for integers up to the early millions; but when faced with, say, "463,563,640,863,759," then the reverse process becomes conscious, even with the partitioning of the component digits into threes to aid the reading.

BIBLIOGRAPHY AND FURTHER READING

The list does not attempt to be exhaustive. Many general books on the history of numbers are indifferent or worse as scholarly sources.

Benoit, P., and others 1992. (Eds.), *Histoire de fractions, fractions d'histoire*, Basel, Switzerland: Birkhäuser.

Bostock, D. 1974, 1979. *Logic and arithmetic*, 2 vols., Oxford: Clarendon Press.

Cajori, F. 1928. *A history of mathematical notations*, vol. 1, La Salle, Ill.: Open Court. [Extensive survey of symbols for numbers and operations.]
Cohen, H. F. 1984. *Quantifying music. the science of music at the first stage of the scientific revolution, 1580–1650*, Dordrecht, Netherlands: Reidel.
Crossley, J. N. 1980. *The emergence of number*, Yarra Glen, Victoria, Australia: Upside Down A Book Company. [A refreshing survey of the development of numbers, including complex numbers.]
Dauben, J. W. 1979. *Georg Cantor*, Cambridge, Mass.: Harvard University Press. [Repr. Princeton, N.J.: Princeton University Press, 1990.]
Dauge, J. F. 1883. *Leçons de méthodologie mathématique*, Ghent, Belgium: Adolphe Hoste.
Davis, M. 1965. (Ed.), *The undecidable*, Hewlett, N.Y.: Raven Press. [A source book.]
Detlefsen, M. 1986. *Hilbert's program*, Dordrecht, Netherlands: Reidel.
Dubois, D. and Prade, H. 1980. *Fuzzy sets and systems: theory and applications*, New York: Academic Press.
Field, H. H. 1987. *Science without numbers*, Oxford: Blackwell.
Fourier, J. B. J. 1831. *Analyse des équations déterminées* (ed. C. L. M. H. Navier), Paris: Firmin Didot.
Grattan-Guinness, I. 1973. "Not from nowhere: History and philosophy behind mathematical education," *International journal of mathematical education in science and technology, 4*, 421–453.
Grattan-Guinness, I. 1994. (Ed.), *Companion encyclopaedia of the history and philosophy of the mathematical sciences*, 2 vols., London and New York: Routledge. [Many relevant articles, especially in parts 1, 2, 5, 6, and 12.]
Hopper, V. F. 1938. *Medieval number symbolism*, New York: Cooper Square. [Repr. 1969.]
Judaica. 1971. Articles on "Alphabet, "Astronomy," "Gematria," "Kabbalah," "Mathematics," "Numbers," and "Sefirot," *passim* in *Encyclopaedia Judaica*, 16 vols., Jerusalem: Keter. [See also various biographical articles.]
Klein, J. 1968. *Greek mathematical thought and the origin of algebra*, Cambridge, Mass: MIT Press.
Livingstone, E. 1986. *The ethnomethodological foundations of mathematics*, London: Routledge and Kegan Paul.
Lounsbury, F. L. 1978. "Maya numeration, computation, and calendrical astronomy," in *Dictionary of scientific biography*, vol. 15, 759–818.
McClain, E. G. 1978. *The myth of invariance*, York Beach, Maine: Nicolas-Hays.
Moore, G. H. 1982. *Zermelo's axiom of choice*, New York: Springer.
Piaget, J. 1952. *The child's conception of number*, London: Routledge and Kegan Paul. [French original 1941.]
Rado, R. 1975. "The cardinal module and some theorems on families of sets," *Annali di matematica pura ed applicata, (4)7,* 135–154.
Reisch, G. 1503. *Margarita philosophica*, Freiburg, Germany: Schott.
Rotman, B. 1987. *Signifying nothing: The semiotics of zero*, London: MacMillan.
Schaar, M. 1851. "Note sur la division ordonnée de Fourier," *Bulletin de l'Académie Royale des Sciences de Belgique, 18,* 144–157.

Schmandt-Besserat, D. 1992. *Before writing: From counting to cuneiform*, vol. 1, Austin: University of Texas Press. [Volume 2 is a catalog of tokens.]

Smith, S. B. 1983. *The great mental calculators*, New York: Columbia University Press.

Sylla, E. 1984. "Compounding ratios: Bradwardine, Oresme, and the first edition of Newton's *Principia*," in E. Mendelsohn (ed.), *Transformation and tradition in the sciences*, Cambridge: Cambridge University Press, 11–45.

Tirthaji, Maharaja Jagadguru. 1965. *Vedic mathematics* (ed. V. S. Agrawala), Delhi: Motilal Banarsidass.

van Dalen, D. 1990. "The war of the frogs and the mice, or the crisis of the Mathematische Annalen," *Mathematical intelligencer, 12*, no. 4, 17–31.

van Heijenoort, J. 1967. (Ed.), *From Frege to Gödel*, Cambridge, Mass.: Harvard University Press. [A source book.]

Willard, D. 1984. *Logic and the objectivity of knowledge: A study in Husserl's early philosophy*, Athens: Ohio University Press.

Yeldham, F. A. 1926. *The story of reckoning in the Middle Ages*, London: Harrap.

Yeldham, F. A. 1936. *The teaching of arithmetic through four hundred years (1535–1935)*, London: Harrap.

CHAPTER THIRTEEN

What Was and What Should Be the Calculus?

Calculus teaching rarely involves the history of the subject. As a result, the various conflicting traditions of terms, notations, and ideas are conveyed in only partial and ill-digested forms. This chapter contains in ¶1–¶6 a brief survey of the developments of the calculus from Newton and Leibniz to Lebesgue and Zermelo (that is, from the 1660s to the 1900s). The final section, ¶7, contains a miscellany of points on educational questions.

> I presume that few who have paid any attention to the history of the Mathematical Analysis, will doubt that it has been developed in a certain order, or that that order has been, to a great extent, necessary—being determined, either by steps of logical deduction, or by the successive introduction of new ideas and conceptions, when the time for their evolution had arrived. And these are causes which operate in perfect harmony. Each new scientific conception gives occasion to new applications of deductive reasoning; but those applications may be only possible through the methods and the processes which belong to an earlier stage [. . .].
>
> Now there is this reason for grounding the order of exposition upon the historical sequence of discovery, that by so doing we are most likely to present each new form of truth to the mind, precisely at the stage at which the mind is most fitted to receive it, or even, like that of the discoverer, to go forth to meet it. Of the many forms of false culture, a premature converse with abstractions is perhaps the most likely to prove fatal to the growth of a masculine vigour of intellect.
>
> —*George Boole (1859, preface)*

The original version of this essay was first published in I. Grattan-Guinness (ed.), *History in Mathematics Education*, Paris: Belin, 1987, 116–135. Reprinted by permission of Société Française d'Histoire des Sciences et des Techniques. I have reorganized the material a little, mainly in ¶4.

1. What Did Newton and Leibniz Introduce into the Calculus?

1.1. *The Innovation*

"Newton and Leibniz invented the calculus," the histories and textbooks agree in announcing, and they point to the "fundamental theorem," the principle that differentiation and integration are inverse processes, as the main source of advance:

$$d/dx \int f(x)\, dx = \int df(x)/dx\, dx = f(x), \tag{13.1}$$

to express the matter in modern symbolism. Yet this is only part of their contribution, and the second part at that. For these two mathematicians first realized that the task of the calculus was to find *new functions or relations* of the variables from the given function or relation: $df(x)/dx$, or some analogue, for differentiation, and the indefinite integral function $\int f(x)\, dx$ for integration. Without these concepts (13.1) degenerates into the falsehood

$$0 = f(b) - f(a) \text{ (say)} = f(x). \tag{13.2}$$

Calculus prior to Newton and Leibniz was concerned with finding particular values of functions (such as points of optimization, or tangent points given by double-root conditions) or evaluating some particular (classes of) areas and volumes (see Whiteside [1961], and Pedersen in Grattan-Guinness [1980, 10–48]).

1.2. *Newton's Theory*

Newton's theory, evolved in the 1660s and cast in a fairly definitive form by the 1690s, took the "fluxion" $\dot{x}$ of a variable x as its basic differential concept. It was conceived intuitively in terms of the rate of change of x against that of some underlying temporal variable t:

$$\dot{x}(t) = \text{“}\lim_{\sigma \to 0}\text{”}(x(t + \sigma) - x(t)/\sigma); \tag{13.3}$$

but the foundations of the process of taking limits were unclear, and the scare quotes around "lim" must be taken seriously. Indeed, as is well known, the famous controversy surrounding Berkeley's *The Analyst* (1734) was centered partly on this question (Cajori 1919).

The rates of flow of variables were relative to each other; double all of them in a given problem, say, and you do not change the problem. So given an increment σ on t, increment $\sigma\dot{x}$ was taken on $\dot{x}$; alternatively, ratios $\dot{y}/\dot{x}$ of flows would be used. Newton's phrase "prime and ultimate ratios" referred to these ratios and expressed his assumption that the limiting value reached from below the given point equaled that achieved from above (in modern language, that the left- and righthand derivatives were equal).[1] Among these flows, the case where $\dot{x} = 1$ was of special importance, for it provided a base reference for the others. We shall return to this point in ¶2.

Newton's version of the indefinite integral was his "fluent" $\overset{|}{x}$ of the variable x. Thus the inversion principle (13.1), had he ever written it explicitly, took the form

$$\overset{|}{\dot{x}} = \dot{\overset{|}{x}} = x. \qquad (13.4)$$

Among his purely mathematical motivations, Newton was especially attracted to the term-by-term differentiation and integration of power series (to use a modern form of expression). He was the first great mathematician to affirm the prominence of power series in mathematics, especially for their capacity to express the equation of curves; his calculus techniques then found new curves from differentiation and integration.

The most important source for Newton's writings on the calculus today is the edition (Newton *Papers*) of his mathematical manuscripts prepared by D. T. Whiteside. Volumes 1 and 2 (1664–1670) contain the first papers, and of the later writings, volume 4 (1674–1684) contains some particularly interesting texts.

1.3. Leibniz's Theory

Leibniz's theory, developed during the 1670s, was quite different, for he took a variable x not to (some kind of) limiting value but down to a differential case dx, an infinitesimal increment on x and of the same dimension as x. So dx, itself a variable, could have a second-order increment taken on it to yield ddx, and so on. Again, the rates of change had a degree of freedom within them, and were really related to some base variable; for him "dy/dx" literally meant a ratio of differentials $dy \div dx$, and "d^2y/dx^2" was $ddy/(dx)^2$. He sometimes allowed for uniform increments where dx was constant and so $ddx = 0$. This category corresponded in role to Newton's case of $\dot{x} = 1$.

Leibniz's integral was the sum $\int x$, an infinitely large value of the variable x, and again of the same dimension as x. But the object to which he gave prime attention was, of course, the area $\int y\, dx$, which he conceived as the (indefinite) summation ($\int$) of areas of rectangles y high and dx wide (if y and x were taken as linear variables).

1. {Maybe Newton's prime ratio corresponds not so much to our righthand derivative, where increment $h \to 0$ through positive values, but involves the less clear notion $0 \to h$.}

The inversion principle for Leibniz took not the form (13.1), which *seems* to be written in his notations, but

$$d\int x = \int dx = x; \tag{13.5}$$

the calculus was a theory of the dimension-preserving operators d and $\int$ acting on variables such as x. The insight to the principle came to him from studying what he came to call "the characteristic triangle" of the difference quotient, and noting that the slope of the tangent $dy \div dx$ was a quotient given by a difference in values of the ordinates, while the area $\int y\, dx$ was provided via the sum of a product involving those ordinates.

Leibniz's calculus is not satisfactorily available in English, although there is an edition (Child 1920) of his early mathematical papers. The genesis of his theory is described in detail in Hofmann (1949) and in Bos (1974). Its status vis-à-vis non-standard analysis is appraised in ¶7.3.

2. Euler and the Differential Coefficient

Of these two traditions the Leibnizian secured much the greater success. The reason usually given is Newton's "crabbed notation," but this explanation is not convincing, since the notation can be successfully mastered in a few minutes (indeed, the dot notation has always been used in dynamics). Leibniz's system has some advantage for efficacy, with its intuitive and pliable notations of little bits (or lots) of variable; but the main reason for its superiority, it seems to me, lay in the quality of his successors, the Bernoulli family and Euler.

The calculus gained much of its importance from application to geometry, and to mechanics, when differential equations, ordinary and partial, were formed and solved. ("Differential equations" referred then, literally, to equations relating differentials: how absurd it is that this expression is still retained today!) Euler was the most productive of Leibniz's followers; and he not only gradually expanded the scope of the subject but also refined its foundations in an important respect.

A calculus problem involved several variables $x, y, z, \ldots$, with their differentials $dx, dy, dz, \ldots ddx, \ldots$ varying together in some degree of indeterminacy. It then followed that expressions involving differentials above the first order (the radius of curvature, for example) depended for their form on the particular conditions applicable to their calculation (for example, which differential was constant); but these conditions not only acted as unwelcome extra luggage but also might not apply in another context.

Euler regularized the practice of specifying uniform differentials and defining other differentials from them; and by this means, he eliminated the indeterminacy. Let us take the simplest case of two variables x and y, with the differentials related as follows:

$$dy = p\,dx; \text{ and specify also that } ddx = 0. \tag{13.6}$$

p, like y, is a function of x; so, as in (13.6),

$$dp = q\,dx, \qquad \text{and so on.} \tag{13.7}$$

Then the second-order differential ddy is calculated as follows:

$$ddy := d(dy) = d(p\,dx) = dp\,dx + p\,ddx = q(dx)^2 + 0 = q(dx)^2, \tag{13.8}$$

so it is composed of two determinate quantities q and $(dx)^2$ (Euler 1755, arts. 124–130).

The p, q, . . . were the "differential coefficients" of y with respect to x {(as the textbook writer S. F. Lacroix [¶3] will name them)}, and thus an important component of functionality came into the calculus, for the variable x that satisfied the relation $(13.6)_2$ was playing the role that we would call the "independent variable," with y functionally dependent on it (Bos 1974, 66–77). Euler sometimes spoke of the calculus as a process of "reckoning with zeros," but in no way was it *founded* on a theory of limits. For example, the integral was *defined* as the inverse, so that $f(x)$ was the integral of p, p of q, and so on.

Euler and others developed the calculus enormously during the eighteenth century; the extension to multivariateness with its *partial* differentials and differential equations was particularly important. (An example of the method is given in ¶7.4.) He also contributed more than anyone else to the doctrine of functions, series, and integrals, and so provided many specific results that could be applied to given problems in pure and applied mathematics.

There are many writings on Euler; the collective volume (Euler 1983) is the most comprehensive single source to date,[2] and several volumes in the edition *Opera omnia* of his works have valuable articles by the editors.

2. {The Euler tercentenary year 2007 led to several new historical works, especially from the Mathematical Association of America; but the description given here still stands up.}

3. Lagrange and the Derived Function

3.1. The Primacy of Algebra

During the late eighteenth century Lagrange challenged the Eulerian tradition by offering his own approach to the calculus, which was to serve instead of both Leibniz-Euler differentials and also theories of limits. To him the rigorous way was to algebrize everything and calculate by means of formal rules. Specifically, he claimed as a proven fact that "every" function had a Taylor-series expansion

$$f(x + i) = a_0(x) + a_1(x)i + a_2(x)i^2/2! + \dots \qquad (13.9)$$

(and so followed Newton in assigning a principal status to power series); the idea was to define the "derived functions" as the coefficients $a_1(x)$, $a_2(x)$, . . . and to obtain them by purely algebraic manipulations of the expressions. He introduced the notations $f'(x), f''(x), \dots$ for his "derived functions." The integral, once again, appeared as the automatic inverse; $f(x)$ was the integral of $a_1(x)$, $a_1(x)$ of $a_2(x)$, and so on. He developed this approach in the 1760s, and gave it a definitive presentation in his *Théorie des fonctions analytiques* (1797), based on his teaching at the Ecole Polytechnique in Paris (¶7.5).

3.2. Reactions

The idea is a nice one but lacks some conviction. For example, Lagrange argued that no term of the form $ui^{m/n}$ could occur in the expansion of $f(x + i)$, but he did not categorically ban multivaluedness; he also rather slipped in the prime notation (1797, 7–14; see also ¶7.4). Other questions arise. First, can "any" function be cast in the Taylor series to start with? Among his followers, Arbogast (1800) tried harder than anyone else to produce algorithms for generating power series; but even his account is not exhaustive, even relative to the functions handled at that time. Second, can the computations involved be carried out without recourse to limits or infinitesimals? Lagrange showed only that a few cases could be done, although his efforts and those of his followers (minor figures such as Brisson, J. F. Français, and Servois) introduced the fascinating new algebras of functional equations and differential operators (Grattan-Guinness 1979; {Panteki 1992}). This is a neglected but important chapter in the history of algebra, since it expunged forever the notion of algebra as the symbolism of only number and quantity.

But where does this leave the calculus? In France, then the mathematical capital of the world, a muddled situation is evident, with all three traditions (limits, differentials, Taylor's series) receiving support. Lacroix, the principal textbook writer of the period, presented all three traditions in his large treatise on the calculus, which came out around 1800 and then in a second edition of the 1810s (Lacroix Calculus; {see now Caramalho Domingues 2008}). Lazare Carnot vacillated over the status of limits and the type of infinitesimals advocated in the various editions of his reflections on the calculus (Carnot *Calculus*: see Yushkevich 1971). The governing body of the Ecole Polytechnique, the Conseil de Perfectionnement, decided under Laplace's stimulus in 1812 to halt the tendency to teach limits (which had probably been encouraged by Ampère's teaching [1806]) and reintroduce infinitesimals; but the course then given by Poinsot (1815) managed to draw on all three traditions at once. He regarded the "differential function" $dy \div dx$ as the "last ratio" of the difference quotient $\Delta y \div \Delta x$ because those differences decreased to the differential "quantities neighboring" orthodox values; but he also used functional equations to calculate differential functions of certain simple functions by purely algebraic means.

However, the situation was soon to change in Paris. Upon the Bourbon Restoration one Augustin-Louis Cauchy was appointed to the staff of the school in 1816, to share the analysis with Ampère for over a decade. During this period he inaugurated the new subject of mathematical analysis.

4. Cauchy and the Primacy of Limits

4.1. Cauchy's Publications

The two most important textbooks of Cauchy are his *Cours d'analyse* (1821) and the *Résumé* of his calculus lectures (1823). The *Cours* contains no calculus, but it is the fundamental work, for in it he refined the theory of limits to a new degree and used it as the basis of mathematical analysis, the new umbrella subject containing not only the calculus but also the theory of continuous functions, the convergence of infinite series, and various related theories. The calculus was presented in the *Résumé*, with 20 lectures on the differential calculus, followed by 20 lectures on the integral calculus. Instead of the encyclopedism of Lacroix or the eclecticism of Poinsot, here was a strict order of material, with a format consisting of basic definitions, theorems laid out in a "logical" order, and the same notations used throughout.

Cauchy's *Résumé* displays his style in analysis in a remarkable way. His mathematics seems to me often to display his personality—namely, the desire for order and system not only in mathematics but also in the political and religious life of the Bourbon Catholic France in which he then flourished. Now, the original printing of the *Résumé* reveals the following feature. Each of the 40 lectures is printed on four pages, or one sheet, perhaps for distribution to the students lecture by lecture. But further, each lecture finishes *exactly on the bottom line of the last page.* The same is true of two later additions, of 12 and 4 pages exactly. And this occurs in a book founded mathematically on the theory of limits.

Cauchy's real-variable analysis is described in Grattan-Guinness (1970) and in Grabiner (1981), both with extensive references to further writings. He carried his ideas through from the real to the complex variable, and further into complex Fourier-integral solutions to linear partial differential equations. But it would take us too far afield to describe here these details; {see Smithies (1997)}.

4.2. *Cauchy's Refinement and Use of Limit Theory*

One of Cauchy's major refinements to the theory of limits was to allow that the passage of the variable to the limiting value A need *not* be monotonic, that A be defined as the approached value rather than the arrived value, and that A might not exist in the first place. I proposed the term "limit-avoiding," a contraction of the more accurate but also more cumbersome phrase "limit-value avoiding," to characteristic analysis as founded by Cauchy and to be refined by the Weierstrassians (¶5; Grattan-Guinness 1970, 55–58). The term emphasizes that the basic definitions and procedures depend on manipulating quantities and *differing* in value from A, and thus avoiding it. Some versions of analysis are contaminated by "limit-achievement," where these limiting values themselves are employed; for example, Newton and Poinsot.

The way in which Cauchy's calculus develops may be summarized as follows. In the *Cours* he presented his (rather unclear) notion of an "infinitesimal" as a variable that "decreases indefinitely in such a way as to go down below any given number" and so "has zero for limit" (1821, 19): it would have helped if he had explained clearly what "any given number" referred to. On this basis he defined the following key ideas for the differential calculus (1823, lecture 4):

$$\text{Derivative: } f'(x) := \lim_{i \to 0} ((f(x+i) - f(x))/i), \tag{13.10}$$

$$\text{Differential: } df(x) := \lim_{\alpha \to 0} ((f(x+\alpha h) - f(x))\alpha), \ h \text{ finite}, \tag{13.11}$$

if these limits existed at all. Much of the *Résumé* was an articulation of these two notions, defined for any finite order and for functions of several variables as well as of one variable. The account also included this statement of the relationship between the two (lecture 4). Putting

$$i = \alpha h \text{ yielded } df(x) = hf'(x) \tag{13.12}$$

from (13.10) and (13.11); and the particular function

$$f(x) := x \text{ furnished } dx = h \text{ and thus } df(x) = f(x)\, dx, \tag{13.13}$$

from $(13.12)_2$. Hence *dy/dx* was again a ratio of differentials, but with differentials defined in the new way by (13.11) and capable of taking finite values, as in $(13.13)_2$ (Taylor 1974). "It is for this reason that one sometimes gives to the derived function the name of *differential coefficient*," explained Cauchy, in words that might not have helped the students too much. "To *differentiate* a function, is to find its differential."

One of Cauchy's major innovations into the calculus was that he defined the integral independently of any notions of the differential calculus. For a continuous function $f(x)$ over $[x_0, X]$ he formed the sum

$$S := \sum_{r=1}^{n} f(x_{r-1})(x_r - x_{r-1}) \text{ where } x_n = X, \tag{13.14}$$

and introduced finer partitions of the original x_r to show, with the help of various lemmas of the *Cours* on intermediate values of a collection of values such as the x_r, that a limiting value would be achieved, that it would be independent of the sequence of partitions used to achieve it, and that the same value would be obtained if arbitrary points p_r of the subintervals $[x_{r-1}, x_r]$ were used instead of the lower limits x_{r-1} as in (13.14) (lecture 21). Yet even this brilliant mathematical analysis contained an oddity, for in explaining the notation "$\int f(x)dx$" for the limiting value he stated that "the letter $\int$ substituted for the letter Σ indicates, no longer sum of products similar to the expression but the limit of a sum of this type," and yet he invoked $(13.13)_2$ to write the integral as

$$\text{“}\int hf(x)\text{” as well as “}\int f(x)\, dx.\text{”} \tag{13.15}$$

I have no idea what $(13.15)_1$ refers to; fortunately he used $(13.15)_2$ in later exegesis, although the status of *dx* in it must now be unclear.

4.3. *Cauchy on the Fundamental Theorem*

Among the results that Cauchy now produced, the inversion principle was, of course, crucial. Now at last could it be a genuine fundamental *theorem* in the calculus, prefaced by sufficient conditions on the function. After providing a somewhat shaky proof of the first mean value theorem in lecture 22, he gave the fundamental theorem in lecture 26 in the form

Theorem 4.1. If $f(x)$ was finite value and continuous over $[x_0, x]$ and $[x, X]$, then respectively

$$\text{“}\frac{d}{dx}\int_{x_0}^{x} f(x)dx = f(x) \text{ and } \frac{d}{dx}\int_{x}^{X} f(x)dx = -f(x).\text{”} \tag{13.16}$$

Curiously, he passed over the other half of the theorem, involving ∫ *d*/*dx*.

Study of this theorem, its conditions, and its relationship to other key results of the calculus (and mathematical analysis in general) was to become a major preoccupation of the analysis during the nineteenth century (Voss 1899; see also ¶7.5 and ¶7.6 for the example of Taylor's series). Indeed, many other theorems came to be examined and refined, for, in fact, how rigorous was Cauchy's mathematical analysis?

5. The Refinements of the Weierstrassians

The next stage of refinement was created largely by Weierstrass and the followers of his lectures at Berlin University from the late 1850s to the 1880s. The history is complicated, for various reasons (especially the fact that Weierstrass himself published very little); basically the analysis as left by Cauchy and his contemporaries was the main motivation, but nice problems were also found in Riemann's posthumously published essay (1867) on trigonometric series.

I summarize the refinements under six main headings. I have necessarily to consider changes in analysis as a whole; obviously the calculus was affected as an important part. For more details, see Pringsheim (1899), Grattan-Guinness (1970, ch. 6), Dugac (1973; where various pertinent manuscripts by Weierstrass are transcribed), {Bottazzini (1986; which also covers complex-variable analysis) and Jahnke (2003)}. Birkhoff (1973) is "a source book in classical analysis" containing English translations of some pertinent papers, but often with very anachronistic terms and notations.

5.1. Multivariate Techniques

Cauchy's analysis is fine when one limit is taken at a time: $\lim_{x\to a} f(x)$, say, or $\lim_{n\to\infty} S_n$. But with more than one variable moving at once—as in the difference quotient (13.10), say, or in trigonometric series—substantial extra problems of rigor emerge. The Weierstrassians gradually brought in the necessary refinements of uniform, nonuniform, and quasiuniform convergence (Hardy 1918).[3]

3. Interesting examples are Stokes (1849) and Seidel (1848) on "infinitely" and "arbitrarily slow convergence," in which the convergence of infinite series of functions is appraised by means that involve taking the suffix value *n* to its achieved value and therefore messing up the system. Thus these limit-achieving definitions are not any mode of uniform or nonuniform convergence, in my view, which is contrary to the opinion of most historians of analysis.

An interesting historical detail concerns Cauchy's possible motivation to introduce uniform convergence over an interval (1853). Seidel and Stokes are sometimes cited as sources, but Professor I. Domar (Uppsala) has recently brought to my attention Björling (1847), where a partly contaminated version of uniform convergence is introduced, in an explicit criticism of Cauchy's *Cours* theorem on infinite series. Cauchy is not likely to have stumbled across this publication of

5.2. Symbolism

These refinements of techniques required much more careful use of symbolism in order that the functional relationship between variables (including incremental variables) could be clearly indicated and thus lapses of rigor avoided. (This is the form of symbolism that is sometimes called "epsilontics," for its use of ε, δ, and other symbols.) Peano played an exceptionally important role here, bringing in a uniform symbolism, which helped him to create a form of mathematical logic (Grattan-Guinness 1986).

5.3. Real-Line Structure

From time to time Cauchy and his contemporaries mentioned properties of the real line in proving certain theorems. However, only around 1870 were definitions or irrational numbers offered: by Cantor, Dedekind, Heine, Méray, and Weierstrass himself (Pringsheim 1898).

5.4. Point Set Topology

The word "system," referring to collections of mathematical objects, is common in mathematics from the late eighteenth century on, but there was little or no set theory developed at this time. However, with the refinements just described, and certain problems emerging out of Riemann's paper (1867) on trigonometric series, set topology had to be developed. Cantor was the principal figure in this area, bringing in a theory that eventually blossomed into his general theory of sets (Dauben 1979).

5.5. Integration

Riemann (1867) included a study of the Cauchy integral, furnishing a version of the so-called Cauchy-Riemann sums and including some implicit set topology and measure theory. These notions were taken up by the Weierstrassians to develop first a contact theory in the sense of Cantor, Jordan, and Peano, and later a fully fledged "measure theory" due to Borel, Lebesgue, Vitali, and Young (Hawkins 1970).

---

the Uppsala Academy, but he almost certainly noticed the French translation of a related paper by Björling published in Liouville's *Journal des mathématiques pures et appliquées* the year before his own paper appeared (Björling 1852). On this part of the story, see Grattan-Guinness (1987).

5.6. Limit Theory

Limit theory received some refinements of its own during this period. Of particular importance was the distinction between the upper limit and the least upper bound, and related distinctions. This led Dini, for example, to refine Cauchy's definition (13.10) of the derivative into upper and lower left- and righthand derivatives (Dini 1878, art. 15). Some details in proofs were tidied up, systematically using positive values of quantities, for example, and distinguishing $<$ clearly from $\leq$.

6. Zermelo's "Bomb" in the 1900s

With all these refinements being introduced and developed (sometimes, it must be said, at incredible length and tediousness), mathematical analysis seems to have reached a climax of rigorous perfection. And yet there was another stage to go through in the 1900s, when Zermelo announced (1904) that there was need of an "axiom of choice" to prove Cantor's theorem that every set can be well ordered. A great dispute arose (Moore 1982), both over the nonconstructive character of the axiom (for it permits the formation of a set by means of infinite selections made without a rule of selection), the different forms of the axiom (denumerable versus nondenumerable selection, and simultaneous versus successive choice), the modifications made to proofs to allow for these selections (a few proofs were in fact cleansed of them), and so on.

Most of the discussion belonged to mathematical analysis and set theory in general and not specially to the calculus, so I shall not describe it in detail here. But there were issues especially in measure theory, where it arose. An interesting case is Lebesgue. Despite his active participation in the discussion in the 1900s, when he introduced the phrase "Zermelo's bomb," he never seems to have been either happy with the axioms or clear over their role in his formulation of the denumerable additivity of the measure function (Moore 1983). This note of uncertainty is a suitable one on which to close this survey of the checkered history of the calculus.

7. Some Educational Questions

Desirous to outline the whole of the remarkable panorama of the calculus from Newton to Lebesgue, I have not had space to present more than the basic principles and features of each version. Study of the practice of each would reveal still more clearly differences between them (as well as some points of common view), all of which are well worth teaching. In this final section I present a few of the numerous educational lessons that the history of the calculus can teach its teachers. The examples that I have chosen are appropriate for the teaching of the calculus to mathematics students at university level, for that is where my own experience lies; but other types of calculus teaching could no doubt profit from historical data also. The first three cases are of a general character; the last three, particular results.

7.1. Different Traditions

The history described above is *confused,* with different approaches being followed with different terms and notations—and worse, the same terms and notations for different concepts; for example, Cauchy's importation of "$f'(x)$" and "derived function" in ¶4 from Lagrange's very different conceptions in ¶3. Traces of these different strands appear in textbooks of our time; but instead of explaining to the (untutored) reader that these various lines do exist, the authors usually present only the ones that they happen to know and omit the others. Thus if students look at several texts that follow different lines, they will be bewildered.

Take, for example, the notation dy/dx. Is it a whole symbol; or is it a ratio of differentials defined in some way such as Cauchy's at (13.11); or is it the operator d/dx, as used in the new algebras mentioned in ¶3? All these interpretations have a distinguished history; so all should receive the appropriate attention, and not just the favored one. Similarly, $\int y\, dx$ should have its origins explained (¶7.3).

7.2. Levels of Rigor

Mathematicians tend to think that mathematics is rigorous (when executed in a way they favor) or not (when otherwise). But the history of mathematics teaches very clearly that in fact *rigor comes in levels,* which therefore must be specified before the candidate piece of mathematics is appraised. A particular clear sequence of examples are provided in ¶4–¶6; the axioms of choice constituted a deepening of the level of Weierstrassian rigor, just as the Weierstrassians deepened that of the Cauchy period.

Few modern textbooks show any understanding of this important point. The usual opening chapter serves up an incoherent mixture of Cauchian- and Weierstrassian-level limit theory, with bits of point set topology thrown in for good measure. However, often measure theory itself does not appear later: only the Cauchy-Riemann integral, which does not need much set topology. Multivariate refinements are not given the attention they deserve; apropos of ¶5.1 in particular, modes of convergence of infinite series of functions are handled only by the introduction here and there of uniform convergence over an interval in order to prove a few results.

For students who want to *understand* anything about mathematical analysis, and the calculus in particular, I direct them to the older textbooks written by authors who also took some interest in the history of the calculus and learned from it. In English, Carslaw (1930), Knopp (1951), and Whittaker and Watson (1927) are especially recommended. Fortunately, the first and last of these books are, or have been, available in paperback—perhaps a sign of consumer resistance to the modern product.

A scan of popular calculus and analysis textbooks of recent decades revealed an absurd confusion of partial explanations of the various senses of $dy/dx, f'(x) \int dx$, and so on. For example, it is not hard to find the derivative presented with dy/dx interpreted as a whole symbol but also the differential dx defined later on. The Award for Unintended Humor went to one of the most widely used texts, Protter and Morrey (1964), which defines derivatives solely in terms of the notation $f'(x)$, notates the definite integral as $\int_a^b f(x)dx$ by "explaining" that "The dx is just a juxtaposition of letters which are considered inseparable," and then defines the differential and gets to (13.12) . . . (pp. 77, 90–91, 184–185).

7.3. *The Place of Differentials*

The Cauchy-Weierstrass approach, based on limits, has of course become normal teaching fare in courses on (pure) calculus or analysis. However, lectures on mechanics, astronomy, mathematical physics, and engineering often draw on Euler's form of the differential calculus (¶2), because of its intuitive malleability in the construction of differential models of the physical phenomena under discussion. Cauchy was criticized at the Ecole Polytechnique during the 1820s by de Prony, one of the graduation examiners, on this point (Grattan-Guinness 1981, 686):

> The instruction, in pure analysis and in analytical geometry, has appeared to leave a little more to be desired in practice and facility for the applications to formulae and general theories: some quite good students, having proved formulae well, find themselves in embarrassment when it is a question of resolving particular cases to which they relate. It is most important never to isolate abstract notions from considerations of detail; the latter always precede the others in the natural developments of ideas, and it is necessary to conform to this law of nature, above all when it is a case of instructing young people for whom the theory must be an instrument of practice.

Plus ça change. . . . In the case of calculus, teaching the hegemony of limits has led to a most unfortunate and quite unnecessary educational schizophrenia: pure calculus is wall-to-wall epsilontics and ignores the Eulerian differential form of the calculus, which applied calculus courses, with very good reason, often use as a staple (compare §1.1).

The situation has changed somewhat with the introduction of calculus courses using infinitesimals based on nonstandard analysis, usually in heavily diluted forms (see, for example, Keisler 1976). But one can use the traditional form, as described in ¶2, with much success as long as the probity of the differential is frankly admitted (see ¶7.4).

I may add that the claims made by some nonstandard analysts that the history of the calculus needs rewriting because of the existence of this subject (and therefore, of course, as a Royal Road to it) are absurd. "I think that Lazare Carnot has nothing to do with nonstandard analysis, and that nonstandard analysis has nothing to do with Carnot," Professor Yushkevich said to me recently after we had sat through a lecture of this type, and of course he is right (see also Bos 1974, 81–84, and Edwards 1979, 341–346). It is precisely because nonstandard analysis is such a fine mathematical achievement that it cannot be transplanted into the past without causing historical distortion there (see ch. 2 on history and heritage).

7.4. Forming a "Real" Differential Equation

Here is a sample of a typical way in which the Eulerian calculus allows the formation of an equation relating differentials: Fourier's derivation of the equation of heat diffusion in the 1800s. It is historically very important, since it was the first major equation of mathematical physics to be found outside mechanics.

Figure 13.1 shows a thin homogeneous bar, diffusing heat within itself with conductivity K and into the environment (assumed to be at temperature 0) with conductivity h. Let Y be the temperature at point B distant x along the bar, and let the bar be sliced into equal differential slices of thickness dx (so that x is an independent variable).

Fourier defined K relative to temperature gradient, in that the loss of heat from x_0 to x_1 was given by Newton's law of cooling as

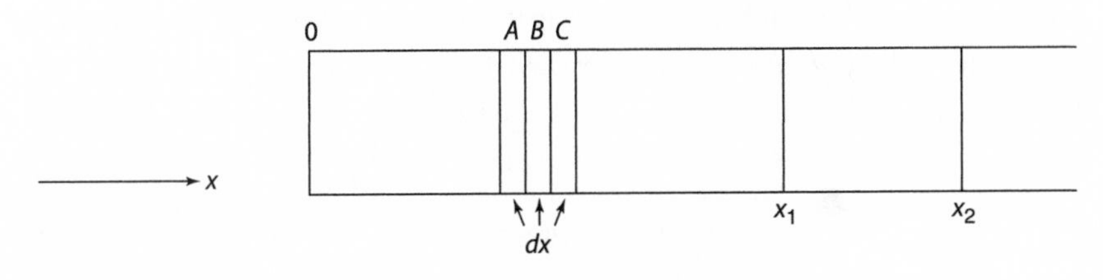

Figure 13.1

$$-\left(\frac{K}{x_1 - x_0}\right)(y_1 - y_0), \text{ or } -K\left(\frac{y_1 - y_0}{x_1 - x_0}\right), \text{ which became } -K\frac{dy}{dx} \quad (13.17)$$

for a slice such as that around B. Then the net gain of internal heat at B was given by the difference between the outflow to C (at temperature y') and inflow from A over the time interval dt:

$$-K\left(\frac{dy'}{dx} - \frac{dy}{dx}\right)dt, \text{ or } Kd\left(\frac{dy}{dx}\right)dt. \quad (13.18)$$

During the same time heat flowed into the environment to the amount

$$-h(y - 0)\, dx\, dt, \text{ or } -hy\, dx\, dt. \quad (13.19)$$

This heat exchange led to a rise in temperature dy in B over the time dt, which calorimetry expressed as the amount of heat

$$CD\, dx\, dy, \quad (13.20)$$

where C was the specific heat of the bar and D its density. Then the heat balance equation for B,

$$\text{(internal heat flow)} - \text{(external heat flow)} = \text{(net heat gain)}, \quad (13.21)$$

became, from (13.17) to (13.20),

$$K\, d(dy/dx)\, dt - hy\, dx\, dt = CD\, dx\, dy, \quad (13.22)$$

or, bearing in mind that dx and dt were constant differentials,

$$ddy/(dx)^2 - hy = CD\, dy/dt \quad (13.23)$$

(Fourier 1807, arts. 19, 23, with some inessential modification).

A few comments are worth making. First, I mentioned in ¶2 that the Leibnizian calculus was dimension preserving; when executed in physical problems, further care had to be taken to define physical constants correctly so that the expressions would be homogenous (all second-order in [13.21], for example). In fact, in Fourier's first attempts, and in those of Biot, who anticipated him, mistakes were made. Second, expressions such as dy/dx were ratios evaluated *at* a particular face of B, not as the limiting value of the difference quotient as the face was approached. It is important to note the difference between (13.22) and the modern diffusion equation

$$K\partial^2 y/\partial x^2 - hy = CD\, \partial y/\partial t \quad (13.24)$$

and its derivation by limit-taking methods: failure to observe this difference can lead to hopeless misunderstandings of the method (Grattan-Guinness 1975, part 2). Third, since both *dx* and *dt* take constant differentials, it is important to remember that the increments *dy* and *ddy* in the argument are partial differentials, which in fact these notations do not indicate. Various modifications to the notation were used in the multivariate differential calculus to indicate partial differentials; for example, Cauchy would have written $d_x y$ and $d_t y$ here to indicate the relevant variables (1823, lecture 8).

7.5. Taylor's Series with Remainder

An interesting feature of the calculus is the proof of Taylor's series, preferably with a remainder term. The history of this topic (on which see Pringsheim 1900) is educationally instructive. Lagrange argued that, given an increment *i* on *x*, then "for any function *fx*,"

$$f(x + i) = f(x) + iP, \text{ where } P = \infty \text{ when } i = 0. \tag{13.25}$$

Similarly, for the new function *P*,

$$P = p + iQ, \tag{13.26}$$

and so on, so that the series was generated by substitution:

$$f(x + i) = f(x) + pi + qi^2 + \dots \tag{13.27}$$

To find the relationship between *f* and the functions *p*, *q*, . . . , he wrote

$$f(x + i + o) = f(x + (i + o)) = f((x + i) + o) \tag{13.28}$$

and expanded each term by (13.27), slipping in prime functions such as f' and p' without explanation. Infinite values for *f* were allowed for by inverse powers of *i* in (13.27). No *clear* remainder term emerged (Lagrange 1797, 7–14, 38–39).

Lagrange's colleagues attempted to improve on this reasoning. Poisson (1805) assumed the more general expansion

$$f(x + i) = f(x) + i^a p(x) + i^b q(x) + \dots, a < b \dots, \tag{13.29}$$

and imitated the argument above to prove that the indices took the successive values 1, 2, 3, . . . and that the primes could be brought in as indicating the "primitive functions" to which the functions related. Ampère (1806, 162–167) took the definition of the difference quotient as the identity

$$f(z) = f(x) + p(x, z)(z - x) \tag{13.30}$$

and carried out various substitutions and successive differentiations to produce the series with a differential remainder form; in a later paper (1827) he differentiated

$$P(a) := (f(x + h) - f(x + ah))/(1 - a) \tag{13.31}$$

successively with respect to a and obtained a similar result.

The proof by integration by parts on $\int_0^h f(x + z)dz$, sketched in d'Alembert (1754, 50) was refined by Fourier and de Prony around 1800 and turned up in Cauchy's *Résumé*, with acknowledgment to de Prony (Cauchy 1823, lecture 36). Cauchy was, of course, more careful than his predecessors on the conditions required on f, and he used the second mean value theorem to convert the remainder term from its integral to its differential form. In the second addition to the book he found the differential form directly by applying the first mean value theorem successively to the first, second, . . . nth partial sums of the series.

These records from the past make admirable educational material for this area of the calculus, since questions of rigor and generality are well exposed. The relationships between the series and the mean value theorems is also brought out, especially if uniformity conditions are applied to the difference quotient when treated as a function of both its base and incremental variables.

7.6. Taylor's Series at All?

At the end of lecture 38 of the *Résumé* Cauchy mentioned that $\exp(-1/x^2)$ did not take a Taylor expansion when $x = 0$, and so killed Lagrange's foundations of the calculus completely. The result was not important to him, since he never espoused Lagrange's approach anyway; but it had more serious consequences for any parts of analysis founded on power-series proofs. This is clear in the paper (Cauchy 1822) in which he announced this and other counterexamples and explored certain consequences.

Poisson was a fervent believer in Lagrange's approach, so he had to reply to Cauchy. In a note (1822) he took the function

$$x^{-m} \exp(-1/x^2), \text{ where } m > 0, \tag{13.32}$$

and argued that if $m \rightarrow \infty$ such that

$$m > (\ln x)x^2, \tag{13.33}$$

then $x^{-m} \rightarrow \infty$ more rapidly than $\exp(-1/x^2) \rightarrow 0$ as $x \rightarrow 0$, so that "there is always a part [of the Taylor's series] which vanishes with x, and another part of which the values are infinite for $x = 0$," thus somehow saving the faith. While the argument is weak even by Poisson's sloppy standards, students would have to work a while to see what is wrong with it.

They may also gain consolation from this historical detail. Euler had found that all the differential coefficients of the function exp $(-1/x)$ (which was another of Cauchy's counterexamples) were zero when $x = 0$; but he failed to realize the consequences for Taylor's series, for his concern was with numerical quadrature of integrals (Euler 1768, art. 327). Lacroix also missed them when describing Euler's procedure in his treatise on the calculus (*Calculus*, 2nd ed., vol. 2 [1814], 143–144). Cauchy may well have found his counterexamples in ignorance of these earlier near misses.

8. Prospects

One reason why the history of mathematics can be useful for mathematical education is that "education imitates history," as I have put it on occasions. Of course there is no need, or value, in pursuing all historical details; but history-satire, in which some kind of imitation of the historical record is effected in teaching, has considerable possibilities (see Grattan-Guinness 1973, and ch. 9).

However, those who advocate such views today do so against a long background of indifference or even hostility to history—a situation that, ironically, itself has a history. One cannot expect the current wisdom to be much affected by arguments that by and large it dismisses in principle. For example, the last time I spoke in Britain on the general question of using history in mathematical education (both for students and for teacher training) was in May 1973, when I addressed an audience of professional mathematicians and mathematics teachers belonging to the Manchester Branch of the Mathematical Association. In question time afterward one of the professionals refuted my views with a deft counterexample that had apparently escaped me: "There is no point in teaching mathematics along the lines you suggest, because fewer theorems could be covered in the given time."

Some of the teachers present seemed not entirely convinced, although they were too shy to say anything. Not so four years later, when I addressed a large group of mathematics teachers (and no professionals) in Sydney, Australia, in November 1977 on the calculus from Leibniz to Cauchy. At one point I started to explain the different interpretations of dy/dx (¶7.1); but I found, to my and their astonishment, that many of my listeners *had never been shown* the notation dy/dx when they learned the calculus.

So we have a tension between the teachers and the professionals. Some of my Sydney audience quite openly expressed to me their fury over the inadequacy of the mathematics education that they had received at college (both over the case of *dy/dx* and many other things). Meanwhile, back at Manchester another professional had comforted me after the lecture at the refutation of my views. "We enjoyed your lecture very much," he claimed, "but, of course, we can't take it seriously."

BIBLIOGRAPHY

Many histories of mathematics contain chapters and sections on the history of the calculus and mathematical analysis, and there is no special need to list them here. Of the histories devoted to a detailed account of the period that I describe above, Boyer (1949) is still a good guide, although it seriously fails to note the difference between uni- and multivariate analysis when describing in chapter 7 "The [sic] rigorous formulation." Edwards (1979, chs. 8–12) covers the material in more detail than Boyer and includes exercises for the reader quite often; but he only brings out definitions of irrational numbers (¶5.3) as the Weierstrassian improvement. Both books contain useful accounts of ancient and medieval calculus techniques.

A general history of this period is Grattan-Guinness (1980, chs. 2–5); in addition, chapter 1 in that volume discusses the formulations immediately prior to Newton and Leibniz, and chapter 6 describes the evolution of general set theory and mathematical logic. A teaching syllabus that I have followed using the book is outlined in Grattan-Guinness (1977; §9.1); the presentation included reading classes, where we read out passages from the primary literature together and discussed the content. {See also now Bottazzini (1986), Caramalho Domingues (2008), and Jahnke (2003).}

Some textbooks have attempted to break out of the rigid mold of the epsilontic orthodoxy. However, it would be too substantial a task to attempt a survey here of the different presentations (compare §9.2.1). Indeed, a substantial (and worthwhile) doctorate thesis could be prepared on the history of twentieth-century textbooks at various levels.

Ampère, A. M. 1806. "Recherches sur quelques points de la théorie des fonctions dérivées . . . ," *Journal de l'Ecole Polytechnique, (1)6,* cah. 16, 148–181.

Ampère, A. M. 1827. "Démonstration du théorème de Taylor . . . ," *Annales des mathématiques pures et appliquees, 17,* 317–329.

Arbogast, L. F. A. 1800. *Du calcul des dérivations,* Strasbourg, France: Levrault.

Berkeley, G. 1734. *The analyst,* London: Towson. [Also in *Works* (ed. A. A. Luce and T. E. Jessop), vol. 4, London and Edinburgh, 1951, 53–102.]

Birkhoff, G. 1973. (Ed.), *A source book in classical analysis,* Cambridge, Mass.: Harvard University Press.

Björling, E. F. 1847. "Doctrinae serierum infinitorum exercitationes," *Nova acta Societatis Scientarum Upsala, 13,* 61–86, 143–186.

Björling, E. F. 1852. "Sur une classe remarquable des séries définies," *Journal des mathématiques pures et appliquees, (1)17,* 454–472.

Boole, G. 1859. *A treatise on differential equations*, 1st. ed., London: Macmillan. [Last ed. 1877; repr. New York: Dover, n.d.]

Bos, H. J. M. 1974. "Differentials, higher-order differential and the derivative in the Leibnizian calculus," *Archive for history of exact sciences, 14*, 1–90.

Bottazzini, U. 1986. *The higher calculus*, New York: Springer.

Boyer, C. 1949. *The history of the calculus and its conceptual development*, New York: Columbia University Press. [Repr. New York: Dover, 1959.]

Cajori, F. 1919. *A history of the conceptions of limits and fluxions from Newton to Woodhouse*, Chicago: Open Court.

Caramalho Domingues, J. 2008. *Lacroix and the calculus*, Basel, Switzerland: Birkhäuser.

Carnot, L. N. M. *Calculus. Réflexions sur la métaphysique du calcul infinitésimal*, 1st ed. 1797, Paris: Duprat; 2nd ed. 1813, Paris: Courcier. [Several reprints; no satisfactory English translations.]

Carslaw, H. S. 1930. *Introduction to the theory of Fourier's series and integrals*, 3rd ed., London: Macmillans. [Repr. New York: Dover, n.d.]

Cauchy, A. L. 1821. *Cours d'analyse* . . . , Paris: de Bure. [Also in *Oeuvres complètes*, ser. 2, vol. 3, Paris: Gauthier-Villars, 1897]

Cauchy, A. L. 1822. "Sur Ie développement des fonctions en séries . . . ," *Bulletin des sciences par la Société Philomathique de Paris*, 49–54. [Also in *Oeuvres complètes*, ser. 2, vol. 2, Paris: Gauthier-Villars, 1958, 276–282.]

Cauchy, A. L. 1823. *Résumé des leçons* [. . .] *sur le calcul infinitésimal*, Paris: de Bure. [Also in *Oeuvres complètes*, ser. 2, vol. 4, Paris: Gauthier-Villars, 1898, 5–261.]

Cauchy, A. L. 1853. "Note sur les séries convergentes . . . ," *Comptes rendus de l'Académie des Sciences*, 36, 454–459. [Also in *Oeuvres complètes*, ser. 1, vol. 12, Paris: Gauthier-Villars, 1900, 30–36.]

Child, J. M. 1920. (Ed. and trans.), *The early mathematical manuscripts of Leibniz*, London: Open Court.

d'Alembert, J. le R. 1754. *Recherches sur différens points importans de système du monde*, vol. 1, Paris: David.

Dauben, J. W. 1979. *Georg Cantor* . . . , Cambridge, Mass.: Harvard University Press. [Repr. Princeton, N.J.: Princeton University Press, 1990.]

Dini, U. 1878. *Fondamenti per la teorica delle funzioni di variabili reali*, Pisa: Nistri. [German trans: Leipzig, Germany: Teubner, 1892.]

Dugac, P. 1973. "Eléments d'analyse de Karl Weierstrass," *Archive for history of exact sciences, 10*, 41–176.

Edwards, C. H., Jr. 1979. *The historical development of the calculus*, New York: Springer.

Euler, L. 1755. *Institutiones calculi differentialis*, St. Petersburg: Academy Press. [Also in *Opera omnia*, ser. 1, vol. 10.]

Euler, L. 1768. *Institutiones calculi integralis*, vol. 1, Leipzig: Teubner, 1913; St. Petersburg: Academy Press. [Also in *Opera omnia*, ser. 1, vol. 11.]

Euler, L. 1983. *Leonhard Euler 1707–1783. Beiträge zu Leben und Werk*, Basel, Switzerland: Birkhäuser.

Fourier, J. B. J. 1807. "Théorie de la propagation de la chaleur dans les solides," manuscript, *passim*, in I. Grattan-Guinness in collaboration with J. R. Ravetz, *Joseph Fourier 1768–1830 . . .* , Cambridge, Mass.: MIT Press, 1972.

Grabiner, J. 1981. *The origins of Cauchy's rigorous calculus*, Cambridge, Mass.: MIT Press.

Grattan-Guinness, I. 1970. *The development of the foundations of mathematical analysis from Euler to Riemann*, Cambridge, Mass.: MIT Press.

Grattan-Guinness, I. 1973. "Not from nowhere: History and philosophy behind mathematical education," *International journal of mathematical education in science and technology, 4*, 321–353.

Grattan-Guinness, I. 1975. "On Joseph Fourier: The man, the mathematician and the physicist," *Annals of science, 32*, 503–514.

Grattan-Guinness, I. 1977. "A history of mathematics course for teachers," *Historia mathematica, 4*, 341–343.

Grattan-Guinness, I. 1979. "Babbage's mathematics in its time," *British journal for the history of science, 12*, 82–88.

Grattan-Guinness, I. 1980. (Ed.), *From the calculus to set theory 1630–1910: An introductory history*, London: Duckworth. [Repr. Princeton, N.J.: Princeton University Press, 2000.]

Grattan-Guinness, I. 1981. "Recent researches in French mathematical physics of the early nineteenth century," *Annals of science, 38*, 663–690.

Grattan-Guinness, I. 1986. "From Weierstrass to Russell: A Peano medley," in *Celebrazioni in memoria di Giuseppe Peano*, Turin, Italy: Department of Mathematics, University, 17–31. [Also in *Revista di storia della scienza, 2* (1985: publ. 1987), 1–16.]

Grattan-Guinness, I. 1987. "The Cauchy-Seidel-Stokes story on uniform convergence again: Was there a fourth man?" *Bulletin de la Société Mathématique de Belgique, (A)38* (1986), 225–235.

Hardy, G. H. 1918. "Sir George Stokes and the concept of uniform convergence," *Proceedings of the Cambridge Philosophical Society, 19*, 148–156. [Also in *Collected papers*, vol. 7, Oxford: Clarendon Press, 1979, 505–513.]

Hawkins, T. W. 1970. *Lebesgue's theory of integration*, Madison: University of Wisconsin Press. [Repr. New York: Chelsea, 1975.]

Hofmann, J. E. 1949. *Die Entwicklungsgeschichte der Leibnizschen Mathematik*, Munich: Oldenbourg. [English trans: *Leibniz in Paris 1672–1676*, Cambridge: Cambridge University Press, 1974.]

Jahnke, N. H. 2003. (Ed.), *A history of analysis*, Providence, R.I.: American Mathematical Society.

Keisler, H. J. 1976. *Elementary calculus* and *Foundations of infinitesimal calculus*, 2 vols., Boston: Prindle, Weber, and Schmidt.

Knopp, K. 1951. *Theory and application of infinite series* (trans. R. C. H. Young), London and Glasgow: Blackie.

Lacroix, S. F. Calculus. *Traité du calcul différentiel et du calcul intégral*, 3 vols., 1st ed., Paris: Duprat, 1797–1800; 2nd ed., Paris: Courcier, 1810–1819.

Lagrange, J. L. 1797. *Théorie des fonctions analytiques . . .* , 1st ed., Paris: Imprimerie de la Republique. [Also in *Journal de l'Ecole Polytechnique, (1)3* (1801), 1–277.] 2nd ed., Paris: Bachelier, 1813. [Also in *Oeuvres*, vol. 9, Paris: Gauthier-Villars, 1881.]

Moore, G. H. 1982. *Zermelo's axiom of choice . . .*, New York: Springer.

Moore, G. H. 1983. "Lebesgue's measure problem and Zermelo's axiom of choice . . . ," *Transactions of the New York Academy of Sciences, 412*, 129–154.

Newton, I. *Papers. The mathematical papers of Isaac Newton* (ed. D. T. Whiteside), 8 vols., Cambridge: Cambridge University Press, 1967–1981.

Panteki, M. 1992. "Relationships between algebra, differential equations and logic in England: 1800–1860," doctoral dissertation, C.N.A.A., London.

Poinsot, L. 1815. "Des principes fondamentaux, et des règles générales du calcul différentiel," *Correspondance de l'Ecole Polytechnique, 3*, 111–131 [including a sequel paper].

Poisson, S.-D. 1805. Démonstration du théorème de Taylor," *Correspondance de l'Ecole Polytechnique, 1*, 52–55.

Poisson, S.-D. 1822. "Remarques sur les intégrales des équations aux différences partielles," *Bulletin des sciences par la Société Philomathique de Paris*, 81–85.

Pringsheim, A. 1898. "Irrationalzahlen und Konvergenz unendlicher Prozesse," in *Encyclopädie der mathematischen Wissenschaften*, vol. 1, part 1, 47–146 (article IA3).

Pringsheim, A. 1899. "Grundlagen der allgemeinen Funktionentheorie," in *Encyclopädie der mathematischen Wissenschaften*, vol. 2, part 1, 1–53 (article IIAl).

Pringsheim, A. 1900. "Zur Geschichte des Taylorschen Lehrsatzes," *Bibliotheca mathematica, (3)1*, 433–479.

Protter, M. H., and Morrey, C. B. 1964. *College calculus with analytic geometry*, Reading, Mass.: Addison Wesley.

Riemann, G. F. B. 1867. "Über die Darstellbarkeit einer Function durch eine trigonometrische Reihe," *Abhandlungen der Gesellschaft der Wissenschaften zu Göttingen, 13* (1866–1867), mathematische Klasse, 87–132. [Also in *Gesammelte mathematische Werke*, 2nd ed., Leipzig, Germany: Teubner, 1892, 227–271.]

Seidel, P. L. 1848. "Note über eine Eigenschaft der Reihen, welche discontinuerliche Funktionen darstellen," *Abhandlungen der Akademie der Wissenschaften zu München, mathematisch-physikalische Klasse, 5* (1847: publ. 1849), 381–393.

Smithies, F. 1997. *Cauchy and the creation of complex function theory*, Cambridge: Cambridge University Press.

Stokes, G. G. 1849. "On the critical values of the sums of periodic series," *Transactions of the Cambridge Philosophical Society, 8*, 533–583. [Also in *Mathematical and physical papers*, vol. 1, Cambridge: Cambridge University Press, 1880, 236–313.]

Taylor, A. E. 1974. "The differential: Nineteenth and twentieth century developments," *Archive for history of exact sciences, 12*, 355–383.

Voss, A. 1899. "Differential- und Integralrechnung," in *Encyclopädie der mathematischen Wissenschaften*, vol. 2, part 1, 54–134 (article IIA2).

Whiteside, D. T. 1961. "Patterns of mathematical thought in the later seventeenth century," *Archive for history of exact sciences, 1*, 179–388.

Whittaker, E. T., and Watson, G. N. 1927. *A course of modern analysis*, 4th ed., Cambridge: Cambridge University Press. [Paperback reprint 1965.]

Yushkevich, A. P. 1971. "Lazare Carnot and the competition of the Berlin Academy in 1786 on the mathematical theory of the infinite," in C. C. Gillispie (ed.), *Lazare Carnot savant*,

Princeton, N.J.: Princeton University Press, 149–168. [Previously unpublished manuscript by Carnot on pp. 169–268, including notes by Yushkevich.]

Yushkevich, A. P. 1976. "The concept of function up to the middle of the nineteenth century," *Archive for history of exact sciences, 16*, 37–85.

Zermelo, E. 1904. "Beweis, dass jede Menge wohlgeordnet werden kann," *Mathematische Annalen, 59*, 514–516. [English trans. in J. van Heijenoort (ed.), *From Frege to Gödel . . .* , Cambridge, Mass.: Harvard University Press, 1967, 139–141.]

PART 3

BYWAYS IN MATHEMATICS AND ITS CULTURE

CHAPTER FOURTEEN

Manifestations of Mathematics in and around the Christianities

Some Examples and Issues

Mathematics and Christianity both have very long histories; but the links between them have not been widely studied, and then mostly for the effect of the latter on the former. This paper reviews the converse influence, where mathematics played roles in the formation and canonical texts of Christianity and in its artefacts (especially buildings), leading to passages and information that are to be interpreted figuratively rather than literally. The main branches of mathematics are arithmetic and geometry with trigonometry. Pre-Christian sources are found especially in Egyptian and Greek mathematics, and the latter link suggests some historical relationships to movements that were to lead to Freemasonry. Finally, the method of crucifixion is appraised as an exercise in mechanics.

1. Aims and Preliminaries

1.1. "Faith" and "Religion"

I use the word "faith" as an umbrella term to encompass all kinds of general belief systems, worldviews, and (so-called) myths referring to the creation and origins of the universe, and usually accompanied by ethical and axiological requirements. When a faith crystallized into a (fairly) formal doctrine, and its related community into a society, then I shall refer to a "religion."

The origins of both mathematics and faiths lie deep in antiquity; and since little was known then in any domain of knowledge, it is not surprising that they have long exhibited connections. Although various specific situations and individuals have been studied, the *general* influences of each upon the other are not widely examined,

The original version of this essay was first published in *Historia scientarum*, (2)11(2001), 48–85; also in *Revista Brasiliera de Historia da Matematica*, 1(2001), 21–56. Reprinted by permission of the editor of *Historia scientarum* and the editor of *Revista Brasiliera de Historia da Matematica*.

and most work has dealt with a faith influencing mathematics. This paper presents a suite of significant cases of an example of the converse influence, where mathematics of some kind affected the content and texts of Christianity;[1] note is taken of pre-Christian origins. Some cases concern its early history and the formation of the Bible; others relate to developments in medieval Europe, especially in church design and architecture. Lack of direct evidence has rendered very obscure the histories of possible connections, and they cannot be resolved here.

The common factor among faiths is as *authoritative sources of knowledge over* other sources, including competing faiths, concerning basic tenets and attendant commentary and justification.[2] Mathematics has also been so extolled: it may not be accidental that Euclid's *Elements* is divided into numbered books and propositions and the Bible into numbered chapters and verses, giving each work the appearance of an encoded body of law. (The origin of the *explicit indication* of the numbers is not clear for either work.) However, there is an important difference. Mathematics was extolled as a source of correct (supposedly "logical") *reasoning* (Euclid especially important here), while by definition a faith supplied truths *without the need for reasoning*. The latter was supplied by theology to satisfy the demand made by lack of belief: Paul even stated a version of the liar paradox without mentioning its logical status (Titus 1:12–13). In 1699 John Craig(e) tried to determine mathematically the decline in level of belief by its successive transmission among adherents.[3]

The two domains of knowledge also overlap in specific notions, such as that of 1, which expresses some unity in monotheistic religions and also serves as a special unit in arithmetic of which numbers themselves are multiples, from 2 onward (in particular, in Euclid *Elements*, 7:1–2). Other topics of common concern include continuity and infinity. The tricky issue of cognitive and noncognitive relations between Christianity and mathematics will be considered in ¶7. Some assumptions need to be aired now.

1. For valuable indications of Muslim numerology and sacred geometry, see McClain (1981). The utility of the Quran as a source for the history of Christianity does not seem to be well recognized.

2. Proofs of the existence of God are exercise less in mathematics than in logic chopping (Whittaker 1946). Kurt Gödel (1906–1978) proved a theorem in modal logic asserting the existence of an object possessing all "positive" properties, in a manuscript published in Gödel (1995, 388–404, 429–437). He was reluctant to interpret it theologically; on his theism, see Wang (1987, esp. pp. 215–218). My own logical stance is, God save us from religions, especially the aggressive ones.

3. See Nash (1991), with an English translation of the original Latin text. A good interpretation in terms of likelihood theory is provided in Stigler (2000, ch. 13).

1.2. Mathematics and Methodology

I take the most relevant parts to be arithmetic (especially numerology and gematria, and the use of digit strings such as 1-1-1), music, geometry, trigonometry, and mechanics, together with astronomy and the measurement of time. Usually the mathematics is very elementary to modern eyes, or existed at the level of maybe intuitive thinking rather than formal(ish) theory; but it should not be derided, for all topics back then were in early states of development, and the cultural impact could have been very considerable. Indeed, I assume that in antiquity very close interactions obtained among these domains of knowledge.

Important also is semiotics, the theory of relationships between signs and their referents. Signs carry information of some kind, an important service in a community where many members are illiterate. Here the possibility of *multiply interpreting* a sign must be emphasized.[4] For example, and indirectly a Christian one in involving crosses, the flag of the United Kingdom, designed after the legislative union with Ireland in 1801, is a mesh of vertical, horizontal, and oblique strips in red and white with sections in blue because it is a semiotic superposition of the Scottish, English, and Irish flags.

Also involving multiplicity is a methodology of *reinforcement.* The normal modern manner of studying a historical situation is to find *the* (single) explanation of it, and to treat alternative explanations as *rivals.* However, this is not the way to tackle ancient thought, when everything was early on and multiple readings and connections admitted. Rival explanations were *companion explanations*: the significance of, say, some number or shape could have been seen in *several* explanatory contexts. This methodological point can apply to *any* historical enterprise, but with especial force in contexts such as the one that we now treat.

1.3. Literal versus Figurative Interpretation of Texts

This last point relates to this hypothesis, not merely concerning mathematics: ancient religious texts contain *important figurative or metaphorical elements,* maybe drawing in double meanings of words (and hence leading to multiple interpretations), and were then so understood; but later such passages have been misinterpreted by ad-

4. For a nice survey of symbols in general in Orthodox Christian practice relative to their possible origins, see Rees (1992). The semiotic side of Judaism is admirably conveyed in Goodenough (1953–1968).

herents as making *literal* claims. The hypothesis has been greatly strengthened for Christianity by some of the texts from biblical times that have been discovered and deciphered in recent decades, for they explicitly use a method that is not mentioned in biblical texts, or else has been excised at some stage. The key word, in Hebrew, is "pesher," which translates as "interpretation of dreams"; it appears in the Bible in its own right (for example, Genesis 40:5, Ecclesiastes 8:1). An appropriate verse of text was supplemented by a pesher explaining the events or person involved, or the intention of the analogy or metaphor stated;[5] thus they contain some of the secrets that theology seeks.

Two nonmathematical examples in Christianity exemplify figuration. First, Jesus is said to have had a virgin birth; but this property often goes with the job as prophet (note also Buddha, Indra, Zoroaster, Dionysius, . . .) and may involve a semantic blur between "virgin," "young woman," and "maiden" ("bᵉthûlâ" and "almah" in Hebrew, and "Παρθενος" in Greek) to "virgo" in Latin and thence into European languages (Thiering 1992, ch. 8). In any case, it is asserted of Jesus only in Matthew 1:23 (note the ambiguous "Joseph the husband of Mary, of whom was born Jesus" in Matthew 1:16) and in Luke 1:27, with the latter contradicted by the genealogy through Jesus's father Joseph related in 3:23–38. The Old Testament mentions such a birth at Isaiah 7:14; it may have inspired the claim made for that of Jesus.

Second, Jesus was named as "the son of God" (for example, Matthew 14:33); but such definite descriptions had been standard modesty among Egyptian pharaohs and other mighty precursors, and so do not entail a literal reading for *any* such figure, though (at least) one of them might have been granted that Office. (Among the many proper names for Jesus, "Jesus" presumably was not one of them; for it is [Latinized out of] Greek, with parentage in the Hebrew/Aramaic "Jo[e]shua.") Other cases of mistranslation surely include Moses and the slaughter of the "first-born" ("bekhori" in Hebrew) instead of the "chosen" ("bechori" or "bechiri"; Velikovsky 1953, 26–29). In addition, the possibility must be admitted of tampering with biblical texts, especially in the New Testament, during the time of its formation from the creation of Christianity by Paul in the +first century (that is, the first century A.D.) to its official establishment under Emperor (and non-Christian) Constantine in the +fourth century (¶5.1, ¶8.1).[6]

5. The method is described in Thiering (1992, ch. 4), quoting examples from a version of Habakkuk. Page references are to the 1993 reprint, where for some reason fig. 17 is omitted from p. 521.

6. An excellent introduction to biblical paleography is provided in Parker (1997), especially on the various extant versions of the Gospels. A similar service for the historical and geographical context of early Christianity is provided by Toynbee (1969).

2. Some Christian Principles

2.1. Orthodoxy and Apocrypha

There have been two main streams of Christian faith, commonly known as "Orthodox" and "Apocryphal" (the latter referring not only to the addenda so named to the Old Testament); I use these names, but without affirming or denying either connotation. The content of the Apocryphal tradition has become much clearer since 1991, when the Huntington Library made publicly available copies of the original texts of the Dead Sea Scrolls of the special committee, largely composed of Dominicans and Jesuits, after their discovery in 1945 and 1956. However, the significance and even content of many texts await evaluation after decades of slowness over publication.[7] A happier history attends the Nag Hammadi texts, discovered also in 1945 in a village in Upper Egypt, which were transcribed and translated much more quickly.[8] In both cases, especially the second, some texts are consonant with the Orthodox tradition, but several not. Both traditions record activities of a Jewish sect with Jesus as a or the main figure, but major differences surround the accounts of his life and role. The three main ones are that in the Apocryphal line: (1) he was born normally, not by virgin birth; (2) he was a rabbi, with Mary Magdalene as his wife, not celibate; and (3) he only fainted on the cross and was taken down alive to recover.[9]

Various subsequent differences are worth noting. The Old Testament retains a much higher level of importance in the Apocryphal line than in Orthodoxy. Again, the Apocryphal denial of Jesus's death spares it from the Orthodox paradox of worshipping the Jew Jesus and also leading for centuries the promotion of anti-Semitism in Europe and its colonies.

The most striking difference concerns Mary Magdalene. In Apocryphal texts her importance equals that of her husband as a leader; but she appears very little in the New Testament, although she witnesses a major event, the appearance of Jesus after the resurrection. Orthodox Christians often know her as a prostitute, more rarely aware that this assertion has *no* biblical authority: Luke 7:37 does not provide the supposed evidence, referring merely to some woman "who was a sinner." The long

7. For a valuable though rather polemical account, see Baigent and Leigh (1991).

8. Robinson (1977). Most texts pertain to the New Testament; a valuable concordance of the third (1988) edition of them consonant with Orthodoxy is provided in Evans, Webb, and Wiebe (1993), which also has a fine bibliography of recent commentaries.

An English translation of older and newer Apocryphal texts is provided in Elliott (1993). Eight influential English translations from William Tyndale (1535) to the Revised Standard Version (1960) can be readily compared in Weigle [1962].

9. The most detailed account to date following this tradition is given in Thiering (1992). Many further references are also given there, and an astonishingly detailed chronology of events is claimed.

and continuing endurance of this appraisal shows the strength of the differences between the two traditions. Again, the important category of Mariolatry may well involve muddles between her and Mary the mother of Jesus as well as disagreements between Catholics and Protestants over the status of the latter Mary. Among other instances of heresy, Leonardo Da Vinci made his position semiotically clear in his famous *Last Supper*: the person next to Jesus seems to be a woman; her and his arms suggest an "M"; and there is no wine on the table (that is, no death by crucifixion).

2.2. *Some Place for Freemasonry*

The historical situation is complicated further by Freemasonry and especially its precursors. While a faith rather than a religion in the sense of ¶1.1, it draws substantially on the same historical background, and its mathematical components are most relevant to our concerns. It takes as its historical origin the death of Hiram Abif, a pharaoh linked to biblical figures such as Joseph, and apparently the builder of the Holy of Holies in the Temple of Solomon in the –sixteenth century (that is, the sixteenth century B.C.). One of the main Masonic symbols is the twin pillars of Boaz and Jachin in front of the temple,[10] with a semicircular keystone over them forming a Trinity. The configuration may also have been a symbol of the Sun waxing up one pillar, rising to its zenith, and waning down the other one; views akin to Sun worship are prominent in Freemasonry. The Masonic extolling of pillars has left residue in the social practice of speaking of a person, usually a man, as "a pillar of society"; cognates include "upright" and "orthodox."

This faith seems to have had ancient connections with Gnosticism, a tradition of religious pre- and post-Christian thought that then became entangled with it in some way. Although Gnostics seem to have revered Saint Paul, the compliment was not returned; on the contrary, early Orthodox Christians were their greatest opponents and destroyers of their writings, especially Irenaeus (+second century), who therefore is a major source for stating the positions that he attacked. One reason may be that Gnostic sources convey a much more personal conception of faith, where the believer finds God through his own contemplation and meditation, and so bypasses the "professional" teachers or leaders extolled in the more socially oriented Orthodoxy. Another Gnostic heresy, shared by some other Apocryphal traditions, seems to have been the claim that Simon was crucified, not his supposed brother Jesus. When

10. 2 Chronicles 13:7; and I Kings 7:21, where Hiram the King of Tyre in v. 13 is not to be confused with Hiram Abif (see also 5:1 and 5:18).

Orthodoxy became established, the destruction of all alternative texts occurred; although traces survived, the discoveries of 1945, especially at Nag Hammadi, have greatly enlarged the textual basis.[11] Other relevant sects include the Mandaeans, who may be the same sect as the Nasoreans and later were also (mis-?)named "St. John's Christians" for their reverence for that saint (Drower 1937).

Apocryphal Christians regarded themselves as descendants of Jesus and Mary, a status captured in English in the phrase "Holy Grail." It seems to come from Saint Gréal (or Graal) in late medieval French, equally meaningless; for it is a light code for *sang réal*, or "true blood," the supposed descendants from Jesus and Mary. During the High Middle Ages this movement centered on the Templarians, the self-appointed keepers of the temple. In the twelfth and thirteenth centuries they were among the chief proponents of the main waves of the Crusades, back to their homeland to "search for the Holy Grail" by recovering the relics from the Holy of Holies. (There are several examples of such elisions in all main languages: English ones include "a norange" to "an orange," and the Masonic case "Saint Clair" to "Sinclair.") The talk of searching for goblets or chalices is either a misunderstanding of the origin of the phrase, or an alternative Orthodox motivation or excuse for those invasions; it comes from the legends of King Arthur and has neither biblical nor doctrinal authority.

As Saint Gréal exemplifies, the main country for the development of medieval European Christianities is France, perhaps with Belgium and Flanders. Many cathedrals were built there—raising the difficult questions of their financing and design. Stonemasons and their sponsors may have had Templarian backgrounds; some geometrical links will be noted in ¶5.3.

The Catholics tried to destroy the Templars by having many members tortured and killed on October 13, 1307 (seemingly the origin of the phrase "Friday the 13th," with maybe the day chosen deliberately to corroborate the reputation for bad luck of 13 at the Last Supper).[12] However, some of the rest of the movement escaped to Scotland and to Portugal. The place of Portugal, then a new country, is very striking. For example, many churches and cathedrals there are dedicated to Santa Caterina,

11. Current understanding of the tangled relationships between Gnosticism and Christianity in view of the Nag Hammadi texts is provided by Perkins (1993); however, numerology and gematria are barely noted.

12. See Knight and Lomas (1997b) for a plausible circumstantial argument that the Turin Shroud carries the image of Jacques de Molay, the Templarian grand master, who was crucified that day and then wrapped in a shroud (a standard item in the Templarian wardrobe, symbolizing Egyptian wraiths). I now prefer this interpretation of that shroud to the interesting suggestion of protophotography effected by Leonardo Da Vinci, made in Picknett and Prince (1994). This book also contains information on other shrouds.

who was associated even in name with the heretical (French) sect of the Cathars. Other Apocryphal indications include frequent use of the equal-armed cross, and depictions of Jesus and his disciples showing 13 heads but with 12 beards.[13]

Although evidence for the activity of Masonic groups goes back to the fourteenth century, the flowering of Freemasonry began in the early seventeenth century, in Scotland and England, with the latter coming to dominate in Britain and elsewhere.[14] The name "Freemason" seems to have come from "Free-stone mason," a craftsman who worked freestone, types of sand- or limestone that could be delicately carved, and thus of greater refinement than a mere "rough-stone mason." The high point of Masonic influence came in the late eighteenth century, spreading across the aristocracy and upper classes of most European countries.[15] It was especially prominent in France, not only before the French Revolution but also after it; Bonaparte was elected a Mason in 1798 while in Malta (Collaveri 1982), and during his reign the top Masonic post of "Grand-Maître" was held by his brother Jerome, assisted as "adjoints" by his brother-in-law Joachim Murat and the Second Consul Jean de Cambacérès.[16] Masons also played essential roles in the War of Independence and the founding of the United States of America, as is still very evident from the money upward (or downward).[17]

13. A modern Portuguese example of this tradition can be seen on the outer facade of the Igreja de Nosso Senhora de Fatima, built in the 1930s on the Avenida de Berna in Lisbon. I have not found any explanation of this practice in histories of early Portugal. It is especially clear in their former colony of Goa in India, both inside and outside the Basilica of Santa Caterina (personal witness).

14. This clause elides some tricky historical questions on connections between the Templars and the Freemasons, and links with Rosicrucianism and with alchemy. There were also controversies over chronology and purpose between Scottish and English Masonic movements; for example, stone cutting was taught only in Scotland (thanks to Snezena Lawrence for information from archival sources). These differences do not seem to bear on the mathematical issues discussed in this paper; however, different properties are apparently involved in the rituals and signs for the various degrees of Masonic membership.

15. A fine summary of these developments is given in Roberts (1972), and an important social history in Jacob (1991). On links to Christian heresies, especially disputes over the status of dualism, see Stoyanov (1994). For a substantial survey of the relationship between Freemasonry and Orthodox Christianity, with a fine bibliography, see Ferrer Benimeli (1976–1977).

16. See, for a typical year of the period, Calendar 1813 (39–40). Some historians doubt Bonaparte's Masonic membership, but these facts support it.

17. For a good compact summary, see Knight and Lomas (1997a), esp. app. 1. Templarians seem to have already "discovered" America, before Columbus. Are they the real origin of that name, "A Merica" in Portuguese (rather than the Spanish "la Merica" of [*ibidem*, 376]), to the land to the west under Venus as the evening star, which in biblical times was known to the Gnostic Mandaeans as Merikh? (Drower 1937, 56, 396). Maybe Martin Waldseemüller's explanation of his proposed name in terms of Amerigo Vespucci in 1507 was a "cover story"; as a resident in Lorraine, the region of the special cross shown in ¶5.1, he would have known the other one. In any case, why should "America" have been widely adopted just because a mapmaker, albeit an important one, put it on a map and gave this description? And why "America" and not "Amerigo" as the name?

The Templars share some major properties with the Hospitallers, who are better known as the Knights of Malta for their residence on that island from the 1570s until its capture by Bonaparte.[18] Both movements were involved with the Crusades (when they were based at Rhodos) and were led by a "grand master"; another similarity is noted in ¶5.1.[19]

Freemasonry and its precursors rival Orthodox Christianity in several aspects of doctrine; this has aroused not only Christian persecution but also much general suspicion and some ridicule. Further, the obsession of Masons with self-secrecy has both accentuated the antagonism and made their place in history difficult to specify. Their general significance seems to be *very under*rated outside the brotherhood, and maybe also within it.[20]

3. Numbers, Arithmetic, and Numerology

3.1. General Considerations

I start with roles for arithmetic, mainly positive integers (hereafter, "numbers") and their properties of combination and as digit strings. Beyond simply mathematical properties, many cultures have attached faith-full or religious significance to certain numbers, thus building up metaphysical doctrines that are known collectively as "numerology." As well as the utility and intrinsic interest of numbers, a great attraction was that, in this mysterious world and universe of which a culture understood so little, they were (apparent) sources of *certainty,* both with their arithmetical properties and also as *invariants,* that is, categories (or even objects) that did not change while others were varying.

18. Bonaparte then proceeded to Egypt, which the French occupied from 1798 to 1801 (Charles-Roux 1937). Was the choice of country stimulated by Masonic ideals as well as by the desire to gain foreign influence to replace the lands in America that they could not expect to retain and indeed sold to President (and Mason) Jefferson in the Louisiana Purchase of 1803?

19. For a well-documented history of the Hospitallers see Sire (1994); he describes various versions of the cross on pp. 101–110 but not its origins or semiotic significance, and rather glosses over links to Freemasonry. The Valletta Public Library contains a little-known but amazing collection of late medieval books (often badly damaged by bookworm, I noticed), mostly left to the Order by Knights after their deaths.

20. There is a surprising amount of literature on Freemasonry and its history, some by Masons themselves, but its early development seems particularly obscure (to outsiders, anyway). I have benefited by the disclosures made in Knight and Lomas (1997a), for its authors are two Masons, who disclose the revealing text of some Masonic discourses and provide fine insights on several related matters, especially Christianity and the historical priority of Scotland over England. However, its utility is heavily compromised by scruffy references and unsupported claims; for example, *where* is it "on record" that Pope Leo X (1513 to his death in 1521) said that "it has served us well, this myth of Christ" (p. 79)? On this pope, see also note 73.

This position was clearly adopted by Iamblichos (flourished +fourth century), in his (attributed) "theology of arithmetic."[21] He characterized the "monad," "dyad," . . . up to the "decad" in terms of their manifestations, ascribing properties also to their factors. For example, for him the hexad 6 was the product of the first male number 2 and the first female number 3, the number of sides of the cube, and the numbers of super- and subterrestrial signs in the zodiac. While maybe unusually enthusiastic, he was a typical Greek adherent to worldviews informed by numerology.

For the hexad Iamblichos also mentioned that 6^2 was 36, an important number, with properties such as

$$36 = 1^3 + 2^3 + 3^3, \text{ and } 3^3 + 4^3 + 5^3 = 6^3 \text{ compared with } 3^2 + 4^2 = 5^2. \quad (14.1)$$

He also noted that

$$36 = 1 + 6 + 8 + 9 + 12, \quad (14.2)$$

for it had long been known from harmony that the arithmetical, geometrical, and harmonic means were definable from the last quartet of numbers.

Surprisingly, neither for the hexad nor the octad did Iamblichos mention that 36 was also the triangle number to 8; that is,

$$36 = 1 + 2 + 3 + 4 + 5 + 6 + 7 + 8. \quad (14.3)$$

These numbers were so called because they could be laid out as triangles of dots, maybe by placing pebbles in the sand. The successive sums of triangle numbers were called "solid numbers." Other collections of number were characterized by layout in regular pentagons and hexagons, certain kinds of rectangle, and so on; collectively they became known as "figurate." The layouts were nested, which may have helped inspire proof by mathematical induction. Rectangular numbers exhibited the important L-shaped sign (see figure 14.1 for an example), of Babylonian origin in literally "completing the square" when that layout is used. Called "gnomon" in Greek, this shape was used in sundials and elsewhere;[22] in the orientation "Γ" it was the Egyptian hieroglyph for the angle or corner of a building.[23]

Greek numerology has received more attention than is now realized; in particular, the fine efforts early in this century of Hermann Usener to stress the importance

21. For a modern annotated English translation, see Iamblichos (1988).

22. Høyrup (1992). On numerology in Celtic religions, especially the influence of 3 and 7, see Graves (1962, *passim*) and MacQueen (1985, chs. 1 and 3 *passim*).

23. Budge (1920, 574b). For a historical introduction to the gradual (mis)understanding of Egyptian script, see, for example, Iversen (1961, ch. 1). The prehistory of cuneiform writing in arithmetical recording is convincingly argued in Schmandt-Besserat (1992).

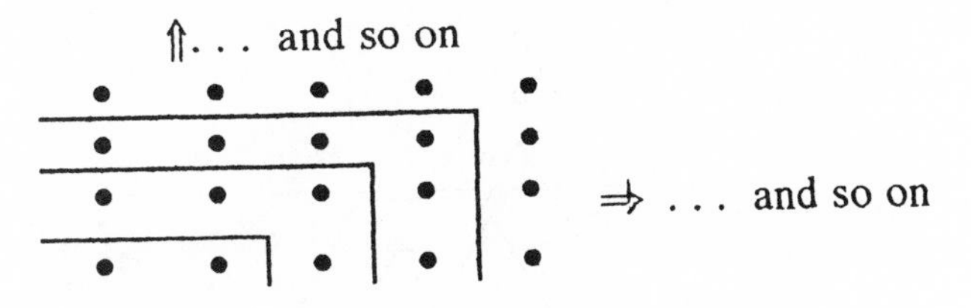

Figure 14.1. An example of the generation of rectangular numbers

to them of 3 are poorly known (Usener 1903), and even less so are the remarkable catalogues of Wilhelm Heinrich Roscher on the remarkable ubiquity of 7, and also of 9.[24] A partial exception to this historical ignorance concerns Plato, to whom we now turn.

3.2. Plato and the Case of 37

Plato discussed "special" numbers in the *Timaeus* and elsewhere, including the nuptial and tyrant numbers, which were related to (14.1) and to various astronomical properties (Young 1924; Erickson and Fossa 1996). As these cases show, not all numerologically significant numbers are small, a feature exemplified also by the successor of 36. One source of importance of 37 is the basic Pythagorean triangle, with sides in the proportion 3:4:5. In the Babylonian division of the heavenly orbits, which later became interpreted in terms of circles and spheres, the angle between the sides 3 and 5 is only 0·13° short of 37°; *if* angles were measured arithmetically in terms of the 360° division, then this value may have been thought in deep antiquity to be close enough to be "meaningful."

In any event, 37 deeply concerned Plato, especially in his account of the 37 "guardians" of the circular city of Magnesia. The main book is the *Laws*, or "nomoi," which also means "traditional melodies." Drawing upon this association, McClain (1978, chs. 8–9) has offered a fine interpretation of the account as an allegory for the ranking of intervals within the musical octave. The guardians are composed of 18 preferred "divine births" interpolated by 19 secondary "new arrivals." Each interval is determined by a ratio of the form $2^p3^q5^r$:7 for low values of the indices, where the denominator 5,240 is repeatedly said by Plato to be the maximal number of "landowners" in the city. The unavoidable complication created by the incom-

24. See especially Roscher (1904). The massive range of contexts of his writings includes Gods, muses, winds, and ages, and many topics in the writings of Hipparchos.

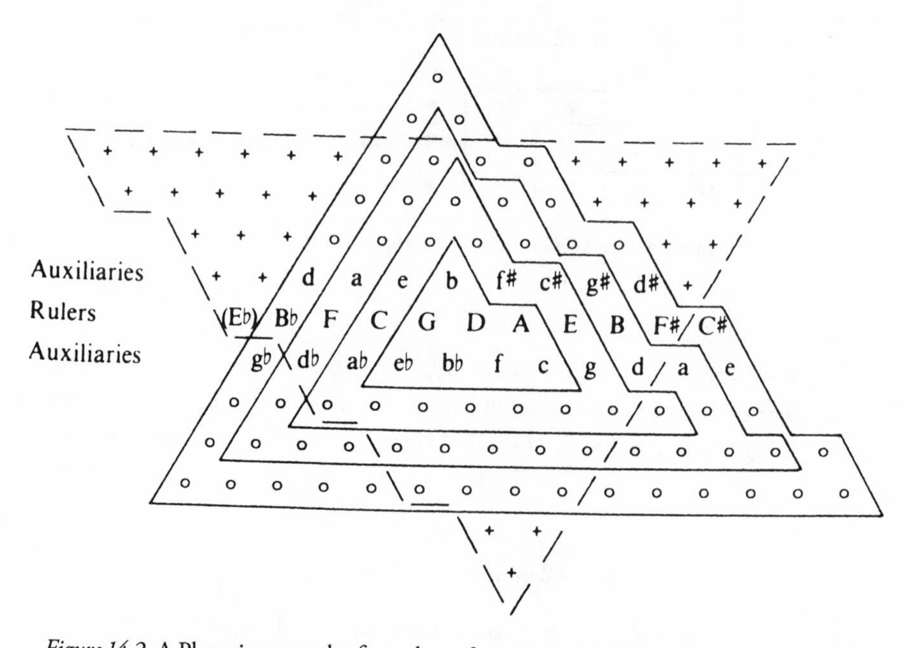

Figure 14.2. A Platonic network of numbers, from E. G. McClain, *The Pythagorean Plato* (York Beach, Maine: Nicolas-Hays), 29

mensurability between 2:1 octaves and 3:2 perfect fifths takes the form here of the "diaschema," the ratio 1:73, which is connected to the number of extra days needs to make an accurate calendar (365 = 73 × 5 = 360 + 5). A detailed analysis is offered by McClain, including the angular location of the guardians within a circle representing the octave. Figure 14.2 shows a case for $3^p \times 5^q$, where each lattice is generated from its base at the bottom lefthand corner with powers of 3 up the oblique columns and of 5 along the horizontal rows, and ratios marked by the musical letters oriented around D as tonic; the hierarchy of "rulers" and "auxiliaries" are determined by the prominence of the corresponding ratios (p. 29). He also shows that if the Pythagorean triples are generated by

$$2pq,\ (p^2 - q^2) \text{ and } (p^2 + q^2), \tag{14.4}$$

then the ratio 2:1 that produces the basic 3:4:5 triangle is 37th in order of magnitude of the ratios produced by his system (pp. 118–119). (For a related but little-known property of this triangle, see ¶5.3.) While Plato does not seem to express a faith or religion as such, clearly a major philosopher was presenting a metaphysical worldview centered on numbers, with music and astronomy in close attendance. I note the importance of 37 in Greek Christian gematria in ¶6.1.

In a companion work, McClain (1976) has used similar connections, not only in Greek sources but also Hindu ones, to argue that some Platonic and other statements concern the number of gods in a faith.[25] The numbers of gods and goddesses *chosen* for assertion in polytheisms would form a difficult but worthy historical topic of study.

4. Numerological Examples in the Christianities

The rest of this paper reviews examples of mathematics (as specified above) in the various Christian traditions. The historical origin of some of those not in the Bible may lie in medieval Europe, but their relevance is not thereby lost. I write the numbers in the customary notation, but they are expressible also in Greek numeration, so the significance might be ancient. One later imposition was the calendar, an interesting system of the +sixth century lacking year 0 (and so causing the premature "millennium fever" recently): the dating of Jesus's birth is now regarded as a few years later. As Peter Griffiths has pointed out to me, the Old Testament has only one explicit naming of a year: "the 15th year of the reign of Tiberius Caesar" (Luke 3:1).

Many of the numbers are work-a-day, such as distances and weights; they are usually expressed in the system to base 10, while many of the more religious ones are related to the system to base 12 (for example, Jesus's disciples, and the tribes of Israel).[26] But the distinction is not sharp, as we shall now see.

4.1. The Important Case of 40

A further property of 37 brings us to Orthodox Christianity:

$$37 + 3 = 40 \text{ and } 37 \times 3 = 111. \tag{14.5}$$

While (14.5) can also be expressed in Greek numerals, the time of its appearance in Orthodox Christianity is not known. But its importance includes 111 as the Trinity number via the digit string 1-1-1 in a Christian cultural setting using arithmetic to

25. On the connections, often overlooked, between musical intervals and ancient symbols of animals in many faiths, see Schneider (1946).

26. For a list of instances, see Bauke (1999). On the many traces of 12 in the Middle Ages, see Ulff-Møller (1993).

An interesting Spanish figure is Ramon Llull (1232?–1316), who used 2, 4, and 8 quite prominently in his great "Art"; some of the results embody combinatorics. Several main writings are translated in Llull (1985); his work is well surveyed and cataloged in Platzeck (1961–1964). The important Arabic influence on him is well captured in Hernández (1977).

base 10, from the thirteenth century onward in Europe. Another such number is 123, as 1-2-3: I am not aware that 41 gained status through the equation analogous to $(14.5)_2$.

This case also involves 40, which is well known in Christian numerology.[27] The Old Testament mentions 40 years for the children of Israel eating manna (Exodus 16:35) and in the wilderness (Numbers 14:33); also of peace in Israel (Judges 3:11, 5:31) followed by disaster (13:1), and similarly for the Egyptians (Ezekiel 29:12). Forty days are involved in Moses twice on Mount Sinai;[28] interesting variants are the length of the Flood after 7 days and nights of rain (Genesis 7:4, 17, and elsewhere), and in 33 days for purifying a mother after 7 days of uncleanness following the birth of a son (Leviticus 12:1–6; 14 + 36 = 50 days are required for a daughter). Other connections include a tradition of taking the length of human pregnancy to be 40 weeks or 10 lunar months—rounded-up figures, maybe deliberately.

The *reasons* for the popularity of 40 are not clear. Do they lie in (14.5), or in 8 × 5, or in the 2:3 ratio of 60, or by reinforcement in all these contexts and others?

4.2. 3s versus 2s

The triangle as a sign represents the Trinitarianism of Orthodox Christianity, which is based on another triangle number, 3 = 1 + 2. Indeed, it ends a run of 3s in Orthodox Christianity, according to which, after birth before 3 shepherds and 3 wise men, Jesus lived for 3 × 10 years in obscurity, apart from an address to the Pharisees in his 3 × 4th year (only in Luke 2:41–52); then he ministered for 3 years a doctrine including the Trinity of Father, Son, and Holy Ghost prior to dying in a 3some on a cross but coming back to life 3 days later and being seen 3 times (the last one is noted in ¶4.3).

The sect to which Jesus belonged lived at Qumran, a name that means "vault"; this may be another sense of the Masonic Trinity of the keystone over twin pillars (¶2.2). By contrast, the cognitive content of biblical Trinity is very hard to understand; its main purpose presumably lies in supplying the Holy Ghost as the originator of the supposedly virgin birth. Among its consequences, Jesus is not a Christian,

27. Roscher (1909); some Mandaean uses, also involving 42, are noted on pp. 98–99. A prolific but less impressive author of that time on biblical numerology was the Belgian C. Lagrange: the main works are 1900, 1921.

28. Exodus 24:18, 34:28 (a book with 40 chapters); Deuteronomy 9:9–11 (compare vv. 18, 25). There is a related tradition of 39 as one less, especially in punishments; see, for example, Deuteronomy 25:3 and 2 Corinthians 11:24.

since such believers worship Jesus, which he did not do![29] Its parentage in the earlier Trinity is clearly hinted by Paul when twice he described Jesus as a "corner stone" (Ephesians 2:20, 1 Peter 2:6).

In Orthodox Christianity 3 is the main number. The New Testament contained 27 books, doubtless deliberately though maybe not cognitively: 3^3, 3 to the cube, proclaiming the faith of the Trinity as the measure of God across the three dimensions of space, "as high as heaven" and "deeper than hell," "longer than the earth, and broader than the sea" (Job 11:8–9). There may have been a criterion behind the number of books in the Old Testament; the Jewish historian Josephus (+first century) stated that there were 22 of them, like the letters of the alphabet: 5 of Moses (Genesis to Deuteronomy), then 9 of later prophets, and 4 of hymns and precepts.[30] Later some books were divided into smaller ones up to the eighth century to produce the 39 that we now have in the Protestant version (or 41 in the Catholic version, which also contains Maccabeus 1 and 2). The definitive biblical structure was set in the Greek edition of the mid-fifteenth century edited by Robert Stephens.

By contrast, the Apocrypha advocates dualities; the literally binomial digit string 1-1 denoting Jesus and Mary Magdalene, and the corresponding number 11 and its powers. A major source for Christian conflict was whether the Blessing should be given with three fingers or with two. During the High and late Middle Ages it was involved in the schisms between Rome and Kiev. When Christianity was chosen for Russia by the occupying pagan Viking forces in the eleventh century, Trinitarianism was (and is) advocated with especial fervor, as one sense of the word "orthodox." Russian sacred paintings often depict just the heads of three saints.

This "contest" between 2 and 3 leaves 1 in a peculiar position: the unit from which numbers are defined as multiplicities (for example, Euclid's *Elements* 7:1–2), its place in numerology is assured. Yet in Christianity it smacks of Unitarianism, ecumenism, and such heresies, despite clear support for it offered by Paul in Corinthians 8:6.

29. On the status of worshipping Jesus in the context of apparent monotheism in Judaism, see the interesting desimplifications, drawing on a very wide range of sources, proffered in Barker (1992, esp. ch. 11).

30. For discussion of possible identification of books see MacQueen (1985, 9–10).

4.3. 7, and Only Some Multiples of 10

Of especial importance is 7;[31] probably under the Greek influence noted in ¶3.1, it is favored also in other faiths, including Gnosticism. It appears as 6 + 1 days at the start of the Bible, and crops up in various contexts thereafter. I noted in ¶4.1 its role in 7 + 33. A quite early manifestation is the North African Saint Augustine of Hippo, a founding father of the Church in the +fourth century. His advocacy of numerical relationships became influential; in particular, he advocated 12 (= 3 × 4) as the second 7 (= 3 + 4). He was probably influenced by Revelation, where the two numbers combine (for example, 12 and multiples in chs. 7, 14, and 21); his own *The City of God* has 22 books. This connection may also lie behind the practice at the inauguration of a cathedral of giving three knocks at each of its four sides; note the similar placing of trios of oxen in the four compass directions (I Kings 7:25).

Among arithmetical links between 7 and 3, this one is quite striking:

$$7^3 = 343 \text{ “=” } 3(3+1)3 = (111 \times 3) + (3 \times 3) + 1. \tag{14.6}$$

Note also that in binary arithmetic 7 relates to the digit string 1-1-1 (compare [14.5]).

Another category involving 7 is the sacraments, which were instituted in the twelfth century, one for each type of mystery. While the notion lacks biblical authority, the instances were created from passages recording Jesus's actions.

A beneficiary of this attention to 7 is 70; its appearances include the Jews' years of captivity (Jeremiah 25:11–12) and weeks to terminate sin (Daniel 9:24), and the number of Jesus's new disciples (Luke 10:1). 70 links 7 with 10, another major number whose rise in general importance is quite shrouded. The usual explanation for its role as the basis of founding arithmetic in many (though not all) cultures is of counting on fingers and thumbs; but direct evidence is scarce, especially that of philological connections or similarities in ancient languages. Even for counting itself, a much more efficient method is to work to base 12 by counting the segments of the fingers on *one* hand by the adjoined thumb. Note also, as ancients surely did, that

$$3 + 4 + 5 = 12 \text{ while } 3 \times 4 \times 5 = 60, \tag{14.7}$$

31. On the frequency of multiples of 7 in the number of words in a few passages in the Greek Bible, see Panin (1934); he optimistically took the property as proof of the existence of the “Master Mathematician of Creation.” On the importance of 7 in the 13 Jewish Sabbath songs found in the Dead Sea Scrolls, see Newman (1984). On its place in Christianities, see Hopper (1938, esp. pp. 78–88).

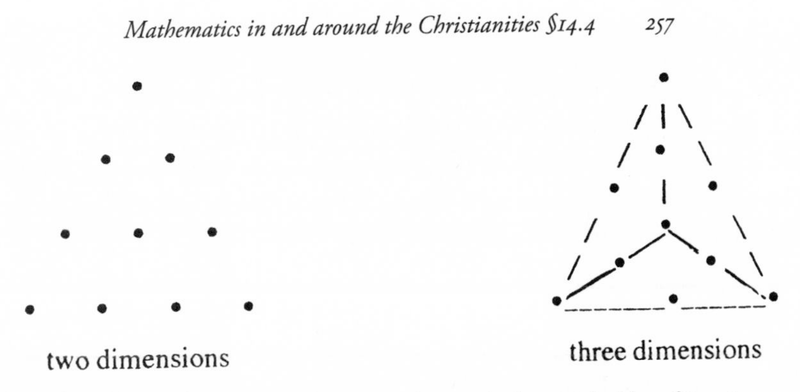

Figure 14.3. The Pythagorean tetraktys, and the tetrahedron with sides of 3 units

with 60 relating to the division of the circle and thereby via astronomy to that of time into minutes and seconds. The number 60 appears in protocuneiform writings; the division is usually attributed to the Babylonians (Høyrup 2002).

One property that would have raised the status of 10 is as the triangle number of 4, the Pythagoreans' tetraktys, symbolic in both two and three dimensions (figure 14.3). But overall the preference for 10 over 12 as the basis is a most curious feature of the history of mankind. The Bible is equivocal; the number of disciples of Jesus shows 12 again, but 10 features mostly through multiples such as 40 and 70, as well as 150 as the number of the Psalms, and population statistics rounded to 100s in Numbers (a book so called for this reason).[32] However, 60 is strikingly scarce in Christianity, despite the importance coming to it from astronomy.

Some of the numbers mentioned above are triangular. An enlightening example of a larger one occurs in a pesher in the New Testament when Jesus appears before the disciples after his Resurrection for the third time and Peter catches 153 fish (John 21:11). The number 153 was chosen in this figurative passage because it is the triangle number of 17, which was the number of Jewish sects, orders, and societies at that time; and the triangle semiotically expresses the spread of the faith as a widening shape. Another property for this number is that $\sqrt{153} = 12{\cdot}36$ is an excellent approximation to the number of synodical months (that is, the time between successive new moons) in the year.

The fish was an early symbol for members of the faith. There may well be various explanations, and it is undecidable historically to determine the original motivation, *if* there was one. There is the Greek acrostic "ΙΧΘΥΣ" (fish), reading in full

32. I have not spotted any numerological features of the large numbers used, or connections to those given in McClain (1976).

"Ιησου Χριστος Θεος Υιος Σωτρ" (Jesus, Christ, Son of God, Savior). Another possible source was the practice of washing converts in salt water,[33] and yet another that at that epoch the vernal equinox, long revered as an important astronomical "timekeeper," lay in the astrological house of Pisces (Bauval and Hancock 1996, 377). I note a geometrical image of fish in ¶5.2.

4.4. Medieval Numerologies

Especially from the High Middle Ages, various numerological systems were developed within the Christianities, with some numbers serving different roles. The modern student is likely to be confused, though luckily a superb anatomy is available.[34]

Among roles for 7 in medieval Christianity was the number of offices within a cathedral or major church, from the bishop or priest down to the doorkeeper. However, various changes were made in organization in the late Middle Ages, to the great confusion of commentators, who therefore were placed "at sixes and sevens." This is the origin of the phrase (Reynolds 1979), showing consternation at the loss of an invariant!

During that period Freemasonry and its precursors began to become prominent, especially with the Crusades. Trinitarianism is central there also, although the triangle "△" came to be used only from that time.[35] It was used more explicitly by Freemasons, who append it to their signature. It takes several semiotic interpretations: together with a horizontal base it is the Egyptian hieroglyph for "pyramid" (Budge 1920, 113b); it represents the Trinity of two pillars and keystone, the locations of the three blows to the body of Hiram Abif that caused his death (Knight and Lomas 1997a, esp. ch. 11), and perhaps also the unification of Upper and Lower Egypt sometime late in the -fourth millennium.[36] In Hebrew the components of this Trinity symbolize respectively "judgment" and "right(eous)ness" joined by "peace." Masons also date years from -4000, their supposed origins of Egyptian

33. Thiering (1992, 447); on another source for "fish" in the Hebrew language, see pp. 515, 517. On Jewish fish symbols of the period, see Goodenough (1953–1968, vol. 5 [1956]).

34. Hopper (1938); among its treasures, and relating to ¶4.3 on 10s, is the curious liking for 50 in Irish numerology (pp. 209–211). Valuable for examples from literature are MacQueen (1985) where, however, the discussion of Plato in ch. 20 was already eclipsed by McClain (1976, 1978), and Surles (1993), where amazingly MacQueen is not cited. See also Meyer and Suntrup (1987).

35. Apparently the Catholic tradition saw the triangle as a parallel transmission from 1st to both 2nd and 3rd sources, while Greek/Russian Orthodoxy read it serially as 1st → 2nd → 3rd (information from Michael Baigent).

36. Whether this unification was political and geographical on land or a symbolic one of land and sky is an interesting question raised in Bauval and Hancock (1996, 234–235).

culture. In addition they developed an elaborate numerological system, drawing on both Christian and Egyptian sources: 3, 6, 18, 33, 42, and 55 (as the triangle number to 10) are particularly important.[37]

One practice, of ancient though uncertain origin, is that of *doubling*: "Think of a number and double it" refers to an operation advocated since the Early Middle Ages[38] and still known as a phrase, at least in Britain. Some important numbers are doubles of others; in particular, 18 and 36. The latter has been noted; the former was significant to the Egyptians for the property that

$$18 = 6 \times 3 = 6 + 3 + 6 + 3; \qquad (14.8)$$

the area of this double square equaled its perimeter. Knowledge of (14.8) is attributed to them by their Greek chronicler Plutarch (+first–+second century) in his *De Iside et Osiride.*[39] His claim is believable, because it "unites" addition and multiplication; physically evidence is provided by the $((3 \times 3) + (3 \times 3))$ pillars supporting the temple of Hathor at Dendorah. Plutarch also credits them with knowing the other figure with this property, the 4×4 square; and this is the arrangement of columns in the main hall of the site KV5 in the Valley of the Kings, currently being excavated.

Doubling also links to the (general) double cube, which seems to have been the preferred shape of sacrificial altars,[40] and later in Freemasonry of a Masonic lodge. Further, doubling 36 gives 72, which was important maybe from deep antiquity for the property that precession takes 72 years to traverse 1° in the division of the circle.[41] Perhaps for that astronomical reason, it is prominent in various religions, including the Septuagint as the number (not 70) of the supposed Greek translators of the Old Testament.[42] In further doubling, 72 may help underpin the importance assigned to

37. For a summary of this system and further references, see Grattan-Guinness (1992), and chapter 16. The number 33 has resonances in both Christianities; its introduction into Freemasonry is attributed to King Frederick the Great in the 1760s (Jacob 1991, 156).

38. An early surviving source for emphasizing doubling and other simple ratios is "De arithmeticis proportionibus," dated to around the +eighth century and attributed to Bede or at least under his influence: for the text and concordance of versions, see Folkerts (1972).

39. See ch. 42 of *De Iside et Osiride* in, for example, Plutarch (1970, 185).

40. On the role of ritual in this and other ancient arithmetical properties, see Seidenberg (1962a, 1962b). There may be some kinship to the different classical problem in Greek geometry of doubling the cube.

41. On the profound possible consequences of the ancient awareness of 72, with various ancient buildings oriented around stellar arrangements dating to −10,500, see Hancock (1998). His study amplifies the claims of ancient knowledge of precession in De Santillana and von Dechend (1970), a work that has also helped to inspire the recent interpretation of Homer's *Iliad* in terms of stellar constellations made in Wood and Wood (1999).

42. On various manifestations of 72 in Celtic numerology, see Graves (1962, 235, 275, 286). Cases from Christianity and other religions of both 70 and 72 are noted in Endres (1951, 226–228). Neither author was aware of the link to precession.

144 in Egyptian numerology, and (thereby?) the biblical manifestation in the 12 × 12 × 1,000 "children of Israel" in their 12 tribes sealed on their foreheads as servants of God (Revelation 7:4–8, in a book strewn in 7s).

Finally, a variant on squaring the cube, used in Masonic numerology, is given by $19^2 = 361 = 360$ degrees in the ecliptic plus a point. From this, 19 is taken to be a number for the Sun as the circling point, being illustrated with 19 or 38 rays. {Another role for 19 comes from the Metonic cycle (as we call it) of the Sun and the Moon; the Mayans seem to have known it (Sullivan 1996, 173).[43]}

5. Sacred Geometry and Semiotics

5.1. Interpreting the Crosses

Crucifixion was a severe and therefore popular form of punishment and murder during the Roman Empire, and many shapes of cross are described. No biblical authority is available for the one used for Jesus's case;[44] the preferable shape is considered in ¶8 as a problem in mechanics.

The historical evidence for the occurrence of the crucifixion of Jesus (on the sixth day) is not clear. For example, before the establishment of the Talmud in the early centuries of the Common Era, the Hebrew word "taloh" did not distinguish "to hang" from "to crucify," construing the latter in terms of hanging alive.[45] This conflation may reflect the practice of hanging a body on a tree or cross *after* execution by other means, for consumption by birds of prey (Genesis 40:19). (This latter feature renders implausible the Apocryphal claim that Jesus was taken down from the cross alive.) The biblical accounts of the surrounding events are also suspect, on both textual[46] and historical grounds; and another semantic blur between "cross" and "stake" looms out of the Greek word "σταυρος."[47]

43. {James North (London) is studying the importance of 19 for Francis Bacon. Manifestations include an emphasis on the 19th letter of the Greek alphabet, tau (T), for example as in "the Great Ins'tau'ration"; and the significance of Isaiah 19:19 (an altar and a pillar in Egypt) and John 19:19 (Pilate puts Jesus's title on the cross [¶8.1]).}

44. For a nice short survey, with valuable references, see Hengel (1977).

45. See Cohn (1972, 209–212), especially the distinction between "tselev" in (post-)Talmudic Hebrew and in Aramaic.

46. For example, the giving of a vinegar-soaked sponge on hyssop (Greek, "hyssopo") to Jesus on the cross (John 19:29) is surprising, since the sponge would fall off such a bush; much more sensible is that the sponge was held up on a javelin (hysso), or "reed" as in Matthew 27:48 and Mark 15:36. For other examples, see ¶8.1.

47. An account of crucifixion under the Romans is given in Cohn (1972, esp. ch. 8), although staking is not mentioned. Broadly accepting the Orthodox version, he strongly queries many details such as the timing of the events around the feast of Passover, defends the role of the Jews, points to inconsistencies in attitudes of Jesus and other figures, and notes the lack of non-Christian

Concerning the shape of the cross, works of art came to show the † version; but they cannot provide authoritative evidence, since even the earliest surviving cases were produced long after the supposed event.[48] But it is striking that no crucifixion seems to be shown until about the sixth century. A revealing source for this feature is the mosaics and paintings in the catacombs in Rome, and in the early churches there and in Ravenna.[49] Instead, early art displays a semiotic cross based on superposing the first letters "chi" and "rho" of "christos" as the monogram ☧; in semiotic style this sign also doubled as "pax" in Latin (and since the sixteenth century has left a residue in the word "Xmas"). It was often accompanied to left and right by the letters α and ω, the metaphor attributed 3 times to Jesus as "alpha and omega, the beginning and the ending" (Revelation 1:8, 21:6, and 22:13), though maybe of pre-Christian Hebrew origin. One wonderful example of the sign from the late +fifth century shows it set within 3 concentric circles and surrounded by a crown of 12 doves (Wilpert 1917, vol. 3, pl. 88). Other mosaics show both kinds of cross, and also the eight-point star ✱.[50]

A symbolic feature of most crosses is that they are figures with 5 points.[51] Their semiotic status is enhanced by being fertility symbols, for in a long pre-Christian tradition, vertical lines were male and horizontal ones female, and you can work out the rest for yourself. Now the Orthodox cross † shows the dominance of male over female;[52] but the Apocryphal tradition shows the equal-armed cross +, representing sexual equality—a semiotic distinction of great importance, which however seems

accounts of the event. For a survey of Roman jurisdiction over Jews, basically believing but astonished by Pilate's reported treatment of Jesus and not describing methods of punishment, see Morrison (1890, esp. pp. 148–152).

48. On this curiously neglected topic, a nice introduction is provided in van der Meer (1957). Much literature exists on *later* illustrations; a good survey is given in Thoby (1959), with two supplements to 1963. As a convention I represent all crosses without seriphs, as in this paragraph of text; but seriphal versions were also admitted.

49. See especially Wilpert (1895, 1917), especially the eye-opening volume 3 of the latter work: there is only Jesus before Pilate (pl. 99, sixth century) and the crucifixion (pls. 179–180, seventh and eighth centuries). Among more recent discoveries in Rome, note Ferrua (1991). A fine illustrated survey of early British crosses is provided in Thomas (1981, esp. chs. 3–6 *passim*): ☧ is very prominent.

50. For a different early source, see Mahr (1932); the equal-armed cross seems rather more common.

51. See Browne (1658), esp. ch. 1 (out of 5); dedicated to Nicholas Bacon, the text exhibits clear Masonic allusions.

52. Knight and Lomas (1997a, 316) claim that, as an ancient Egyptian hieroglyph, † means "savior"; but this is not corroborated by Budge (1920, 653b), where that word is given a sequence of signs, and words involving "God" make much use of sequences of the sign "+" (pp. 402–405). The hieroglyph ☥ denotes "life" (p. 112a), as is mentioned in Knight and Lomas (1997a). Budge is still the classic catalog of hieroglyphs, though not always of their meanings. Most analysts of Egyptian grammar and semantics understandably render the hieroglyphs in the modern codings given by letters of the modern alphabet.

to be largely unnoticed today, including by Christians. The latter shape continued in Apocrypha, coming to be connected with the Rosicrucian movement as the Rosy Cross, the circular rose enclosing that cross: ⊕, but sometimes ⊗ using the ⊗-shaped cross.[53] Both orientations were also adopted by the Templars and the Hospitallers (¶2.2), the latter with two corners on each arm giving the eight-point "Maltese cross." Nevertheless, all those movements, especially Freemasonry, were male sexist, although there are traditions of female and co-Masonry; contradictions are not confined to Orthodoxy.

Also in Apocrypha, the Cross of Lorraine, ‡ or sometimes ‡, symbolized in its second horizontal bar the claim that Jesus did not die, with – denoting "not." The former shape also has an Orthodox reading, where the upper bar suggests the notice stating Jesus's kingship of the Jews, which was said (compare ¶8.1) to have been placed on the †-shaped cross. A variant design ‡, the "patriarchal cross," maybe intended as a compromise between the traditions, divides the vertical line into equal thirds by two horizontal bars in such a way that the lower bar forms an equal-armed cross with its vertical component. Russian Christianity uses the similar shape ☦; I have not found a convincing semiotic explanation of it, and two well-known historians unexplain it as a "peculiarity."[54] Some Russians adopted the tradition of regarding it as picturing a footrest (*suppedaneum*); but unless used as help in fixing the victim on the cross as it was hoisted into position or to support the nailed arms (see ¶8.2), it is not likely that he would have been provided with such creature comforts. Presumably Orthodoxy was not being infringed semiotically.

A related shape is the swastika, also known as the "gammadion" as the semiotic arrangement of four Greek capital gammas: Γ. Sometimes appearing with Christian crosses in networks of both shapes,[55] it is *very* multicultural, serving also as a geographical sign for the Navajo Indians, as an example of 9 (its principal number) for the Hindu religion, and in various other cases.[56] The optimism of the righthanded version, increasing the energy of the world, contrasts with the lefthanded "sinister" pessimism—which the Nazis were to adopt as their sign, doubtless consciously, following its use in Germany in an anti-Semitic tradition for some decades.

53. This feature is missed in Yates (1972), although it is exhibited in her figures 15b and 27b! She prefers the origin for "rose" in "ros" for dew (p. 156), which may well be another source.

54. Ouspensky and Lossky (1982, 185); see also the illustrations. The + appears in Russian artefacts; its interpretation is also unclear.

55. A superb Irish example of perhaps the +sixth century, the Shrine of the Bell of Saint Patrick's Will shows quartets of sinistral swastikas and equal-armed crosses enclosing each other (Mahr 1932, vol. 1, pl. 80; the description in vol. 2, 156, makes no mention of the feature).

56. For an anthropological study, see Quinn (1994).

5.2. *Other Shapes in Design*

The shape as sign can be an important part of a religious message. As well as the semiotic cases above, basic shapes such as polygons and circles have played a role in Christian symbolism and in the construction of churches and cathedrals.[57] The transition during the Middle Ages from a round plan (as used by the Saxons, among other cultures) to a rectangular one and then especially to the †-shape, may image a change of religious stance;[58] indeed, the clash of the two Christian traditions may be exhibited here, for the builders of medieval cathedrals and churches took as basic proportions the 2:1 of "ad quadratum" and the 3:1 of "ad triangul(at)um" (Lund 1921–1928) and built the Christianities into their basic design (a possibly relevant theorem is presented in ¶5.3). Further, in contrast to the plan, the front elevation was dominated by two towers; at the church of St. Peter in Caen, the slightly shorter one is even called Tour de la Dame. A major French example is the cathedral de Nôtre Dame in Paris (an intriguing name in the context of Mariolatry mentioned in ¶2.1). The third tower in the roof of the nave was added in the nineteenth century by Eugène Viollet le Duc; not coincidentally a great authority on medieval architecture, perhaps he saw a chance here to affirm Orthodoxy.

Even in †-shaped churches circles are well evident, especially in large "rose" (sic) windows in which various other curves are shown in the interior designs. A well-known type shows a circle surrounded by 6 others of equal radius, giving both a 7 and the regular hexagon and hexagram; in a variant the circles are of larger radius and only arcs of the outer ones are shown. Each case represents a 7 as 1 surrounded by 6, perhaps connoting the supposed creation of the universe in 6 days followed by a day of rest (Genesis 1, 2:2). Much less known is the property in solid geometry, that a given sphere can be surrounded by 12 equal spheres that almost touch it and each other. Was this taken as an image of Jesus with his disciples? Is it also part of the "meaning," whether or not with Christian connection, of the many dodecahedral metal artefacts from Roman times that have been found in Northern Europe?[59]

Another little-known shape is the "vesica piscis" (literally, "fish bladder"), defined

57. A good collection of illustrations is provided in Pennick (1980). For various geometrical figures, and theorems, often not well known and sometimes of religious import, see Heilbron (1998).

58. Of especial interest is the hybrid type called "round church," where the rectangle is completed by a semicircular apse. Few of them survive in Britain, but I happen to live by one of them, which was constructed in the 1120s. For earlier examples, see Thomas (1981, chs. 6–7).

59. A modern planar analogue, often seen but rarely recognized, is the twelve stars in the flag of the European Union. Nothing to do with the number of member countries at any time, it copies a

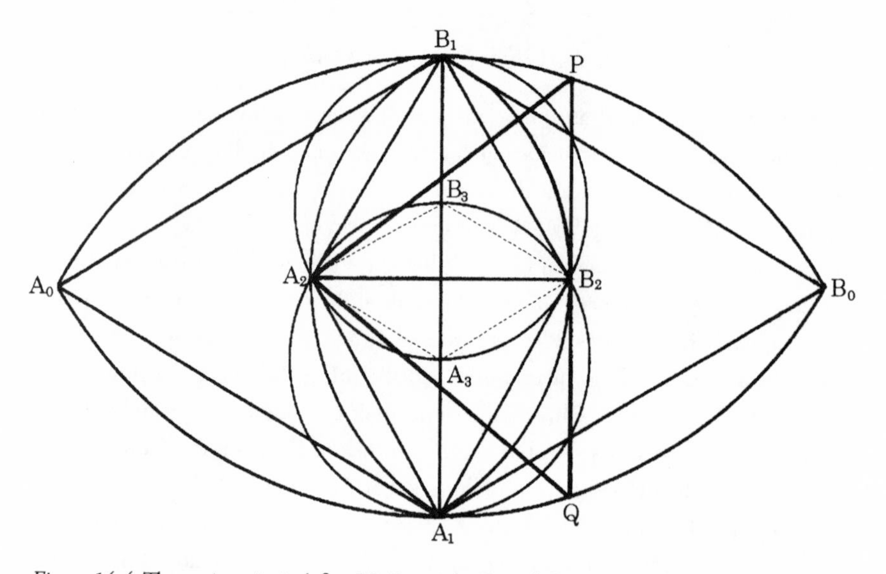

Figure 14.4. The vesica piscis (after N. Pennick, *Sacred Geometry: Symbolism and Purpose in Religious Structures* [Wellingborough: Turnstone, 1980], 11)

by arcs of two circles of equal diameter each passing through the other's center. A beautiful early mathematical example of self-similarity, three stages are shown in figure 14.4: centers A_1 and B_1 generate respectively the arcs $A_0B_1B_0$ and $A_0A_1B_0$, then A_2 and B_2 serve similarly for $B_1B_2A_1$ and $B_1A_2A_1$, and finally A_3 and B_3 produce $A_2B_3B_2$ and $A_2A_3B_2$. Doctrinally it is well equipped with 3s, for each arc is 1:3 of a circle, and each pair of arcs encloses two equilateral triangles; other such constructions include the equilateral triangle A_2PQ.[60] Further, the figure arises in Euclid's construction of the regular hexagon within one of the circles (*Elements*, 4:15), which could have been an inspiration, which Christians would relate to the crucifixion occurring on the sixth day. Finally, each pair of arcs symbolizes a fish, the early symbol for Christians discussed in ¶4.3; the origins of this shape seem to be unknown, but it often features in the design of large church windows and in the decoration of ceilings. As a bonus, an equal-sided pentagon can be constructed with A_3B_3 as base by drawing the circle with B_2 as center and B_2A_3 as radius and using its points of intersection with circles $B_1A_2B_2$ and $A_2B_2A_1$ and the midpoint of A_3B_3.

---

symbolic representation of the disciples in Orthodox Christianity preserved in a church on the rue du Bac in Paris. It seems to have been adopted following the exertions of a French bureaucrat—if so, another striking example of French influence.

60. Pennick (1980, 11–14, 20–22), including its use in another faith.

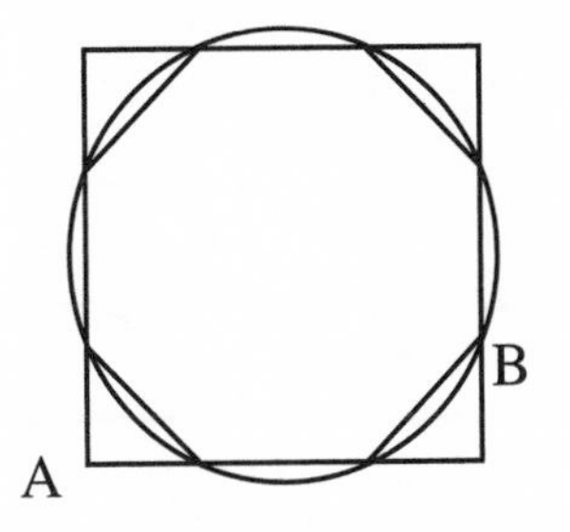

Figure 14.5. Egyptian nonregular octagon, to approximate to π with 256:81

A further favored shape in church architecture is the regular octagon. No obvious biblical authority supports it, although the resurrection of Jesus is presented as occurring on the 7th + 1st day (Matthew 28:1; Mark 16:9; Luke 24:1; John 20:1). The shape may have gained authority from the Egyptian approximation to π given by 256:81. This excellent value arises for the area of the circle by considering the (nonregular) octagon in figure 14.5, which is created by dividing each side into equal 1:3s. Note also that AB is $\sqrt{10}$, also close in value to π. Perhaps in deep antiquity these approximations were thought to be exact, thereby enhancing the status and mystery of geometry by solving (or maybe, with the $\sqrt{10}$ property, *inspiring*) the problem of squaring the circle; however, the Egyptians knew that the ratio 8:9 of the area of the circle to the circumscribed square *was* an approximation. Similarly, a description of a "circular molten sea" assumes that $\pi = 3$ (I Kings 7:23) but maybe figuratively or numerologically while known to be inaccurate mathematically.[61]

The cathedral builders probably laid out the regular octagon by this simple construction, known in antiquity. Take a square, and from each corner lay half the diagonal on each connecting side; then the octagon is given by joining up the resulting octet of points.[62] A similar construction is shown in figure 14.6: at each corner extend the sides by an amount equal to half the diagonal, and join up the octet of endpoints. In addition, semicircles with that radius and centered on each corner of the square intersect in an attractive way that is also used on church windows.

61. On this theme, see Tsaban and Gerber (1998). The symbol π dates from the early eighteenth century, as the first letter of "perimetros" and "periferos" alluding to the ratio of diameter and circumference. The value is also involved in the area of the circle, and in the area and volume of the sphere—*deep* theorems in ancient geometry with various semiotic connections (a later proof is noted in ¶5.5).

62. The important architect Mathes Roriczer (1430s–1490s) publicized this construction in a booklet that is printed and translated in Shelby (1977, 120, 64, 182): the introduction and bibliography of this edition are very valuable. He also gave the construction of the pentagon mentioned in ¶5.2.

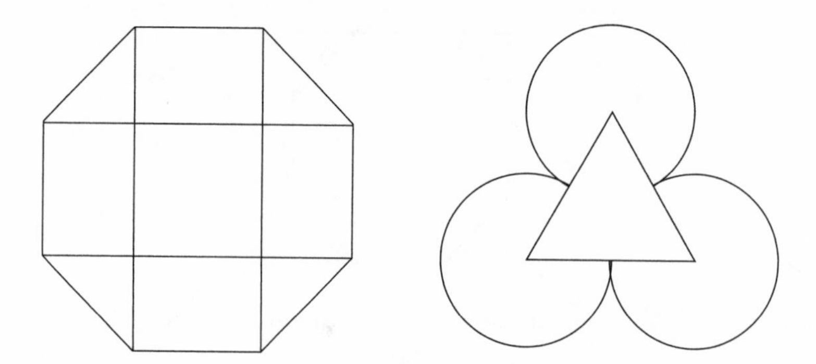

Figure 14.6. The square and octagon as well as the equilateral triangle and intersecting arcs of circles

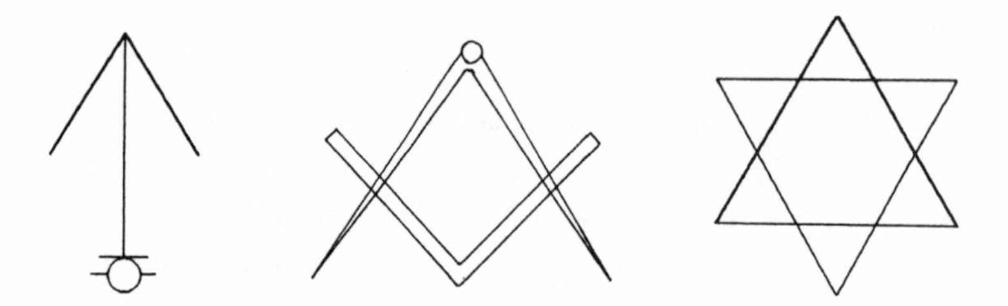

Figure 14.7. Egyptian balance, the Hiram key, and the Jewish cross

There is also a Trinitarian analogue using three arcs of circles of equal radius at the corners of an equilateral triangle. An example is given in figure 14.6; sometimes the triangle is inverted.

Geometrical symbolism in sacred art is very wide, and not only the various crosses. An interesting case is halos: initially circular with pre-Christian Hellenic origins in Sun symbolization and worship, they went elliptical in the late Middle Ages when sensitivity to perspective in art increased. But rare are spirals: no apparent biblical mentions or place in Orthodox pictorial representations, although they gained some currency in Gnosticism.

One of the main signs in Freemasonry is the "Hiram key," a pair of angles (figure 14.7). The lower one is a right angle, for a reason soon to be noted. The upper angle is shown as a pair of compasses, maybe inspired by the Egyptian hieroglyph ʌ, which denoted "to balance" or "to weigh" (Budge 1920, 614a, 622b, 688a). It is usually shown open at around 60°, suggesting the △ symbol again; note also its similarity to the lower part of the "pax" sign in ¶5.1. Further, only a slight modification of the key is needed to obtain from it the Jewish hexagrammic cross (figure 14.7). I have seen this connection forcefully shown in a *church,* in the village of Clare in Suffolk:

a plaque on the wall to the memory of Prince Leopold (1853–1884), the fourth son of Queen Victoria, and master of the Royal Clarence Lodge, with that cross placed immediately below the key. Note also that a hexagrammic cross of two right-angled triangles can be constructed from the regular octagon of figure 14.5.

5.3. An Obscure Theorem about the 3:4:5 Triangle

Given the basic right-angled triangle with sides in proportion 3:4:5, the isosceles triangles set on each corner as shown in figure 14.8 have respective ratios of altitude to half-base of 1:1, 2:1, and 3:1. The theorem is most easily proved by setting ∠ACB or ∠ABC = 2X and relating tan 2X (the ratio of the short sides of the original triangle) to tan X (the ratio of the short sides of each subtriangle) by the formula (in modern terms)

$$\tan 2X = 2 \tan X / (1 - \tan^2 X). \qquad (14.9)$$

It is a remarkably beautiful and simple result, and also numerologically significant in going from 3, 4, 5 to 1, 2, 3; the 3:4:5 triangle may also be understood as the 6:8:10 triangle. Further, ∠RCB ≈ 18° (as noted in ¶3.2, its double ≈ 37°) and ∠QBC ≈ 27°, two numerologically important numbers.

I have no idea of its history, and no mathematician or historian whom I have asked has even seen it before; I found it, in different depictions and without historical account, in Pennick (1980, 83). Given its relationship to the basic building

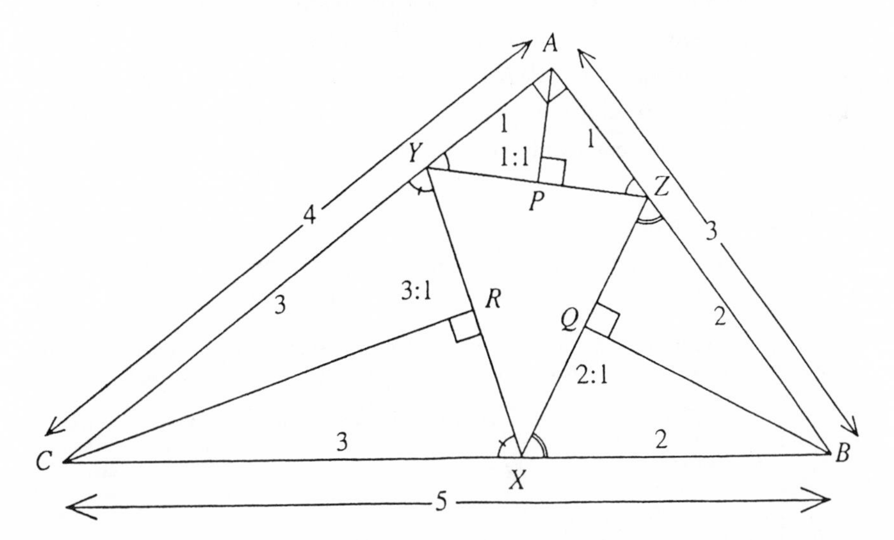

Figure 14.8. The 3:4:5 triangle and its proportions

proportions, it may have been known in the Middle Ages but kept as one of the masons' metaphysical secrets (as opposed to the craft secrets): workbooks for the masons that I have consulted present 2:1 and 3:1 separately, specified relative to the sides of appropriate triangles.[63] It may not be accidental that in Freemasonry the right-angled isosceles triangle (ΔAYZ in figure 14.8) was known as the "Mason's square," half a square as in the lower part of the Hiram key, semiotically showing both 2:1 and also the Babylonian L shape of figure 14.1.[64] But from around the late seventeenth century the name was attached to the short arms of the 3:4:5 triangle (AB and AC in figure 14.8) and was so worn by Masons as an ornament.[65] I have not found the reasons for the change, but this theorem shows how both 2:1 and 3:1 can be encompassed.

The corresponding theorem for *any* ratio n:1 can be established. The construction occurs on the corner C, with the triangle at B leading to a nonintegral ratio p:q where $q \neq 1$. The proof can be related to the generation of the sides of ΔABC in (14.9) {(Grattan-Guinness 2000). Some other little-known triangle theorems, of a secular character, are given in ch. 18.}

5.4. Squares and Word-Squares

Squares arise in Christianity, especially concerning "four angels standing on the four corners of the earth, holding the four winds of the earth," together with "another angel ascending from the east, having the seal of the living God" (Revelation 7:1–2), a passage that may be a source of the tradition of building churches to face east. The more usual explanation for Europe of facing (roughly) toward Jerusalem is later; it was reinforced during the Middle Ages up to around the fourteenth century, for the Catholic Church decreed that maps of or containing the Mediterranean Sea should have Jerusalem at the center with the east at the top. Indeed, this is the *origin* of the word "orientation."

The association of the Earth with a square is very ancient. The famous classical problem in Greek mathematics of squaring the circle, exciting because it cannot be

63. The books of Villard de Honnecourt (1175–1240) were very influential; a modern edition is included in Bucher (1979). The theorem is also absent from, for example, the booklets edited in Shelby (1977).

64. See the important illustration from the +twelfth century in Knight and Lomas (1997a, pl. 21). How are we to understand the decoration of older cathedrals and churches where a sequence of isosceles triangles along the keystone follows the line of the original semicircular keystone? Note also that the dramatic and intimidating interiors of Britain churches were eliminated by the destructions of the Reformation period.

65. For a well-known illustration of this triangle worn by Mason Mozart and his brothers, see Robbins Landon (1991, pl. 26).

effected by using only rule and compass, surely also carries a "religious" message in the impossibility of unifying the square Earth with the circular heavens. Further, the latitude of Gizeh, at exactly 30° north, recalls another Greek classical problem, of the impossibility of trisecting any angle by ruler and compass; 3 arises again.

Squares with crosses can be made out of letters that form words. Inscriptions were very widely used in the Roman Empire, with names and events summarized in rectangular blocks of (capital) letters. Two semiotic variants are symmetric letter-squares of five five-letter Latin words.[66]

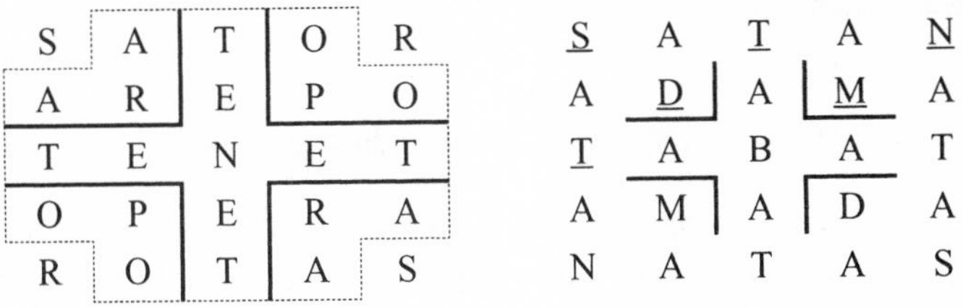

S	A	T	O	R
A	R	E	P	O
T	E	N	E	T
O	P	E	R	A
R	O	T	A	S

S	A	T	A	N
A	D	A	M	A
T	A	B	A	T
A	M	A	D	A
N	A	T	A	S

Figure 14.9

The first one, apparently dating from the +first century and found in various places in the Roman Empire (including Pompeii), uses 8 letters. Seemingly of Orthodox Christian inspiration, the two instances of TENET form an equal-armed cross; for reasons argued in ¶8.1, the choice of a word using *T*s may not be accidental either. In addition, ATO on two edges and OTA on the other two are the acronyms of "alpha/omega tenus omega/alpha," with "tenus" meaning both "up to" and "down to." The text sort of means "the planter turns the motion of the heavens with his plough," but it served mainly as an anagram: leave *N* in the center, and around it arrange both as a horizontal row and as a vertical column

A P A T E R *N* O S T E R O. (14.10)

Again there is an equal-armed cross and "alpha" and "omega," now enclosing "our father"; further, the 13 letters of each line symbolize the disciples surrounding Jesus, who is represented by *N*, the 13th letter of the Latin alphabet.

In the second word-square, of Templarian origin, 7 letters are used. The array of 12 *A*s forms the equal-armed cross surrounded by 8 stars, like the Maltese cross

66. See, for example, Endres (1951, 51–55). A British potsherd showing part of the first one, starting with ROTAS, is shown in Thomas (1981, pl. 1); his interpretation on pp. 101–102 misses the equal-armed cross, "tenus," and the place of "n" in the Latin alphabet.

(¶2.2); it serves also as a heraldic device.[67] The six noncentral consonants as underlined in figure 14.9, including both *T*s, serve as the acronym for the Templarian phrase "Solomonis Templum Novum Dominorum Militiae Templariorum."[68]

Among other anagrammic decodings, one of Gnostic inspiration reads "rosa sarona patet Peter et ero" (that is, "the Rose of Sarona," a Gnostic name of the blood of the saints, maybe inspired by the temple at Sharona on the Nile river) "is open to Peter and culprit" for his deprecations and 3 denials of Jesus (Matthew 16:22–23, 26:69–75; Endres 1951, 56). But this reading seems too contrived to be the motivation for creating the letter-square; it may have been put forward later as an alternative. Its content, of course, is not affected.

5.5. The Tarot and Judaism

Other signs of Masonic origin include the practice of a Crusader marking his role by being portrayed on his tomb with crossed legs—akin to the equal-armed cross. Another manifestation is the two crossed thighbones in their symbol of skull and crossbones, later and better known in the pirates' flag.

The Templars may also be the, or a, source of Tarot cards ("Ta," "Ro," Egyptian for "the royal road"; Cavendish 1975). Clear cases include the Hanged Man card with his crossed leg, and the High Priestess sitting between the pillar of B(oaz) and J(achin); this latter is sometimes replaced by the Female Pope, presumably in allusion to the status of Mary Magdalene. The Major Arcana has 22 cards, which relate to the Hebrew sefirot tree of life as follows. "Sefirot" is Hebrew for "ten numbers," and the tree is composed of 10 circles (or spheres) representing the aspects of God laid out in 3 vertical "pillars." These circles are joined by 22 "paths" (figure 14.10), each one corresponding to a letter of the Hebrew alphabet ("alef bet," meaning "learn wisdom," and maybe supporting the *A B* in the letter-square above).[69] Further, in the Talmud the 22 letters are construed as the initial alef and then 3 7s denoting respectively the mysteries of grace, mercy, and strict justice, and are presented alternately as male and female. In ¶4.3 we saw 22 also as the number of

67. Heraldry is a fine example of semiotics, especially in combinations and layouts of the same sign. Its development in the High and late Middle Ages seems to have been closely entwined with the Christianities, Gnosticism, and Freemasonry.

68. On the history of this word-square, see especially Atkinson (1951); as with Thomas (1981), his interpretation is incomplete. Girolamo Cardano (1501–1576) presented it in a medical text as if already well known (Cardano 1557, ch. 44).

69. Cavendish (1975 49–51); this and further connections to the Templars are made in Knight and Lomas (1997a, ch. 4), maybe too categorically. On the importance of tree motifs, and also wings, and their possible Babylonian origins, see Parpola (1993). On the sefirot and its role in the Kabbalah from its rise in the thirteenth century, see Scholem (1955) and Idel (1988, esp. ch. 6).

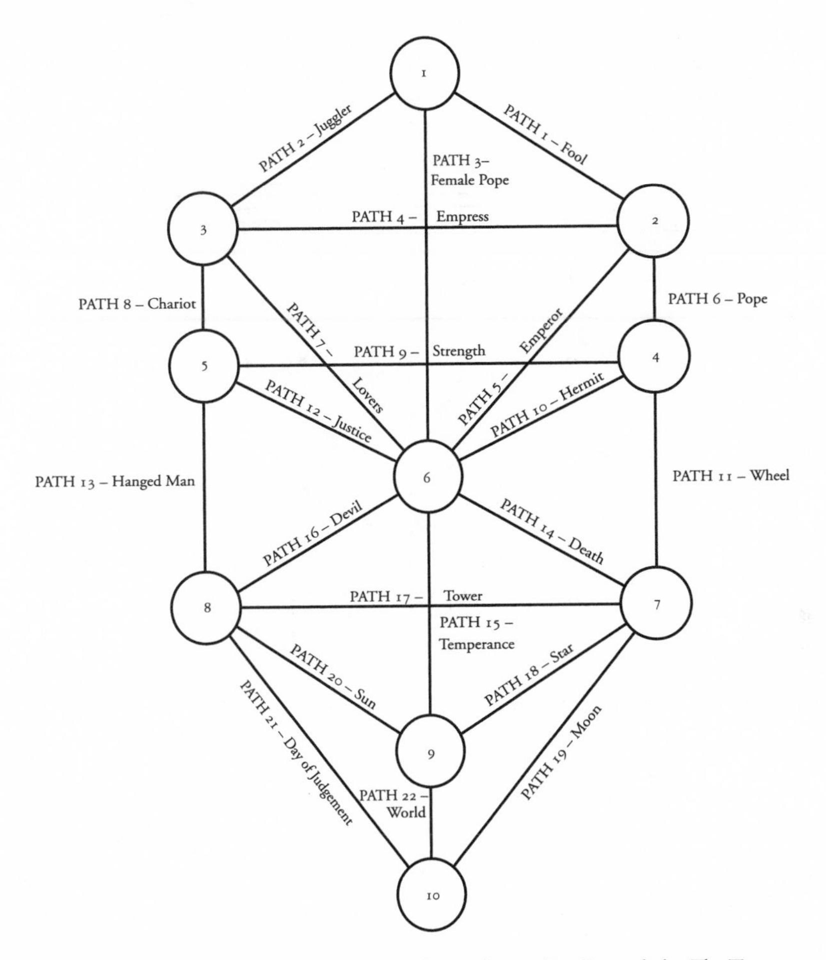

Figure 14.10. The 22 links between the 10 sefirot spheres (R. Cavendish, *The Tarot,* [London: Michael Joseph, 1975], 54)

chapters in Revelation; I Kings also has 22, but it exists as a book following one of the subdivisions of Old Testament books mentioned in ¶4.1, so deliberate intention there is less clear.

Lying at the heart of Jewish mysticism, the sefirot admits multiple interpretations. One with semiotic import is as a symbol of the infinite in a sequence of concentric circles (why?), each one cut at its top (figure 14.11). It also connects to mathematics, since it provides a beautiful intuitive argument for this profound theorem concerning the circle:

$$\text{Area} = (\text{Diameter} \div 2) \times (\text{Circumference} \div 2): \tag{14.11}$$

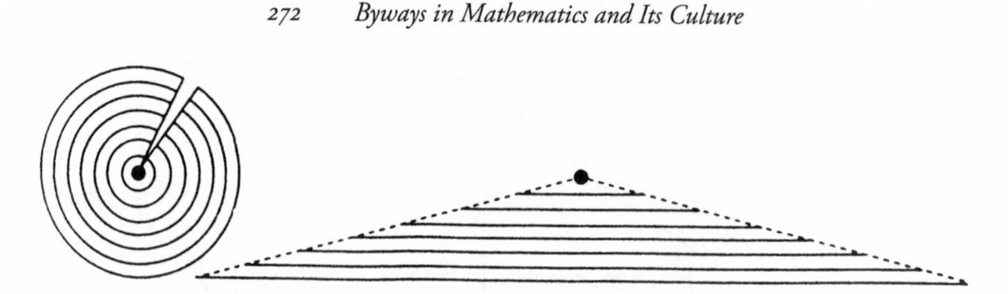

Figure 14.11. The Jewish sefirot sign, and its bearing on π

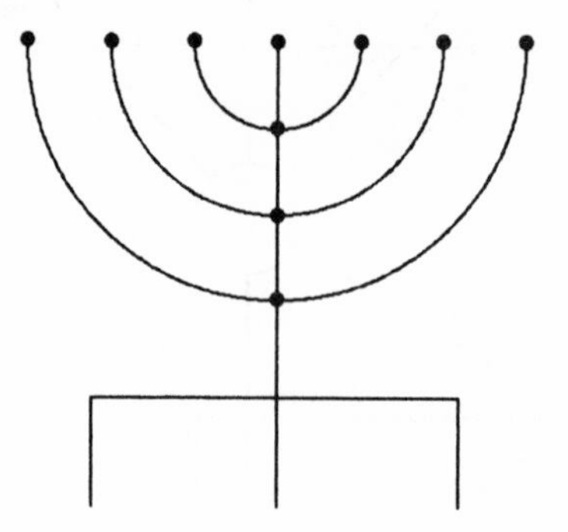

Figure 14.12. The extended menorah

unrolling the circles into parallel straight lines produces an isosceles triangle. The history of this illustration is quite curious; as far as I can find, *both* the symbol and the theorem date from around the +twelfth century, the theorem appearing in an encyclopedia produced by Abraham bar Hiyya.

Further, if 3 such circles are taken and cut across the horizontal diameter and a vertical central line inserted, then we have the Jewish menorah candlestick, with 7 branches and 10 points. The extended version (figure 14.12) seems to be ancient; it may represent the Jewish association of the menorah with kingdom, power, and glory, a triad that also ends the so-called Lord's prayer (Matthew 6:13).[70]

70. Thiering (1992, 489), though not with the full figure, which can be seen in some of the mosaics in Wilpert (1917). Among relevant literature, see Scholem (1974) and Blumenthal (1978).

6. On the Role of Gematria

6.1. Possible Origins

To Jews seems also to be due the origins of "gematria," to use the Latin name that was applied to it later: when a language has developed an ordered alphabet, assign a number to each letter and thus find the number of a word or phrase as the sum of its constituent letters.[71]

A stimulus may have been the ancient practice, presumably done to save space on the writing material, of not separating words in written texts (which are called "uncial"). It may also have arisen as a natural extension of the use of letters of the alphabet also as numerals, and the standard classical systems assign 1 to the first letter, 2 to the second, and so on up to 10, then in 10s up to 100, then in 100s. Hebrew runs to 300, the so-called greater canon in Latin to 500, and Greek to 900 since special signs were used for 6, 90, and 900 (Scholem 1971).

Once established, a role for gematria was to serve as a code, where a "dangerous" word or phrase was replaced by another one with the same gematriac sum.[72] Among more everyday uses, a graffito at Pompeii is a gematriac declaration of love; such human relations, whether friendly or not, are still evident in the phrase about "knowing your number."

An example early in the Old Testament concerns Abraham's desire to rescue Lot by deploying 318 men; for this number is that of the Hebrew version of the name Eliezer, of his chief steward (Genesis 14:14,15). The use of gematria in the New Testament was probably encouraged by the alpha-omega metaphor attributed to Jesus (¶5.4). The best known biblical case is 666, "the name of the beast, or the number of his name" (Revelation 13:17–18). An attraction of 666 may have been its property as the triangle number of 36, itself triangular to 8 as in (14.3); and a link to 37 in (14.5) is provided by

$$37 \times 36 = 2 \times 666 \qquad (14.12)$$

71. For an excellent historical introduction to this topic, see Dornseiff (1922), esp. pp. 91–118 of a chapter on its appearance in "various domains" after a survey of origins attributed to the Pythagoreans. The role of gematria in the development of alphabets in written languages still has to be determined; much valuable information on alphabets is given in Graves (1962, esp. chs. 8, 10, 11, and 13).

72. I have an unpublished manuscript of 1978 by one J. Hughes of Melbourne, with whom I have not been able to make contact, pointing to many such equalities in English with a Christian context using the standard 1–26 letter code; for example, 123 = The Holy Bible = God story (compare ¶4.1).

(note the doubling of a number again). Among the many decodings, a convincing one takes it to be a criticism of the method of initiation into celibate communities, arithmetically as the sum 60 + 200 + 400 + 6 of Samekh, Resh, and Taw; these letters relate to stages of the method, together with Waw, to make a word.[73] Another candidate is anti-Roman: the sum of the numbers with numerals d, c, l, v, and i.

The number 37 and its multiples seem to play important roles in Greek Christian gematria, for they are the numbers of a disproportionately large collection of key words and phrases. Two of them are "Ιησους" (Jesus) at 888 and "Δεοτος" (God) at 592; when added they give 1480, which corresponds to "Χριστος" (Christ). (Near-)doublings include "Ιχθυς" ("fish," of ¶4.3) at 1219 and "ξυλον" ("cross," from "wood") at 610. Many other pairs of words and phrases have the same numbers, not related to 37.[74]

Another role for gematriac numbers seems to have been to relate pairs of them to important ratios in sacred geometry. Two examples connect to the fish bladder of figure 14.4. First, "Ο λογος του κυριος" (The word of the Lord) at 2213 with "Σωτιρ" (Savior) at 1408 give a ratio that is only 0.6 percent in excess of π:2. Second, 1408 with 1209 from above is just 0.8 percent larger than 2:$\sqrt{3}$, the side to the altitude of the equilateral triangle.[75] An example not involved with 37 is "Ιαννως" (John), which takes the nice Trinitarian digit string 1119; this is also 3 × 373, the number for "λογος."

Of course many such equalities and relationships must arise, but the frequency here of semantic concord is striking, especially involving 37. Instant "rationalist" dismissals need tempering by our very fragmentary knowledge of the formation of alphabets and the practice of gematria in general.[76]

Gematria continued to bear on faiths, at least up to the late Middle Ages, in guiding the phraseology used in sacred texts and the numbers of repetitions of key

73. Thiering (1992, 514); presumably Revelation was written in (demotic) Greek, but this would not prohibit an allusion to Hebrew gematria. For various other candidates for 666, see Graves (1962, ch. 19). In the 1510s Martin Luther's friend the mathematician Michel Stiefel tried to identify 666 with the Latin name of the current pope, Leo X (see note 20)—an exercise not only unsuccessful but also absurd, as undertaken relative to a wrong language.

74. For extensive surveys of these properties, see Bond and Lea (1917) and Lea and Bond (1919–1922, pt. 2, 133–201) on 37. Another remarkable source is a copy of Lexicon (1860) held in the British Library (callmark 012924.d.20); someone spent part of 1948 writing in the margins the number for every word and phrase in the book, and tipped in a manuscript using words for 37 and its multiples; seemingly unaware of Lea and Bond, his list is fairly different from theirs, implying that neither one is complete. A general review of gematria in the Gnostic tradition is given in Hopper (1938, ch. 4).

75. See Bond and Lea (1917, 50–62) on these and other examples (which, however, includes a rather contrived reading of the text for 153 fish).

76. This point bears on the startling predictions found in the Hebrew Bible in Drosnin (1997); however, the lack of vowels in Hebrew must increase the chances of coincidences.

words there. It has also played a role in music, both sacred and secular.[77] Sadly, it has been even less well investigated than has numerology; and since it is even more covert, study is especially difficult. But one context where some progress has been made is in study of the construction of religious buildings.

6.2. Gematria in Architecture

One apparent use of gematria was to build key words or phrases literally into churches. An early example is the original Basilica of Saint Peter at Rome (replaced by the current building at the turn of the sixteenth and seventeenth centuries); measures included not only 666 units but also 888, the number of "Jesus" (Bannister 1968).

For reasons that are not clear, around the thirteenth century the Latin "greater canon" was replaced by a "lesser" one in which simply a = 1, b = 2, . . . i = j = 9, . . . u = v = 19, . . . z = 24, so that smaller totals were produced. One important manifestation of this canon is Chartres Cathedral, on its rebuilding into its present form at the turn of the twelfth and thirteenth centuries. The motto "assumptio virginis beate mariae" (assumption of the Blessed Virgin Mary), takes the gematriac integer 306, which is the distance of an important measure in the building when calculated in Roman feet (the unit commonly used in medieval architecture). Doubling may also be involved, for this number is also twice 153, which was noted in ¶6.2; some other lengths in that cathedral correspond to other phrases. Among other features is the layout in various places by regular hexagons (James 1978, 1981); and the indeed amazing maze obviously has numerological features, some possibly linked to astronomy.[78] The other cathedrals of this period, especially in France, should be analyzed this way.

7. Cognitive and Noncognitive Relations

The last remark can easily be extended: the interrelations between religions and mathematics deserve an extended and systematic survey, which has never been attempted, {though see now the extensive collection of essays (Koetsier and Bergmans 2005) on "mathematics and the divine."} Some tentative general remarks are made

77. On the influence of German cabalistic numerology on J. S. Bach, see Tatlow (1991). The place of gematria in the libretto of Mozart's *Die Zauberflöte* (1791) is explained in Irmen (1991); see also chapter 16.

78. Critchelow and others (1975). Very similar mazes were constructed in other French churches and cathedrals.

here, principally about cognitive and noncognitive relations; that is, where the relation does (not) bear on the *content* of the theory involved, whether on the religious or the mathematical side, or both.

7.1. Some Cautions

A major difficulty in interpreting early cases is that in antiquity every body of knowledge was at an early stage, so that distinguishing cognitive from noncognitive relations between them is not easy to effect (or at least little enlightening evidence is available), and may even be anachronistic. Even distinguishing the religious from the mathematical may be impeded. Rather, we see much *interaction* and *reinforcement* (¶1.2) between Christianity and mathematics (with astronomy), with little textual guidance as to the direction from cause to effect. For example, the appearance of 7 in one realm may have raised its status in the other one, and thus guided the *choice* of 7 as the number of days that God used to create the universe (¶4.3).

Reinforcement also relates to the distinction *motivation* from *justification,* where again the boundary between cognition and noncognition is not at all easy to draw. The repeated manifestations of, for example, 3, 7, and 12 in Christian texts and art are motivated by their numerological importance, and thereby are cognitive within the religion even though they are not essential to the doctrine.

Even in ancient mathematics the distinction may not be clear. Take, for example, the definition of Platonic solids and proof that there are only five of them, attributed to the Pythagoreans and publicized in the addendum to Euclid's *Elements,* 13:18. The theorems are all mathematical, but the motive could well have been more metaphysical, inspired by examples such as the octahedron, which is the Egyptian pyramid placed over its reflection; and ¶5.2 reminds us that the another such solid is the dodecahedron.

In antiquity there seems also not to have been much concern in distinguishing logical relations between two aspects of reality (as perceived at some given time) from epistemological ones between the two domains of knowledge. The eminence of 7 was surely much inspired by the (apparent) presence of 7 planets in the heavens; but arithmetical properties for 7 such as 4 + 3 also play roles. Iamblichos did not make the distinction in his lists of manifestations of numbers from 1 to 10 (¶3.1), and we saw at (14.8) in ¶4.4 epistemological evidence for 18 and 36 but ontological for 72 via precession.

Another important obstacle, especially affecting the epistemological relations, is that in all religions these relations are exhibited without explanation (the original sense of "occult," and here linked to the pesher method of ¶1.2); the believer has to *realize* that they are present. Another pitfall specific to Christianity is that its historical record is so incomplete and contradictory that one cannot confine cognitive relations to biblical authority. This factor is compounded by the minuscule knowledge of the pre- and early history of Christianity that I usually find in believers with whom I have discussed the matters raised in this paper; fortunately, it is not itself pertinent to this research. I have tried to avoid cases where the relevant brand(s) of Orthodox Christianity must be determined, but clashes with the Apocryphal version, such as the two forms of cross, must be treated. I am painfully aware of my own ignorance of Hebrew, which has surely prevented me from rendering proper justice to the Jewish contexts.[79]

7.2. Some Cases

The cognitive place of 3 from the origins of Christianity is clear, with both arithmetical examples such as Trinitarianism and geometrical ones such as 3 nails in the body on (whatever) cross; however, not all 3s need be cognitive. The "†" form of cross is also cognitive (though its historical veracity will be questioned in ¶8.1), as are consequences such as the medieval plan of a church, and the pairs of pillars with (in the High Middle Ages) the semicircular arches. Many other manifestations are not cognitive; for example, the use of halos (¶5.2), or the confusions over 6s and 7s (¶4.3).

But in some contexts analysis is not straightforward. Many cases of numbers, not only 3, seem to be cognitive while the (rounded or truncated) data in Numbers are seemingly not. Numerical illustrations such as the 13 letters of the line (14.10) in ¶5.4 signifying Jesus with his disciples are not cognitive relative to doctrine, though surely so as a sign. Some instances of numbers of chapters and verses of text may be cognitive, in at least some editions of the Bible; in particular, surely the Psalms are not 150 in number by happenstance, and numerology suggests that the German practice of numbering the incipit (where existent) as verse 1 is more authentic than the British one of *not* numbering it (for example, at 68:18 or 68:17 respectively).

Presumably the numbers in the miracles and parables are symbolic rather than literal, since the texts are peshers. For example, feeding the 5,000 with 5 loaves and 2 fishes presumably symbolizes the ordination of many new ministers by the few

79. For examples of further insights from Jewish sources, see Tsaban and Garber (1998).

initiates, perhaps 5 Levites and 2 Essenes; the 12 baskets of leftovers might signify the next team of initiates. The companion story of 4,000 and 7 loaves and "a few small fishes" with 7 leftovers makes the symbolic side clear, since reporter Mark quotes Jesus reviewing the numbers in both cases and rightly wondering "How is it that ye do not understand?"[80]

Fish constitutes a good issue of cognition. In itself the sign is not cognitive; but its choice *was* so guided if the conjecture is correct about the location of the vernal equinox in Pisces when Christianity was being founded (¶4.3). It may also have motivated the fish bladder shape, with its 3-sided triangle (figure 14.4). This is a link between geometry and arithmetic, which we saw also in Peter's count of his catch of fish at 153, cognitively the triangle number to 17 (¶4.3). Finally, this example shows that mathematics can relate to a religion in ways that do not obtain with physical and natural sciences; for the fish bladder has no consequences for natural history or biology.

The case of Christianity and music is simpler. Although music and its instruments are mentioned in the Bible, from its fatherdom in Jubal (Genesis 4:21) right through to the "harpers harping with their harps" before an audience of $(12^2 \times 10^3) + 1$ (Revelation 41:1–2), not enough information is provided with which to form cognitive relations. Thus, beyond the pre-Christian Platonic allusions of ¶3.2, I have not treated the later connections of music with numerology and gematria; once again, they are rich but not well studied (see ch. 16).

Most cases of influence discussed above were positive; but negative cases should also be noted. Especially from the eighteenth century, there was a great swathe of these, freeing mathematics from whatever role, cognitive or otherwise, that religions (usually Christianity) had previously held. But Christian factors still found a place on occasion thereafter (ch.15).

This secularization of mathematics has led historians to ignore religious factors in the development of mathematics, and even more of mathematical elements in religions. However, mathematics introduced figurative or metaphorical meanings into (at least) Christianity that were probably well understood at first, maybe only to initiates, but have been misunderstood later as literal utterances by believers, including initiates.

80. Mark 8:19–21, after the miracle at vv. 1–9. For the case of 5,000, see Matthew 14:13–21, Mark 6:30–44, Luke 9:10–17, and John 6:1–14. For a valuable discussion of these and other miracles in terms of pesher techniques, see Thiering (1992), 32–34, 121–122, and the interpretation of numbers on pp. 489–492.

8. A Case Study in Mechanics: Equilibrium in the Crucifixion Process

8.1. On What Shape of Cross Was Jesus Crucified?

As was noted in ¶5.1, the standard reading of the crucifixion takes, without biblical authority, † as the shape of the cross involved; it is known as "crux immissa" (interestingly, "grafted'). Another possibility is the Y shape, especially if using a tree and two suitably grown branches;[81] indeed, this form of crucifixion is recalled in Acts 5:30 and 10:39. However, in yet another inconsistency, the Gospels agree that a cross was used; maybe another semantic blur is involved, between "tree" and "beam." But both physiological and material mechanics suggest that the T shape gives a *much* more reliable device, so that the shape should be interpreted physically; within Orthodox Christianity it is also called Saint Anthony's cross, after his alleged murder on it.[82] (It is also called the "crux commissa" [criminal], and also the T cross after the Greek letter tau.) Let us consider the reasons for its superiority, as a piece of mechanical history.

The task at hand is to suspend a human body for a long time on a wooden structure. If a natural tree of suitable size and shape was not available, then a pair of hewn planks or stumps would be needed. There seems to be no reliable evidence on the state of forest cover in the region in biblical times; indications of climate and water supply, not least in the Bible, suggest that it was modest, although analyses of geological deposit for that epoch intimate that a somewhat more temperate climate than today's then obtained in the region. Further, contrary to normal later portrayal, the vertical component of the cross need not have been long, requiring still more wood; feet of the victim off the ground was the key factor. Making a cross was not a trivial matter, even with repeated uses of the vertical member; signing out for iii crucifixes for the event may not be an idle joke, given also the Roman obsession with stock control.

A major problem is the manner with which the planks should be joined to be secure enough to hold a victim, always heavy and maybe sometimes writhing, over

81. The cover of Hengel (1977) includes an illustration with this Y shape, sadly without giving provenance. On evidence for a tree, see also ¶8.2.

82. A few painters have used this shape, for example, Tommaso Masaccio's "Trinity" (±1426), analyzed from other points of view in Field (1997, ch. 3).

A more recent example of an instrument with a "meaningful" shape is the guillotine: a circle inside a rectangle onto which descends a triangle. Such semiotics would not have been overlooked in the Terror of the French Revolution.

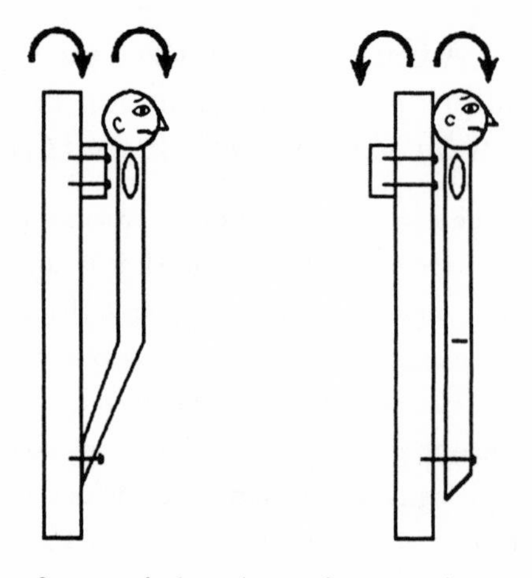

Figure 14.13. Manners of using a †-shaped cross for a crucifixion

the required period. With the T shape the horizontal crosspiece could be fixed behind or in front of the vertical one; the former would give a more stable structure, since the turning moments of the crosspiece and of the body would lie in opposing directions (at the cost of some torque through the crosspiece itself), whereas in the latter case they follow the same direction (figure 14.13). However, neither possibility is technically promising, since the joining of both crosspiece to the post and body to the crosspiece would have to be effected with *horizontal* (maybe sloping) nails, with considerable risk of detachment occurring under the weight carried; the Romans were well aware of the brittleness of iron. Further, in his agony the victim might swing his arms and torso in the plane of the cross itself, a further risk of dismantlement. The X shape, the "crux decussata" (crosswise), is similarly susceptible; but while also known within Christianity as Saint Andrew's cross, for reasons similar to Anthony above, its record may be more semiotic than factual, as was suggested in ¶5.1.

From a mechanical point of view the T shape is far preferable to all these options; for a thick vertical post, maybe a tree trunk, carries atop a horizontal plank fixed with *vertical* nails. The arms of the victim could also be nailed vertically, although the structure might be strong enough to take suspension by horizontal nailing. This preference would apply also to the lifting of the cross into its vertical position after the victim was nailed to it as it lay on the ground (presuming that this was the procedure followed).

If so, then great doubt is cast on the claim that a written description of Jesus as the King of the Jews was affixed to the upper part of the post (a claim occasionally symbolized by the cross of Lorraine in the form ‡); for there is no such part on the T-shaped cross. In fact, only in John 19:19 was it stated to be placed explicitly somewhere on the post, or fixed on top of the horizontal member; in Matthew 27:37 it is "over his head," and in Mark 15:26 and Luke 23:38 it is called a "superscription."[83] Maybe these texts are later additions, or alterations of a report that the description was carried separately on a pole, following a requirement of Roman law that the offense be indicated on the occasion of the punishment. None of the books in the Nag Hammadi collection mentions the description at all.

The authority of the accounts is also questioned by the very active role attributed to Pontius Pilate; for the records of Roman government suggest it as unlikely that the prefect of the Province of Judea would have been so involved in this rather minor event, although many details (such as the description) would have been effected under his authority. It is also very surprising that he pardoned Barabbas, for only the emperor normally held this power. Further, surely Barabbas was no thief but "son of the father," as his name proclaims, and thus a religious leader like Jesus. However, the presence of soldiers is very believable, since as an event the Roman authorities would have licensed the crucifixion.

8.2. How Was Jesus Nailed to the Cross?

The claim that Jesus was nailed at all is also questionable, since the increased agony (to be analyzed shortly) would have greatly reduced the duration of the punishment; lashing with ropes was the normal practice, when the pain would be less but the endurance longer, with asphyxiation competing with loss of blood, effects of wounds, hunger and thirst, and attack by animals or birds, as the days went by.[84] If he was nailed, the tradition that nailing was done through the palms is very doubtful (as is now fairly well understood), even if the T-shaped cross was used. For the tensile strength of the flesh would be insufficient to tolerate the measure of shear, especially if the victim writhed; the palm would rip through the (single?) nail and free the arm, thus sabotaging the punishment. Nailing would be much more rigid if made through the upper arm.

Biblical authority here is weak, for none of the Gospel accounts of the crucifixion mentions nailing at all. Evidence comes only from the case after the Resurrection

83. The history of the description is well rehearsed in Thiede and d'Ancona (2000), though they consider only Orthodox texts and assume that the cross took the † shape.

84. On this and several other issues about nailing, see the valuable reflections in Hewitt (1932); and also Cohn (1972, 219–222).

of doubting disciple Thomas, who wanted to put his own finger through the holes in Jesus's palms (John 20: 20–28; compare the Pauline Colossians 2:14). Further, Thomas would have had to rest content with touch of the injuries, since the flesh of Jesus's hands would already be closing back to its equilibrate condition.

For the feet, a separate nail through each one is far more effective than the one nail though both, for in that case danger from turning moments arises again. This latter tradition, dominant in both doctrine and representation, may again be semiotic and figurative rather than literal: the Trinity evident in the 3 nails forming a triangle. Further symbolism may attend the issue of whether the right leg crossed the left or vice versa. There is also a tradition of nailing effected through the ankles, each one separately. The Nag Hammadi texts (¶2.1) are much more informative on these points. One states of Jesus that "They nailed him to the tree, and they fixed him with four nails of brass."[85] The link between the tree and the carpenter's son was noted in Philip (92) 73.9–15.

With this background we can understand the *kind* of agony that the victim would have suffered (a point of theology that receives little attention).[86] Awful enough to be nailed through feet and upper arms to a cross, the torture was much subtler and terrifying, whether nailed or lashed, for gravity would cause the body to sag, thus reducing the volume around the rib cage and thereby causing difficulties of breathing. When a deeper breath had to be taken, then the victim would have to raise his upper body, but in the process he would throw the weight of his body onto his nailed feet, and so send shattering waves of pain through to the brain. After that shock, he would also have the *mental* anguish of trying but failing to avoid taking the next such breath, with seconds passing like eons. It is likely that the victim would pass out in a few hours, as is believably implied for Jesus in all the Gospels.[87] Then the body would sag further (thereby increasing the turning moment, incidentally), and the victim would eventually die from asphyxiation—unless the process was terminated earlier, as Apocryphal Christianity claims for the case of Jesus.

85. The Second Treatise of the Great Seth 58.24–26; maybe the material was what we call "bronze." See also especially Truth 18, 24, 20.23–27; the Apocalypse of Peter 81.12–20; the letter of Peter to Philip at 139.18–19; and the Interpretation of Knowledge, 5.30–34.

86. A surprising omission is by the Reverend Samuel Haughton, who used mechanical principles to discuss preferred ways of hanging but ignored crucifixion (Haughton 1873, 8–14), drawing on his paper (1866), where he also considered the simultaneous hanging of 12 women as recorded in Homer.

87. Matthew 27:48–50, Mark 15:33–34, Luke 23:44–46, and John 19:29–30. But the tradition that Calvary or Golgotha was on the top of a hill does not gain biblical authority from the relevant passages, located a few verses before those just cited.

ACKNOWLEDGMENTS

I am indebted to various people for advice and opinions, especially Immanuel Velikovsky, David Singmaster, and Jens Høyrup, and several members of the Canonbury Masonic Research Centre (London), and to Nicolas-Hays Company for figure 14.2 and Michael Joseph for figure 14.10.

BIBLIOGRAPHY

Atkinson, D. 1951. "The origin and date of the 'Sator' word-square," *Journal of ecclesiastical history, 2*, 1–18.

Baigent, M., and Leigh, R. 1991. *The Dead Sea scrolls deception*, London: Cape.

Bannister, T. C. 1968. "The Constantinian Basilica of Saint Peter at Rome," *Journal of the Society of Architectural Historians, 27*, 3–32.

Barker, M. 1992. *The great angel: A study of Israel's second God*, London: John Knox Press.

Bauke, D. 1999. "Mathematik in der Bibel," in M. Toepell (ed.), *Mathematik im Wandel*, Hildesheim: Franzbecker Verlag, 24–41.

Bauval, R., and Hancock, G. 1996. *Keeper of Genesis*, London: Heinemann.

Blumenthal, D. R. 1978. (Ed.), *Understanding Jewish mysticism*, New York: Ktav.

Bond, F. B., and Lea, T. S. 1917. *A preliminary investigation of the Cabala contained in the Coptic Gnostic gospels, and of a similar gematria in the New Testament*, Oxford: Blackwells.

Browne, T. 1658. *The Garden of Cyrus, or the Quincunx*, London: H. Brome. [Many later editions.]

Bucher, G. 1979. *Architector: The lodge books and sketchbooks of medieval architects*, vol. 1, New York: Abaris.

Budge, E. A. Wallis. 1920. *An Egyptian hieroglyphic dictionary*, London: John Murray.

Calendar 1813. *Calendrier maçonnique pour 5813*, Paris: Poulet. [Copy in the Warburg Library, London.]

Cardano, G. 1557. *De rerum varietate*, Basel, Switzerland: Petri.

Cavendish, R. 1975. *The Tarot*, London: Michael Joseph.

Charles-Roux, F. 1937. *Bonaparte: Governor of Egypt*, London: Methuen.

Cohn, H. 1972. *The trial and death of Jesus*, London: Weidenfeld and Nicholson.

Collaveri, F. 1982. *La Franc-maçonnerie des Bonaparte*, Paris: Payot.

Critchelow, K., and others. 1975. *Chartres maze: A model of the Universe?* Cambridge: The Land of Cockayne.

De Santillana, G., and von Dechend, H. 1970. *Hamlet's mill*, London: Macmillan.

Dornseiff, F. 1922. *Das Alphabet in Mystik und Magie*, 1st ed., Leipzig, Germany: Teubner. [2nd ed. 1925.]

Drosnin, M. 1997. *The Bible code*, London: Weidenfeld and Nicolson.

Drower, E. S. 1937. *The Mandaeans of Iran and Iraq*, Oxford: Clarendon Press.

Elliott, J. V. 1993. *The Apocryphal New Testament*, Oxford: Clarendon Press.

Endres, F. C. 1951. *Mystik und Magie der Zahlen*, 3rd ed., Basel, Switzerland: Rascher.

Erickson, G. W., and Fossa, J. A. 1996. *A pirâmida Platônica*, João Pessoa, Brazil: Editora Universitária.

Evans, C. A., Webb, R. L., and Wiebe, R. A. 1993. *Nag Hammadi texts and the Bible: A synopsis and index*, Leiden: Brill.

Ferrer Benimeli, J. A. 1976–1977. *Masoneria, iglesia e ilustración: Un conflicto ideologico-politico-religioso*, 4 vols., Madrid: Seminario Cisnerós.

Ferrua, A. 1991. *An unknown catacomb: A unique discovery of early Christian art*, New Lanark, Scotland: Geddes and Grosset.

Field, J. V. 1997. *The invention of infinity: Mathematics and art in the Renaissance*, Oxford: Oxford University Press.

Folkerts, M. 1972. "Pseudo-Beda: *De arithmeticis proportionibus*," *Sudhoffs Archiv, 56*, 22–43.

Gödel, K. 1995. *Collected works*, vol. 3, Oxford: Clarendon Press.

Goodenough, E. R. 1953–1968. *Jewish symbols in the Greco-Roman period*, 13 vols., Princeton, N.J.: Princeton University Press.

Grattan-Guinness, I. 1992. "Counting the notes: Numerology in the works of Mozart, especially *Die Zauberflöte*," *Annals of science, 49*, 201–232.

Grattan-Guinness, I. 2000. "A portrayal of right-angled triangles which generate rectangles with sides in integral ratio," *Mathematical gazette, 84*, 66–69.

Graves, R. 1962. *The white goddess*, 3rd ed., London: Faber and Faber.

Hancock, G. 1998. *Heaven's mirror*, London: Michael Joseph.

Haughton, S. 1866. "On hanging, considered from a mechanical and physiological point of view," *Philosophical magazine, (4)32*, 23–34.

Haughton, S. 1973. *Principles of animal mechanics*, London: Longmans, Green.

Heilbron, J. L. 1998. *Geometry civilized: History, culture, and technique*, Oxford: Clarendon Press.

Hengel, M. 1977. *Crucifixion*, Philadelphia: Fortress.

Hernández, M. C. 1977. *El pensamiento de Ramon Llull*, Valencia, Spain: Editorial Castalia.

Hewitt, J. W. 1932. "The use of nails in the crucifixion," *Harvard theological review, 25*, 29–45.

Hopper, V. F. 1938. *Medieval number symbolism: Its sources, meaning, and influence on thought and expression*, New York: Columbia University Press. [Repr. New York: Cooper Square, 1969.]

Høyrup, J. 1992. "Mathematics, algebra and geometry," in *Anchor Bible dictionary*, vol. 4. New York: Doubleday, 602–612.

Høyrup, J. 2002. *Lengths, widths, surfaces: A portrait of old Babylonian algebra and its kin*, New York: Springer.

Iamblichos 1988. *The theology of arithmetic* (trans. R. Waterfield), Grand Rapids, Mich.: Phanes.

Idel, M. 1988. *Kabbalah: New perspectives*, New Haven, Conn., and London: Yale University Press.

Irmen, H.-J. 1991. *Mozart Mitglied geheimer Gesellschaften*, 2nd ed., Zülpich, Germany: Prisca.

Iversen, E. 1961. *The myth of Egypt and its hieroglyphs*, Princeton, N.J.: Princeton University Press.

Jacob M. 1991. *Living the Enlightenment*, Oxford, Clarendon Press.

James, J. 1978, 1981. *The contractors of Chartres*, 2 vols., London: Croom Helm.
Knight, C., and Lomas, R. 1997a. *The Hiram key: Pharaohs, Freemasons and the discovery of the secret scroll of Jesus*, London: Arrow.
Knight, C., and Lomas, R. 1997b. *The second Messiah*, London: Century.
Koetsier, T., and Bergmans, L. 2005. (Eds.), *Mathematics and the divine: A historical study*, Amsterdam: Elsevier.
Lagrange, C. 1900. *Mathématiques de l'histoire*, Brussels: Kiessling.
Lagrange, C. 1921. *La Bible: un miracle*, Brussels: Kiessling.
Lea, T. S., and Bond, F. B. 1919–1922. *Materials for the study of the Apostolic Gospels*, 2 pts., Oxford: Blackwells. [Repr. Wellingborough: Thorsons, 1979–1985.]
Lexicon 1860. *The analytical Greek lexicon*, new ed., London: Bagster, [1860?].
Llull, R. 1985. *Selected works*, 2 vols. (ed. A. Bonner), Princeton, N.J.: Princeton University Press.
Lund, J. L. M. 1921–1928. *Ad quadratum: A study of the geometric bases of classic and medieval religious architecture*, 2 vols., London: Batsford.
MacQueen, J. 1985. *Numerology*, Edinburgh: Edinburgh University Press.
Mahr, A. 1932. *Christian art in ancient Ireland: Selected objects illustrated and described.* Dublin: Stationery Office.
McClain, E. G. 1976. *The myth of invariance*, York Beach, Maine: Nicolas-Hays.
McClain, E. G. 1978. *The Pythagorean Plato*, York Beach, Maine: Nicolas-Hays.
McClain, E. G. 1981. *Meditations through the Quran*, York Beach, Maine: Nicolas-Hays.
Meyer, H., and Suntrup, R. 1987. *Lexikon der mittelaterlichen Zahlenbedeutungen*, Munich: Fink.
Morrison, W. D. 1890. *The Jews under Roman rule*, London: Putnam's; New York: Fisher Unwin.
Nash, R. 1991. *John Craige's mathematical principles of Christian theology*, Carbondale: Southern Illinois University Press.
Newman, C. A. 1984. "'He has established for himself priests': Human and angelic priesthood in the Qumran Sabbath *Shirot*," in L. H. Schiffman (ed.), *Archaeology and history in the Dead Sea Scrolls*, Sheffield: Sheffield Academic Press, 101–120.
Ouspensky, V., and Lossky, V. 1982. *The meaning of icons*, Crestwood, N.Y.: Boston Book and Art Shop.
Panin, I. 1934. *Biblical numerics*, London: Covenant.
Parker, D. C. 1997. *The living text of the Gospels*, Cambridge: Cambridge University Press.
Parpola, S. 1993. "The Assyrian tree of life: Tracing the origins of Jewish monotheism and Greek philosophy," *Journal of Near Eastern studies, 52*, 161–208.
Pennick, N. 1980. *Sacred geometry: Symbolism and purpose in religious structures*, Wellingborough, U.K.: Turnstone.
Perkins, P. 1993. *Gnosticism and the New Testament*, Minneapolis: Fortress.
Picknett, L., and Prince, C. 1994. *Turin shroud: In whose image?* London: Bloomsbury.
Platzeck, E. W. 1961–1964. *Raimond Lull*, 2 vols., Düsseldorf: Schwann.
Plutarch 1970. *De Iside et Osiride* (ed. J. G. Griffiths), Cardiff: University of Wales Press.
Purce, J. 1974. *The mystical spiral*, London: Thames and Hudson.
Quinn, M. 1994. *The swastika: Constructing the symbol*, London: Routledge.

Rees, E. 1992. *Christian symbols, ancient roots*, London and Philadelphia: Kingsley.

Reynolds, R. E. 1979. "'At sixes and sevens'—and eights and nines," *Speculum, 54*, 669–684.

Robbins Landon, H. C. 1991. *Mozart and the Masons*, 2nd ed., London: Thames and Hudson.

Roberts, J. M. 1972. *The mythology of secret societies*, London Secker and Warburg.

Robinson, J. M. 1977. *The Nag Hammadi library in English*, 1st ed., San Francisco: Harper and Row.

Roscher, W. H. 1904. "Die Sieben- und Neunzahl im Kultus und Mythus," *Abhandlungen der Königlichen Sächsischen Akademie der Wissenschaften zu Leipzig, philologisch-historische Klasse, 24*, no. 1, 126 pp. [Other of his works in this journal will be found at *21*, no. 4 (1903), 94 pp.; *24*, no. 6 (1906), 240 pp.; *26*, no. 1 (1907), 170 pp.; Roscher 1909; and *28*, no. 5 (1911), 155 pp.]

Roscher, W. H. 1909. "Die Zahl 40 im Glauben, Brauch und Schrifttum der Semiten," *Abhandlungen der Königlichen Sächsischen Akademie der Wissenschaften zu Leipzig, philologisch-historische Klasse, 27*, 93–138.

Schmandt-Besserat, D. 1992. *Before writing: From counting to cuneiform*, 2 vols., Austin, Tex.: University of Texas Press.

Schneider, M. 1946. *El origen musical de los animales simbolos en la mitologia y la escultura antiguas*, Barcelona: Instituto Espagñol de Musicologia.

Scholem, G. 1955. *Major trends in Jewish mysticism*, New York: Shocken.

Scholem, G. 1971. "Gematria," in *Encyclopaedia Judaica*, vol. 3, Jerusalem: Keter, cols. 369–374. [Several other related topical articles here, especially by this author.]

Scholem, G. 1974. *Kabbalah*, New York: New American Library.

Seidenberg, A. 1962a. "The ritual origin of geometry," *Archive for history of exact sciences, 1*, 488–527.

Seidenberg, A. 1962b. "The ritual origin of counting," *Archive for history of exact sciences, 2*, 1–40.

Shelby, L. R. 1977. (Ed.), *Gothic design techniques: The fifteenth century design booklets of Mathes Roriczer and Hanns Schmuttermayer*, London: Feffer; Amsterdam: Simons; Carbondale and Edwardsville: Southern Illinois University Press.

Sire, H. J. A. 1994. *The Knights of Malta*, New Haven, Conn., and London: Yale University Press.

Stigler, S. 2000. *Statistics on the table*, Cambridge, Mass.: Harvard University Press.

Stoyanov, Y. 1994. *The hidden tradition in Europe: The secret history of medieval Christian heresy*, London: Penguin.

Sullivan, W. 1996. *The secret of the Incas*, New York: Three Rivers Press.

Surles, R. L. 1993. (Ed.), *Medieval numerology: A book of essays*, New York: Garland.

Tatlow, R. 1991. *Bach and the riddle of the number alphabet*, Cambridge: Cambridge University Press.

Thiede, C. P., and d'Ancona, M. 2000. *The quest for the True Cross*, London: Phoenix.

Thiering, B. 1992. *Jesus the man*, London: Doubleday. [London: Corgi, 1993 printing cited here.]

Thoby, P. 1959. *Le crucifix des origins au Concile de Trente*, Nantes: Belger.

Thomas, C. 1981. *Christianity in Roman Britain to AD 500*, London: University of California Press.

Toynbee, A. 1969. (Ed.), *The crucible of Christianity: Judaism, Hellenism and the historical background of the Christian faith*, London: Thames and Hudson.

Tsaban, B., and Garber, D. 1998. "On the rabbinical Approximation of π," *Historia mathematica, 25*, 75–84.

Ulff-Møller, J. 1993. "Remnants of medieval number usage in Northern Europe," in Surles 1993, 145–155.

Usener, H. 1903. "Dreiheit. Ein Versuch mythologischer Zahlenlehre," *Rheinisches Museum, 58*, 1–47, 161–208, 321–362. [Repr. in book form Hildesheim: Olms, 1966.]

van der Meer, J. 1957. *Early Christian art*, London: Faber and Faber.

Velikovsky, I. 1953. *Ages in chaos*, London: Sidgwick and Jackson. [1973 printing cited here.]

Wang, H. 1987. *Reflections on Kurt Gödel*, Cambridge, Mass.: MIT Press.

Weigle, L. [1962] (Ed.), *The New Testament octapla*, New York: Nelson.

Whittaker, E. T. 1946. *Space and spirit: Theories of the Universe and arguments for the existence of God*, London: Nelson.

Wilpert, J. 1895. *Fractio Panis. Die älteste Darstellung des eucharistischen Opfers in der "Cappella Greca,"* Freiburg im Breisgau, Germany: Herder.

Wilpert, J. 1917. *Die Römische Mosaiken und Malereien*, 2nd ed., 4 vols., Freiburg im Breisgau, Germany: Herder.

Wood, F., and K. 1999. *Homer's secret Iliad*, London: John Murray.

Yates, F. 1972. *The Rosicrucean Enlightenment*, London: Routledge.

Young, G. C. 1924. "On a solution of a pair of simultaneous Diophantine equations connected with the nuptial numbers of Plato," *Proceedings of the London Mathematical Society, (2)23*, 28–44.

CHAPTER FIFTEEN

Christianity and Mathematics

Kinds of Links, and the Rare Occurrences after 1750

This chapter begins with a presentation of eight types of links between mathematics and (a) religion, with especial reference to Christianity. The links are potentially manifest in any period, and collectively the causal connections work in both *directions. Then a review is given of the manifestations of these links between European mathematics and Christianity in the period from around 1750 to the early twentieth century. A noteworthy feature is the* rarity *of their occurrence, and moreover during a time when the links between the natural sciences and Christianity were discussed intensively. Possible reasons for this contrast are sought in the final part.*

> But what matters to me is not whether it's true or not but that I believe it to be true, or rather not that I *believe* it, but that *I* believe it . . . I trust that I make myself obscure?
>
> —*Sir Thomas More on the theory of Apostolic succession, in Robert Bolt,* A Man for All Seasons *(1960, 53).*

1. Introduction

While the links of Christianity to science have been well studied, those to mathematics have gained far less attention, presumably because of the lack of interest in mathematics among both historians and theologians.[1] Much has been missed, however, since several branches of mathematics and religions have interrelating histories going back to antiquity; indeed, mathematics seems to have affected the *formation* of religions in the first place as well as received influence from them (ch. 14), a

The original version of this essay was first published in *Physis*, new ser. 38 (2001), 467–500; three small additions made here. Reprinted by permission of the editor.

1. For three examples among many, beyond some general remarks on mechanics and modeling in astronomy, mathematics is virtually absent from the valuable surveys of issues given in Brooke (1991), Barbour (1997), and Ferngren (2000).

property evident only in astronomy among the other sciences. This article is a foray into some features of links with Christianity.

In the first part of this chapter I propose eight kinds of links between Christianity and mathematics, both cognitive and noncognitive. They correspond to some of the ways in which a human being informed in the mathematics and Christianity of his time might ponder the relationships between them. While basically different, they are not always mutually exclusive. Further, the distinctions between them may not always have been followed or even accepted; some conflations will be noted where appropriate. Applicable to any period, the examples given come from before 1750, when manifestations occurred with moderate regularity.

The second part of the chapter records the occasions between the second half of the eighteenth century and the early twentieth century when European mathematicians invoked a link in their work. However, a main feature is that these occasions were *very few in number,* and rarely influential from one to another, even though far more mathematicians were at work than previously and (as far as I know) most of them professed some version of Christianity. This quietude also contrasts strikingly with the massive contemporary *increase* in concern in the natural sciences, especially in geology, mineralogy, and the rise of Darwinism. The final part of the chapter reviews this striking contrast, and notes some further mathematicians who never deployed such links in their work even though their belief was invoked in relationships between Christianity and other sciences and philosophy, and in a few cases they even participated in theological disputes.

Some limitations to this foray need to be explained. Unfortunately, no attempt can be made to summarize the history of links between mathematics and Christianity prior to 1750, since the literature available is rather sparse and general, and lacking in detail (on the various kinds of link, for example).[2] Again, attention to the relationships of both mathematics and Christianity to logics are restricted to a few mentions and one major figure. Similarly, I confine to a few cases concern with the differences between the numerous versions of Christianity. I also do not consider other religions beyond a few mentions; but I follow some of my historical figures

2. It is not possible to give a representative selection of texts, for they usually occur as passages in works dealing with many other aspects or concerns (biography of a major figure, for example). D. E. Smith waffles in (1921) about analogies between mathematics and Christianity (concerning notions such as infinity and continuity). A selection of relationships from various periods is provided in Henry (1976). For short reviews of links, see Breidert (1999) and Chase (1996). The bibliography (Chase and Jongsma 1983) lists publications on a wide range of themes, including philosophical, logical, and educational issues, but rather few specifically on history; in particular, it confirms the lack of literature (as of 1983) on the period after 1750.

in admitting deism, a position where the existence of some God is asserted but the Christian account is not necessarily affirmed. I postpone these desimplifications to another occasion, along with other questions such as the use made of mathematics, not always perspicuously, in *other* disciplines (including scientific ones) as a source for supporting Christian belief.

A general philosophical point also needs to be noted. When a scientific theory is (apparently) incompatible with experience or experimental findings, then some component(s) of it have to be modified, though the detection of the weakness(es) can be a very complicated process. But if mathematics is involved in that theory, its status is epistemologically more tricky. For example, in the early 1860s William Thomson (1824–1907) estimated the age of the Earth, a study that bore on the relationship between geology and Christianity (Burchfield 1975); but his use of Fourier series involved no link to *mathematics* and Christianity, nor was the legitimacy of the theory endangered. Hence the defense and criticism of mathematical theories, both pure ones and especially when they are used in physical sciences, is a subtler matter than obtains for empirical sciences, so that possible roles for Christianity vary accordingly.

PART 1: EIGHT KINDS OF LINK

Each link is given an abbreviating code, of the form "M***C" or "C***M." The capital letters denote "Mathematics" and "Christianity" respectively, the first one in each code being the source of the link involved and the second its recipient. The asterisks represent three lowercase letters abbreviating the kind of link proposed; thus, for example, "CemuM" names the link where Christianity influences mathematics by being its source of *emulation* concerning rigor and truth.

2. Mathematics Influencing Christianity

MfrmC: Involvement of mathematics in the *formation and development of Christianity.* Numerology was an important source already before Christianity and continued so throughout its formation and prosecution, drawing on ancient reverence for positive integers and their properties (ch. 14). A widespread example is the prominent role given to three in orthodox Christianity; but various larger numbers were also involved (such as Peter catching 153 fish in John 21:11). Other manifestations include the number of gods asserted to exist in a polytheism.

The other main branch was geometry with trigonometry, where high status was given to basic shapes such as the circle, the square, and the regular pentagon, hexagon, and octagon. Later designs include the so-called fish bladder, where two intersecting arcs of circles of the same diameter pass through each other's center and thereby enclose a pair of equilateral triangles (fig. 14.4). Many examples involve the influence of both mathematics and Christianity on architecture. Texts and diagrams of this type exhibit both overt and covert or symbolic levels of meaning: the Bible is a rich case.

Apart from new theories of the infinite (¶13), very few innovations were made after 1750, although some good historical scholarship was achieved.

MexgC: provision of *proofs of the existence of God,* or at least of attributes such as omnipresence and creating the universe, which God should possess. While called "mathematical," they often hinged more on logical reasoning than on mathematics as such (Whittaker 1946). For example, G. W. Leibniz (1646–1716) depended mainly on the ontology of preestablished harmony to show that God had created the best of all possible worlds; he also assumed that syllogistic logic was part of mathematics (Iwanicki 1933). But this kind of link was falling out of favor even in Leibniz's time: for example, and an important one, John Locke (1632–1704) had already asserted "faith and reason, and their distinct provinces" (1689, book 4, ch. 8),[3] a separation which seems to have gained later favor, especially under the further stimulus of the Locke-awakened Immanuel Kant (1724–1804).

3. Emulations

MemuC: *Mathematics emulated by Christianity* as a source for *certainty of theories.* Conversely, CemuM: *Christianity emulated by Mathematics* as a source for *certainty of theories.* The remark on interaction at the head of ¶1 can be expanded here. Christianity and mathematics developed together in antiquity, and both with close connections to astronomy; thus these two sciences exhibit more complicated relationships with religions than pertain with the other sciences, and over a far longer time. In particular, in these three domains of knowledge, primitive humans *began to systematize their scattered scraps of comprehension into theories,* and so brought some order and perhaps *certainty* into their understanding. Principles or assumptions were offered deploying invariant concepts: numbers, for example, or virtue.[4]

3. The histories of logic seem to be very inadequate in treating the bearing on MexgC, but a few eighteenth-century examples are mentioned by R. Kennedy in his introduction to his survey of Harvard philosophy (1995, 1–60).

4. Good insights on the growth of arithmetic and geometry, but not religion, are found in Hallpike (1979, esp. chs. 6 and 7). Deploying anthropological principles nicely informed by de-

Consequences were drawn from these principles; so (presumably) we also see the birth of logic, which often was and is entwined unclearly around mathematics. These developments may also have been interpreted as evidence that humans are (created) *capable* of such achievements.

The presence (or illusion?) of certainty in Christianity and mathematics is an important common feature. There is a striking similarity between Euclid's *Elements* (–fourth century) divided into books and propositions and providing authority by proof, and the Bible (settled +fourth century) divided into chapters and verses and providing authority by decree. It is not known whether Euclid was an influence on the formation of the Bible, and the books of the Pentateuch may have been written still earlier anyway (though the Greek translations [and organization] by the Septuagint dates from the –third century). Maybe both works exhibit a style of presentation common to all authoritative texts; codes of law provide other examples. An interesting later case of MemuC, and of MexgC, is the attempt of the physician and astrologer Jean-Baptiste Morin (1583–1656) to emulate the rigor of Euclid in his proofs of the existence of God (Iwanicki 1936).

A related common feature is that of *idealization*. In Christianity it arises, for example, in ethics. In mathematics it occurs in objects as humble as the straight line, which humans cannot experience directly, not only because of the impossibility of us drawing it exactly straight but also that in being one-dimensional, the line is invisible to us anyway. Extensions of such positions led to general "Platonic" philosophies of mathematics, though usually without a religious import.

The *losses* of certainty in Christianity and in mathematics also show some similarities; invoking theology to fill the vacuum caused by reduction or loss of faith has some parallel with the use of metamathematics to *study* the (lack of) rigor of mathematical theories. A well-known example is the criticism of the calculus made by Bishop Berkeley (1685–1753) in his book *The Analyst* (1734), where he defended Christianity by rightly ridiculing the shaky foundations of the theory, whether limits as mishandled in Newton's fluxional version or the mysterious infinitesimals of Leibniz's alternative differential and integral calculus. Berkeley actually muddled the two theories together in his (secular) solution by compensating errors, which works anyway only for a small class of functions; but his criticisms led to serious efforts to improve the foundations. However, the links to Christianity were not strengthened.[5]

---

velopmental psychology, he also relies on the work of Jean Piaget to an extent that I find hard to endorse.

5. The general history has often been rehearsed, best in Breidert (1985, 7–69), Guicciardini (1989, chs. 3–6 *passim*), and Pycior (1997, epilogue). On Berkeley's solution see Grattan-Guinness (1970).

4. Christianity Influencing Mathematics

Emulation is much concerned with the *(in)validity of arguments* for or against a given position; it should be distinguished from claims of the *truth (or falsity) of propositions,* both of principles and of deduced consequences, which constitute another important similarity between Christianity and mathematics. In particular, until late in the nineteenth century, axioms in mathematical theories were usually regarded as self-evident propositions; hence all the worry over the centuries about proving or replacing Euclid's parallel axiom. The pair following relate to truth claims, the second with cognitive import.

CtgtM: Christianity invoked as the ultimate *target* of (mathematical) knowledge without, however, being involved in its content. God may have made the integers, but the Bible plays no role in, say, finding properties of the primes. This kind of link takes an ontological form when mathematics is said to be used within, say, mechanics and physics to help disclose the "Design of the Universe" and "First Cause," or reveal "God's Book," to quote three metaphors; there was also epistemological versions, which would regard a *theory* as dedicated to God, or expressing His truths. Such positions seems to have been commonly held by European mathematicians, though often not explicitly mentioned—in contrast to the routine though doubtless sincere declaration "In the name of God . . ." at the head of an Arabic text.

CfrmM: Christianity playing a role in the *formation and development* of a mathematical theory cognitively, unlike the kind just described. I note a case in ¶13 involving continuity and infinity.

A very important type of this kind is where the (Christian) God is invoked to *guarantee* the truth, rigor, or generality of at least one part or principle of a mathematical theory. Mechanics seems to have been the branch of mathematics most often involved, especially in connection with cosmology and physics.[6] While not claiming to establish existence as in kind MexgC, this kind goes further than the emulation of CemuM; however, analogies may be deployed, as we shall see in ¶12.

This kind is converse to MfrmC (though I have not noticed a case where mathematics was invoked as a guarantee for Christianity). The two kinds can occur together when, for example, a mathematical theory uses numbers that have already been given status in Christian numerology under MfrmC. Isaac Newton (1642/43–1727)

6. For a selection of medieval examples of this type, see the texts in Grant (1974, esp. pp. 554–568).

seems to provide an example. He adhered to Arianism, a faith heterodox without being apocryphal, holding a seven-point creed that included a Trinity composed of Christ between God and man rather than the Orthodox Father, Son, and Holy Ghost. Numerology is surely in place here: the status of three was noted in MfrmC, and seven also has a very long numerological history, with massive Greek advocacy.[7] These integers have both cognitive and noncognitive effects on his work. It is surely no accident that his *Principia* (1687) and his *Optics* (1704) were written in three books, the latter also in seven parts (and both, especially the *Principia*, organized Euclid-style into numbered definitions and propositions). Further and more important, the mechanics of his *Principia* is based on three laws of mechanics, which he must have known were insufficient for the theory developed.[8] Similarly, in the *Optics* he advocated a spectrum of seven colors in the rainbow and in his rings, relating the colors to the main intervals in the musical scale (Gouk 1988), and seeing also a connection with "Apollo with the Lyre of seven strings."[9] Maybe seven was chosen not only to challenge the sextet of colors asserted by René Descartes; indeed, he confessed himself that the extra color, indigo, was hard to see, and even that "my owne eyes are not very criticall in distinguishing the colours"![10] Again, his theory of the motions of the Moon, which was not based on his mechanics, used a mathematical expression of seven terms (Kollerstrom 2000). This sensitivity to numerology is entirely in line with his deep involvement with alchemy.

5. Social and Institutional Influences of Christianity

CprsM: Christianity affecting the *prosecution or presentation* of mathematics but not its content. There are many examples of positive influence here, such as the structuring of a theory in a book or paper in some way that does not affect its content. Apparent examples in Newton's books were just noted; a similar one appears in ¶8. Another type of case is providing opportunities to develop mathematics at all; it also has a negative version, of preventing its prosecution. Other negative cases include European Christian resistance in the High Middle Ages to studying the pagan Greeks, in any domain of knowledge; in Europe it was gradually overcome,

7. In a series of important but forgotten writings of the 1900s in the Philologisch-historische Klasse of the Abhandlungen der Königlichen Sächsischen Akademie der Wissenschaften zu Leipzig, W. H. Roscher surveyed this literature in great detail: see, for example, 1904 on 7 and 9.

8. See the discussion of Newton manuscripts in Westfall (1971, ch. 8).

9. Quoted from a manuscript in McGuire and Rattansi (1966, 115); also in Cohen and Westfall (1995, 102).

10. Newton to Henry Oldenburg, December 7, 1675, in Newton (1959, 376).

but in Russia the opposition of the Orthodox Church to all sciences was very strong until Peter the Great founded his Academy of Sciences in Saint Petersburg in the 1720s, with the young Daniel Bernoulli (1700–1782) and Leonhard Euler (1707–1783) from Switzerland among the early members. An example from the nineteenth century is recorded in ¶14.

CtskM: Assignment of a *task* or a problem by a religion requiring mathematics for its fulfillment or solution.[11] A well-known Christian example was the construction of calendars, especially for determining Easter.[12] Indeed, the omission of zero from the year numbering when it was established in the +sixth century exemplifies kind MfrmC in that it seems to have been motivated by misidentifying zero with nothing (§12.9). Concerning artefacts, perspective and design were encouraged in Western painting (where purely artistic stimuli were also significant) and in mechanics by the construction of churches and especially cathedrals.[13] Among other types of case is the medieval concern with the possibility of angels flying with infinite speed (although contemporaries such as Nicole Oresme [1323?–1382] urged that science and religion be separated), and then efforts to Christianize the ancient doctrines of the "music" (better, "harmony") of the spheres,[14] which is based on properties of ratios. An unusual later example is the attempt made by John Craig(e) (d. 1731) in 1699 to determine the decline in the level of belief due to its successive transmission among adherents.[15]

Similar cases can be found on other religions; an important one in Islam was calculating the *qibla* angle at a given location for a worshipper correctly to face Mecca (King 1994). Religion is a *necessary* source in this kind; for example, without Christianity there is no Easter to determine. They differ from cases in kind MfrmC where, say, the fish bladder property (¶2) pertains in geometry independently of any religion. Maybe some religious position *stimulated* the interest discovery of such a theorem or property in the first place, though since most come from deep antiquity or secret(ive) contexts, historical evidence is very hard to find.

11. A curious example is the corroborated predictions of the Hebrew Bible described in Drosnin (1997). The absence of vowels in Hebrew must increase the chances of coincidences.

12. Francoeur (1842) gives a detailed catalog of the main properties of the Jewish, Gregorian, and modern calendars, and rules for their determination.

13. Notable recent literature on these topics includes Field (1997), Radelet-De Grave and Benevuto (1995), and Sakarovitch (1998). There are also various relevant articles in Grattan-Guinness (1994, esp. pts. 1–2 and 13).

14. These and related topics are well reviewed in Knobloch (1994). They are now under detailed examination by O. Abdounur (University of Sao Paolo, Brazil).

15. See Nash (1991), which has an English translation of the original Latin text.

PART 2: EVIDENCE AND CASES, 1750–1900S

We come now to written manifestations of these kinds between 1750 and the early twentieth century; nuances appropriate to each context will be made. As mentioned in ¶1, the cases are very rare, and moreover few of them influenced later ones; but they include negative as well as positive influence (that is, reactions against a link as well as those in favor), come from both pure and applied mathematics, and involve both mathematical theories as such and individual mathematicians. They have been organized by a mixture of chronology and kinship of theme.

6. A Turning Point? Newton, Lagrange, and the Stability of the Planetary System

As is well known, one feature of the so-called scientific revolution during the seventeenth century was the attempts to separate science and religion (as already with Oresme). Many scientists and theologians followed the dictum, including after 1750; thus much of the later silence can be explained. However, disputes over geology and mineralogy from the late eighteenth century onward, and then the rise of Darwinism in the nineteenth century, show that the separation was impossible to maintain, either concerning explanations of phenomena or the susceptibility of theories to religious interpretation (Brooke 1991, esp. ch. 3). Further, some students of the physical sciences did not obey the dictum; let us consider one of the most eminent.

An important case of influence of Christianity on Newton's conception of mechanics concerns the stability of the "System of the World." It did not seem to be provable from Newton's laws; the basic path of a planet *P* was determined by the central force from the Sun, but the perturbations off it caused by the forces coming from other planets could be large enough either to send *P* elsewhere in the ecliptic or to give it a quite different inclination. The accidental nearby passage of a comet might produce similar effects.

For Newton these possibilities did not cause qualms: God would arrive to restore equilibrium should danger attend *P*, and hence God exists, CfrmM-style.[16] (This argument might be convertible into a proof of the existence of God of kind MexgC, though for Newton and many others that was already assured.) Indeed, it

16. See Kubrin (1967). I do not (need to) consider other aspects of Newton's religious position in this chapter; a good survey, with further references, is provided in Snobelen (1999).

was thanks to the "Divine Arm" that the planets moved the same way around the Sun, and in nearly circular orbits.

Similarly, Euler, as a good Christian "sav[ing] divine revelation from the objections of freethinkers" (quote from the title of his 1747 work) would have held the same view of stability; he implied as much in his philosophical "Letters to a German Princess" (1768) when seeing "perfections of the highest degree" in the organization of the planetary system (Euler 1768, esp. letters 60–61). However, he also provided the mathematical means to articulate an alternative, secular, position.

Perturbation theory was technically a very cumbersome affair, since inverse powers of distances in various directions and angular orientations have to be taken together (and vectorial mathematics had not yet been invented). To make feasible both this and some other problems in celestial mechanics, Euler realized during the 1740s that if the distances were expressed in polar coordinates, then the sum of their appropriate powers could be rendered as power series of the angles pertaining to P and the other planets, and thence via Cotes-type theorems as a trigonometric series in multiples of these angles (the form $\sum_r a_r \cos rx$; C. A. Wilson 1980). (These series are not to be confused with Fourier series.) In the mid 1770s Joseph-Louis Lagrange (1736–1813) adapted this approach by replacing the angles with new independent variables that allowed him to cast into an antisymmetric form the system of equations of motion, stated according to Newton's second law. Then, assuming also that the planets moved in the same direction around the Sun, he reduced the stability problem for P to a form that, in terms of matrix theory, which this work would help to create later, required proof that all the latent roots of a certain matrix associated with the coefficients of the expansion for P were real numbers, and that their attached latent roots took only real-valued components. He formulated this method for a collection of mass points in general; P. S. Laplace (1749–1827) modified it more specifically for the planetary system.[17]

This analysis still left unresolved a related question: the *locations* of the orbits. A striking presage of the later paucity is evident in Newton: in the first edition of the *Principia* he declared that God had chosen the locations,[18] but he withdrew this *sole* explicit mention of God from the later editions, and the Bode-Titius Law was an *empirical* proposal of the 1770s.

17. For a summary with context, see esp. Grattan-Guinness (1990a, 324–329). Matrices are used here for brevity; Lagrange and Laplace worked on the quadratic and bilinear forms associated with the matrix. On this and the later history of the problem, see Hawkins (1975).

18. See *Principia,* 1st ed. (1687), book 3, prop. 8, cor. 5. For discussion, see I. B. Cohen (1992).

7. God and Non-Newtonian Mechanics in the Nineteenth Century

While Newton's laws soon became the principal means of working in celestial mechanics, they did not enjoy the same prestige within mechanics in general, especially on the Continent, where most major developments occurred after 1750. An alternative theory stressing energy had gained ground, not least through the advocacy of *vis viva* in the 1690s by Leibniz; and from the 1740s a third line came through Jean d'Alembert (1717–1783), with a principle named after him that claimed to reduce dynamical situations to statical ones (Grattan-Guinness 1990b).

In addition, at that time the principle of least action was expounded by Pierre Maupertuis (1698–1759) and Euler.[19] It asserted that in a mechanical situation, the "action" integral $\int_U m\,v\,ds$ of a system of masses m moving at velocity v along the path s over some curve U was minimal relative to all possible curves (Pulte 1990). To the discomfort of others, the principle gave mechanics approach a strongly teleological character, in that the path of least action seemed to be predetermined. Perhaps for that reason, Maupertuis asserted, as a "metaphysical" principle, that God was the guarantee of the principle—a "proof" CfrmM-style of both its truth and its generality. Euler defended "notre Ill. Président" of the Berlin Academy, although he relied more on mathematical concerns and on authorities such as Leibniz in advocating the principle. But in some of his later contributions to mechanics, such as the basic equations of elasticity theory and of hydrodynamics, he made no use of the principle, for he was finding *new* equations, a task for which teleological theories are usually ill suited.

The principal mathematician in the development of this kind of mechanics during the eighteenth century into the "variational" or "analytical" tradition was Lagrange (Fraser 1983). However, as with the stability problem, he made no appeal to Christianity for the generality or truth of d'Alembert's or the least action principles, or of the principle of "virtual velocities" that he adopted later (Grattan-Guinness 1990b, chs. 5–6). This trait not only characterizes Lagrange's mathematics but also locates part of his influence: he wished to secularize mechanics, and indeed mathematics in general. In his lifetime he was called an atheist, although I suspect that if Thomas Huxley's word "agnostic" had been available at the time, it would have been more appropriate.[20]

19. For a good discussion preceding a reprint of many of the main texts, see Fleckenstein (1957).

20. Note the little-known deathbed conversations of Lagrange reported in Grattan-Guinness (1985).

The same judgment applies to Laplace; his famous attributed remark to Napoléon in the early 1800s about not needing the hypothesis of the existence of God in his astronomical theories conforms to agnosticism, although he may have phrased himself thus in order to accommodate the official atheism of his audience.[21] Both then and later he spoke of successful theories as "exact," not true (Gillispie 1998, esp. chs. 24–26).

Striking national differences attend these developments of mechanics, mostly in connection with kind CfrmM (especially guarantee). In Britain, Newton's influence had given celestial mechanics a religious flavor absent from Continental versions; there Newtonian mechanics was less warmly received, with central forces accepted but the inverse square law treated circumspectly (Guicciardini 1999). By contrast, the preference of Newtonian mechanics in Britain gave small place to least action, and thus to the invocation of God. However, as we shall see in ¶9 and ¶10, British applied mathematicians were to contribute much of the modest interest in roles for Christianity in the nineteenth century.

8. Cauchy against the Secularization of Mathematics

By the 1780s France was by far the leading mathematical country, with Lagrange (who moved to Paris from Berlin in 1787) and Laplace among the senior figures. One of the main cultural features of its science is the virtual absence of religious talk, *especially* in spectacular problems such as the stability of the planetary system. The earlier efforts of the French Enlightenment to reduce the status of Christianity are well known;[22] Voltaire's satire of Leibniz in *Candide* (1758) wonderfully encapsulates changes in attitude. The French Revolution of 1789 heightened the status of France still further, especially with the many new and existing institutions for higher education maintained in science and engineering. The Church was sidelined until Napoléon's opportunistic reversion to Catholicism in 1806; but the secular

21. I have not found an authoritative source for Laplace's remark, but the story is quite credible; some short undated manuscripts rather skeptically appraising Christianity are conserved in his *dossier personnel* in the Académie des Sciences. On the context see Hahn (1986).

22. A significant factor may lie in Freemasonry, which enjoyed worldwide influence at that time (for example, in the founding of the United States). Partly because of its obsession with secrecy and much loss of documentation, its historical place is poorly understood; for an important pioneering attempt using archives of various secret societies, see Jacob (1991). The links between mathematics and faith are clear in the study of perspective theory by Brother Brook Taylor (1685–1731), and later the descriptive geometry of Brother Gaspard Monge (1746–1818), where stone cutting was one of his favorite applications. Otherwise, however, the effect of Masonhood upon the work of a scientist may be hard to specify; for example, I have not noticed any particular consequences for Brother Laplace (1749–1827). See also chapter 16.

presentation and development of mathematics was not affected, not even after the Bourbon Restoration of 1816, when King and Church were at center stage (Grattan-Guinness 1990a).

However, then there flowered an extraordinary counterexception. A.-L. Cauchy (1789–1857) was born in the year of the French Revolution, but he became one of its most passionate opponents, affirming Catholicism to a fanatical degree and adhering ardently to the Bourbons from their restoration. Their reign from 1816 to 1830 corresponds exactly to a phenomenal amount of brilliant mathematics from him (Belhoste 1991, esp. chs. 8–11). His most substantial achievements during that period included a reformulation of mathematical analysis and the calculus based on limits, which came eventually to dominate over all other approaches; the invention of the calculus of complex variables and their integrals; and the basic theory of linear stress-strain elasticity theory, with applications to optics. He produced work that fills about 12 of the 27 quarto volumes of his (almost complete) collected works, and most of the best items. They include five textbooks, several pamphlets, and such a production of papers that in 1826 he even started his own journal, *Exercices de mathématiques*, and managed to produce 32 pages per month until near the end of the decade. Then the Revolution of July 1830 threw out the Bourbons; so he followed them into exile for eight years, serving as mathematical tutor to their pretender to the throne.

Cauchy the mathematician, Cauchy the royalist, and Cauchy the Catholic were united to an extraordinary degree: truths in heaven and truth in applied mathematics (a mixture of kinds CtgtM and MemuC), and order in society and order in mathematics, to which altar he consciously brought new standards of rigor, even pioneering the systematic numbering of formulae in papers and books. He presented his version of the real-variable calculus in his *Résumé* of the course given at the Ecole Polytechnique in the important Christian number of 40 lectures, 20 each on the differential and on the integral calculus; in practice, he gave at least 50 lectures each year.[23] One of his theories concerned continuity, where his stress on an unbreakable sequence of values or states followed a Jesuit formulation as presented by predecessors such as Ruggiero Boscovich, S.J. (1711–1787; Baldini 1993).[24]

23. Cauchy (1823). Furthermore, each lecture was printed right to the bottom line of the fourth page of the signature Grattan-Guinness (1990a, 747–749), and moreover as a presentation of a theory based on limits; this feature is lost in the reprint in the *Oeuvres complètes, ser. 2,* vol. 4 (1899). The actual numbers of lectures given by Cauchy, recorded in the archives of the Ecole Polytechnique, are given in my book just cited.

24. While Boscovich doubtless subscribed to the CtgtM link, he does not seem to have used Christianity cognitively in his scientific writings. The same holds also for the mathematical and logical work of the Catholic priest Bernhard Bolzano (1781–1848).

Cauchy's activities during exile included a strange lecture course in Turin on physics, published posthumously (1868), which included passages on man and God. After his return to Paris in 1838, he participated in organizations concerned with Catholic education and refused to take oaths to the ruling authority.[25] His treatment of mathematical analysis was presented in textbooks from the 1840s by F. N. M. Moigno (1804–1884); but this abbot avoided Christian overtones even when describing continuity or infinitesimals (Moigno 1840, esp. intro. and lectures 1, 6). However, after Cauchy's death, he also published the lecture course (Cauchy 1868) on physics.

While Cauchy's Catholicism seems to have impinged cognitively on only a few aspects of his scientific work (CfrmM), it inspired his desire to raise the level of rigor in mathematics (CemuM), and to see applications to the physical world as asserting truths about it rather than merely proposing hypotheses (CtgtM). Another effect of Catholicism *on* him was negative: he chastised his colleagues in the Académie des Sciences for studying natural history and biology, for their objects of study were God's work alone![26]

9. Whewell and the Status of Mechanics

Unlike Cauchy, applied mathematicians in the nineteenth century not only usually avoided appealing to Christianity but also more often spoke of hypotheses in their theorizing. Especially from the 1840s, the certainties of Christianity became more out of reach, with alternatives to established Catholicism and Protestantism flourishing.

In this context William Whewell (1794–1866) is an intriguing British figure. As mathematician, scientist (his word), and Orthodox Christian, with a lifelong career at Trinity (sic) College Cambridge, he affirmed and indeed extended the principles of natural theology, a doctrine of medieval origin where Christianity absorbed aspects of Platonism and Aristotelianism to reveal (the existence of) God through the study of nature.[27] He pursued this position as an alternative to the growingly

25. See the revealing hagiography in Valson (1868, chs. 13–16).

26. See Belhoste (1991, esp. chs. 8–11) and Valson (1868, chs. 12–16). Cauchy also expressed his Christian fervor in poetry, some of which is transcribed in Valson (ch. 11).

{A contrast to Cauchy is provided by the reticence of A. J. C. Barré de Saint-Venant (1797–1880), who wrote a survey of mechanics from the point of view of the conservation of energy, to which he gave a theological interpretation; but he left it in manuscript, publishing only a brief extract in 1842 (Darrigol 2001).}

27. During the century natural theology became an umbrella position; the internal differences lay more in the natural than the physical sciences (Corsi 1988, ch. 15).

popular secular utilitarianism of his day. Especially concerning mathematics, his aim was not only epistemological but also pedagogic: the "liberal education" (his phrase) provided by Cambridge University, in particular the Mathematical Tripos, was for the benefit of its students.

As a philosopher Whewell admitted roles for both empiricism and ratiocination and studied the relationship between contingent and necessary truths. Mathematics was a prime ground for his exegesis, especially mechanics, and he adopted kind CemuM in seeking to demonstrate the truth of mathematical theories in emulation of Christian truths. Mathematics then launched a ladder of certainty up through the physical sciences (especially astronomy) to the natural sciences, and even further to consider man's place in nature and "come to believe in God and the Governour of the moral word" through the use of "Reason or Revelation."[28]

In applying celestial mechanics to natural theology, Whewell hoped to convince, MexgC-style, that the intricacy of the mechanism of the heavens exhibited the existence of a designing God. Curiously, while duly stressing the teleological aspects of natural theology, he preferred to use Newton's laws rather than the more congenial principle of least action (¶7); presumably Trinity College associations took prime place!

Whewell's use of kind CemuM shows that for him Christianity was preeminent: while science could inspire questions, the answers lay only in Christian revealed truth; thus kind CtgtM was also in play. Perhaps this was the reason for his surprising silence over the role of mathematics in Christian architectural design (¶2): his *Architectural Notes on German Churches* (1835) is notable both for his detailed knowledge of architecture and *lack* of concern with the religious significance of geometrical designs, of which he must have been aware.

Whewell granted two levels of utility for mathematics: training the (student's) mind "liberal" style, and more elevated tasks such as studying God's "Forces" and "Laws of force" involved in the motions of the heavenly bodies, as "discovered by Newton." The quotations come from Whewell's colleague Thomas Worsley, master of Downing College, when considering in 1865 the Christian element in Cambridge education; while not citing Whewell, his position was very similar.[29]

28. A typical phrase from Whewell's Bridgwater Treatise on astronomy and physics (1833, 252). Among many publications by him on mechanics, see especially 1835 and 1837. From his massive *History* and *Philosophy* of the inductive sciences he distilled the passages most pertinent to natural theology in his 1846 work. On Whewell see Becher (1980) and Fisch and Schaffer (1991), especially the articles by Becher, Fisch, J. H. Brooke, P. Williams, and D. B. Wilson.

29. Worsley (1865, 75): he also hinted at the stability problem on p. 136. In ch. 2 he treated mathematics, and in chs. 3 and 4, the "cosmical symbol."

One of the very few students of mechanics who *explicitly* advocated similar links was the American Benjamin Peirce (1809–1880), a major figure in a country that was then still of little importance in mathematics. In affirming his belief, he saw mathematical theories as products of the Divine Mind (CtgtM); but even he rarely put his view in print. One example occurs in a 500-page textbook of 1855 on mechanics: a very short passage on the theme that the occurrence of perpetual motion in nature "would have proved destructive to human belief, in the spiritual origin of force and the necessity of a First Cause superior to matter, and would have subjected the grand plans of Divine benevolence to the will and caprice of man."[30] Even here only the rhetoric of CtgtM was involved, and maybe in a general deist sense; nowhere did he invoke any cognitive relations.

Peirce's most explicit religious assertions occurred in a posthumous work on "ideality in the physical sciences," where the first word connoted "ideal-ism" as evident in certain knowledge (1881, 165). However, even though it was "pre-eminently the foundation of the mathematics," he concentrated almost entirely on cosmology and cosmogony with some geology; even types of mechanical equilibrium were not discussed.[31]

10. Ethers and Mathematical Physics

One general connection from earlier times did endure throughout the late eighteenth and nineteenth centuries, though forged more closely with physics than with mathematics. In electricity and magnetism (and, after 1820, in their interactions), theories were still often based on assuming the existence of the ether; and for some figures this ubiquity pantheistically showed God at work.[32] Cauchy affirmed this position at the end of his physics lectures of 1833 (¶8). In Scotland Thomson also assumed the existence of the ether but seemed not to give it any religious connotation. But an allusion occurred in an essay of 1852 "On a universal tendency in Nature to the dissipation of mechanical energy," when he asserted that "it is most certain that Creative Power alone can either call into existence or annihilate mechanical

30. Peirce (1855, 31). The book was based on the Lowell Lectures given at Harvard University in 1879. The American enthusiasm at that time for Protestantism in the sciences is excellently conveyed in Schweber (1999).

31. Peirce was more guarded here on the authority of the Bible than in his earlier writings; for discussion see Peterson (1955, with some inaccurate page citations).

32. On this general theme see Cantor and Hodge (1981), especially the articles by J. Z. Buchwald, D. M. Siegel, and M. N. Wise. The important case of A. M. Ampère is convincingly argued by Caneva (1980); Cauchy also followed it in optics, as had Ampère's lodger A. J. Fresnel (Grattan-Guinness 1990a, ch. 13 and pp. 1036–1040).

energy" (Kelvin 1882, 511); however, he left more developed considerations only in draft (Smith and Wise 1989, 329–332). With P. G. Tait (1831–1901) he published a *Treatise on Natural Philosophy* (editions of 1873 and 1879), very comprehensive in its range and especially concerned with energy; but no links to Christianity of any kind were mentioned.

Similarly, G. G. Stokes (1819–1903), friend of Thomson and Whewell, kept secular his many contributions to mathematical physics, even in ether theory. However, he was much concerned with the status of Christianity; largely following Evangelical principles, he claimed authority for the Bible. In the early 1890s he delivered two series of Gifford lectures on natural theology, but *no* kind of link to mathematics was analyzed: he touched on Trinitarianism twice (MfrmC) but slid quickly away.[33] His main argument for the existence of God was that of design, such as that of the eye (an example sometimes put forward nowadays as evidence of *bad* design!). His appeal to the infinite was unusual: he opposed the doctrine of eternal punishment (normally affirmed by Evangelists) but defended conditional immortality.

Whewell, Thomson, and Stokes were graduates from Cambridge University, where the status of the Mathematics Tripos was strange from our point of view: a major course in an Established institution, with mechanics and some mathematical physics given major roles; nevertheless links to Christianity were rarely mentioned {(Warwick 2003)}. But an exception occurred when James Challis (1803–1882), the Plumian Professor of Astronomy and Experimental Philosophy, meditated at the end of his career on "Cambridge mathematical studies, and their relation to modern physical sciences" (1875). His stimulus was the anonymous book on the *Unseen Universe* (1875) published by Tait and Balfour Stewart (1828–1887), in which they tried to reconcile sciences (especially the physical ones) with Christianity.[34] Challis did not object to their enterprise as such, but he faulted their failure to regard notions such as atom and inertia as absolutes; for him they "must be conceived to have been originally imposed, and to be maintained, together with constancy of form and magnitude by the will and power of the *Creator of the Universe*"—the kind CfrmM of link that Thomson had kept mostly in draft. Challis agreed with his opponents about the existence of the ether, for example, that its pressure was the source of force, but he complained about their construal of forces, and also atoms,

33. Stokes (1891–1893), vol. 1, 164 ("threefold aspect of the Divine body"), vol. 2, 225 (the Trinity). Stokes records several times his Evangelist disquiet over Lord Gifford's stipulation that revelation not be considered by the lecturer; but this cannot have determined his silence over mathematics. For discussion, see D. B. Wilson (1984).

34. See Heimann (1972) and Smith (1998, 247–260). Thomson and Maxwell noted also.

in terms of continuity (the principal notion in their argument), again preferring "the immediate will" of the Creator.[35]

The link of science via ethers continued until the early twentieth century, partly connected to efforts to reconcile science and Christianity through spiritualism (Turner 1974). However, E. T. Whittaker (1873–1956) omitted all manifestations from his *History of the Theories of Aether and Electricity* (editions of 1911 and 1951)—a very striking silence in itself, for he was not only a fine mathematician but also a fervent Christian who wrote at length on religious issues, without, however, invoking mathematics in any particular way.[36]

11. The Place of Probabilities and Statistics

The influence of the Enlightenment on mathematics was most evident in probability theory, especially in dominating France. But the developments after 1750 did *not* buttress Christian belief; on the contrary, discussion of the probability of miracles decreased greatly after 1750 (Daston 1988, 306–342). An important application was to determine the reliability of belief (as formed by judges in tribunals, for example) using reason alone. A major probabilist of that time was the Marquis de Condorcet (1743–1798); he doubted the merits of the interpretation of probability as a measure of belief and sought for it a more empirical basis.[37]

Probability theory continued to develop in the nineteenth century, though rather fitfully and with more opposition to its utility in the social sciences than had obtained during the Enlightenment period. Further, although a frequent argument for the truths of Christianity was the high likelihood of a Designer, redolent of kind CtgtM but not MexgC, probability *theory* does *not* seem to have been invoked to elaborate or quantify the reasoning. The Irish writer James Wills (1790–1868) even sought in 1860 the "antecedent probability" of Christianity, and found it to be "the highest existing probability, the only solution of the known sum of things"; in particular it was "very high" that the Creator must have had a "distinct purpose" or final end in view, with the "human race" as the "main object" (Wills 1860, 163, 39).

35. Challis (1875, 76, 82). On his mathematical work, of moderate significance, see Burkhardt (1908, 1056–1107).

36. See Whittaker (1946; cited in ¶1) on the kind MexgC (proving existence); and (1942), largely on CfrmM (especially guarantee) since science could not account for creation. Sadly, there is no major study of any aspect of his work.

37. See Daston (1988, 211–219) and Baker (1975, esp. chs. 3–4). Condorcet discussed "la persistance de l'âme" only in an undated manuscript published in 1994 (316–326).

However, he proffered no mathematical arguments, Bayesian or otherwise.[38] Similarly, while evolution theory contained elements of statistical *thinking* (for example, in the notion of variation), no mathematical *theory* seems to have been involved in discussing its conflict or reconciliation with Christianity.[39]

A noteworthy link between probability theory and Christianity was of kind CtskM. In a converse to Craig(e) on the decline of belief (¶5), in 1872 Francis Galton (1822–1911) attempted to detect the efficacy of concerted prayer by comparing data from target and control groups of appropriate persons from "the praying and non-praying classes" (Galton 1872).

Various of Galton's reflections on statistics and its data were noted by a distinguished probabilist, John Venn (1832–1923), in his *The Logic of Chance*, which appeared in editions between 1866 and 1888. But Venn did not include the prayer experiment; indeed, the most explicit connection made between Christianity and probability theory was to *reject* CtgtM-style the "old Theological objection" that admission of chance denied the Creator overall control. Again, Christian texts were not cited in the chapters on testimony and on the "Credibility of extraordinary stories."[40] Yet he was a fervent believer; soon after the publication of the first edition of this book, he gave the Hulsean lectures on divinity at Cambridge, choosing as his topic "some characteristics of belief." Again probability theory played only a passing role, and his main concerns lay in the complications of the historical evidence and the vagaries of emotion.[41] Venn's other main research interest lay in logic, to which we now turn. England again is very prominent.

12. Boole and the Ecumenism of the Algebra of Logic

The secular status of probability theory is evident also in the contributions made to it by George Boole (1815–1864); and his case is especially striking because he *did* invoke his religious stance when developing logic. He formed an algebra to symbolize logical reasoning, in which he conceived of a universe of discourse 1 and split it into complementary parts x and $(1 - x)$. Part of his inspiration seems to have been his youthful awareness of Jewish monism, and in his mature development of the theory,

38. The general histories of nineteenth-century probability theory of which I am aware do not discuss links with religions at all; but this may reflect a lack of concern by the historians themselves.

39. Among the available literature, I am guided for Britain by Moore (1979, esp. ch. 9).

40. Venn (1888), 236–238, on objection; note also p. 260 on the Creator and the distribution of stars, and chs. 16 and 17.

41. Venn (1880): note (only) the appendix on Pascal's interest in probability theory and the impossibility of proving the existence of God by reason alone (pp. 115–126).

he linked it to ecumenism, that is, a single unifying Christian doctrine affirmed over and above the xs and $(1- x)$s of the numerous disputing brands. He held this position from exactly the time, the late 1830s and the 1840s, when the dissenting faiths were rising very rapidly in prominence in Britain. It lay in tune with Unitarian truth, construed as an invariant relative to all means of human examination, which gained considerable popularity at this time. It furnished further links via logic to truth; for example, he invoked his universe again in writing "$A = 1$," which asserted that the proposition A is true for all time.[42] Thus he exemplified kind MemuC in the analogy between ecumenism and his logical universe; and if the Jewish influence was formative, he also exhibits kind CtgtM (guarantee).

In his later years Boole became very enthusiastic for the presentation of logic by the French theologian A. J. A. Gratry (1805–1872), who stressed claims such as God as the source of truth (kind CfrmM), and topics such as nullity versus unity, universal laws of thought, and the exercise of the human mind (Grattan-Guinness 1982). In the late 1850s he and his wife became profound admirers of Frederick Denison Maurice (1805–1872), who in 1853 had been dismissed from his post as professor of divinity at King's College London for advocating ecumenist instead of party-line Trinitarian Christianity; Boole had Maurice's portrait placed by his deathbed.

However, as usual with our figures, Boole rarely mentioned religion in his writings; in particular, he did not analyze the mind itself as a product of God's creation, and his logic was concerned only with the products of thought, not with the processes of their formation. But in his *The Laws of Thought* (1854), he analyzed passages from the writings of Samuel Clarke and Baruch Spinoza, which surely did *not* just happen to deal with properties of the one and only God. Again, in his final paragraph he saw the "juster conceptions of the unity, the vital connexion, and the subordination to a moral purpose, of the different parts of Truth, among those who [. . .] profess an intellectual allegiance to the Father of Lights" (Boole 1854, ch. 13 and p. 424), this phrase a code term for his kind of dissenter to refer to the unifying Godhead.

This aspect of Boole's logic was not durable; his widow seems to have been its only follower.[43] Among the others, Venn did *not* adopt it in his strong advocacy of Boole's logic (Venn 1881). Indeed, in the strong revival of logic in Britain during the nineteenth century, he seems to be the only significant figure using cognitive rela-

42. On this theory of truth relative to mathematics, see Richards (1988, esp. ch. 1). Earlier British concerns with truth in the context of algebra(s) are noted in (Pycior 1997, 11).

43. See Laita (1980). On manifestations of Christianity in Boole's (unpublished) poetry, see the examples and discussion in MacHale (1985, ch. 12).

tions with Christianity. The founding work of the revival was the *Elements of Logic* by the Reverend Richard Whately (1787–1863) when it first appeared in book form, in 1826; it was in its fourth revised edition in 1832 (the year after he was appointed archbishop of Dublin) and reached its ninth in 1848. But he drew on Christianity only for a few biblical passages for logical analysis; he never invoked God as the creator of the human mind itself and of its powers of reasoning, with which logic was then often taken to be closely linked.[44]

13. Cantor, the Transfinites, and the Creator

Boole invoked Christianity in a theory of a very foundational character. The same appeal was made by Georg Cantor (1845–1918), the principal creator of set theory, including a (for him, the) theory of the actual infinite. His "Mengenlehre," as he came finally to name it, grew out of mathematical considerations; in fact, it was partly inspired by Cauchy (and, following him in Germany, Karl Weierstrass) using the theory of limits to increase the level of rigor in mathematical analysis (¶8).

At first, in the early 1870s Cantor had called his transfinite numbers $\infty, \infty + 1, \ldots$ "infinite symbols" as he had no basis for these mathematical objects, especially concerning ". . . ." But in the early 1880s he found a grounding for his symbols to make them ordinal *numbers* "$\omega, \omega + 1, \ldots$," in a theory in which he also developed their arithmetic; and he became aware, and indeed deeply informed about, predecessors who had affirmed the actual infinite, often in a Christian context. Especially drawn to Giordano Bruno and (like Boole) to Spinoza, his metaphysical framework included a central Christian component,[45] for he saw his theory of the infinites as explicating and clarifying Christian doctrines of infinitude (kind MfrmC). The audience for this aspect of his theory increased after 1879, when Pope Leo XIII issued an encyclical requiring Catholics to pay much more attention to the development of science, and various philosophers and theologians of that persuasion corresponded with him.[46]

Cantor refuted the common belief that the actual infinite was God's work, not man's. But he was also aware of a third kind of infinitude, the "absolute infinite" of the supposedly greatest cardinal number *G*, and found that two paradoxes fol-

44. Whately (1826). The book was based upon an article published in the *Encyclopaedia Metropolitana* in 1823, which did *not* gain much attention. On his work, see Neil (1862), Van Evra (1984), and Corsi (1988, esp. chs. 7 and 14).

45. Heuser-Kessler (1991); Bandmann (1992, pt. 1).

46. Dauben (1979, esp. ch. 7). {See now also the edition (Tapp 2005) of Cantor's correspondence with theologians.}

lowed—both $G = G$ but also $G > G$, and $G > 2^G$ but also $G \leq 2^G$. (The corresponding version of the first paradox also awaits the supposedly greatest ordinal number.) Cantor invoked Christianity (kind CfrmM) to avoid *this* kind of infinity: it "appears to me in a certain sense as an appropriate symbol of the Absolute" (1883, 205), and if humanity played with it, so much the worse for them. By the late 1890s he had decided that humans should stick to "ready" (*fertig*) sets, for which it "is possible that [. . .] *all its members as being together* without contradiction, thus to think of the set itself as a *composed thing for itself*."[47]

As with Boole, these aspects of Cantor's theory were not durable. However, they arose again in 1904 when Bertrand Russell (1872–1970) argued against the need for an axiom of infinity in set theory, on the grounds that infinite sets could be produced by deploying mathematical induction. He was replying to the claim made by the American mathematician and Christian Cassius J. Keyser (1862–1947) that alleged proofs of the existence of such sets begged the question. Keyser had good mathematical points to make, but he saw the actual infinite as beyond human capacities and belonging to God's realm (kind CfrmM). His exchange with Russell occurred in the *Hibbert Journal*, a "quarterly review of religion, theology and philosophy" (and deserving to be *far* better known by historians of science and of religion).[48] By 1906 Russell admitted that an axiom was needed after all, though as a fervent atheist he had changed his mind because of mathematical reasons similar to Keyser's handled without the theological overlay (Grattan-Guinness 1977, 24–25).

Later in the decade Keyser (1908–1909) described in the *Journal* many features of Cantor's theory in a (rambling) "message of modern mathematics to theology," akin to MfrmC in informing Christianity and matching Cantor in ambition though not in exposition. More important, shortly earlier he had arranged for Benjamin Peirce's son, the polymath Charles S. Peirce (1839–1914), to contribute to the *Journal* in 1908 a variant on kind MexgC in the form of "A neglected argument for the reality of God"; it was based on the assumption that "musement" upon God could persuade the muser of his reality by means expressible within Peirce's triadic (but not necessarily Christian) metaphysics.[49]

A superior version of Keyser's type of interest was made in 1914 by another

47. For commentary see Purkert and Ilgauds (1987), 150–159, with letters on pp. 224–231; quotation on p. 226.

48. Keyser (1904, 1905); in between, Russell (1904). The journal also contains, for example, regular articles by Oliver Lodge on science and spiritualism; for context, see D. B. Wilson (1971).

49. Peirce (1908). Much later Kurt Gödel (1906–1978) proved in 1970 a theorem in modal logic showing the existence of an object possessing all "positive" properties, but he was reluctant to interpret it theologically as a case of MexgC (Gödel 1995, 388–404, 429–437).

mathematician and believer, the Russian Pavel Florensky (1882–1937). He included in a dozen letters on theology and belief not only set theory but also mathematical logic and a little probability theory, in contexts such as infinity, paradoxes, and identity. The Revolution three years later stifled his enterprise, and 20 years after that decided on his murder.[50]

PART 3: REVIEW

14. The Variety of Links, the Paucity of Cases

In the first part of this chapter a taxonomy of links between mathematics and Christianity was proposed, with influence passing in each direction. The purpose was to bring via the attendant distinctions some clarity to a complicated web of connections that has received little analysis.

In the second part of this chapter, cases were described in mathematics where some kind of role was assigned to Christianity between the mid-eighteenth and early twentieth centuries. Most belong to the kinds CfrmM (formation), CemuM (emulation) or CtgtM (exemplar or guarantee of certainty), and some come from CtskM (task) and CprsM (presentation). Theories with a large foundational component were involved relatively often (Cauchy, Boole, Cantor). The affirmations were made either covertly or by allusion (Boole) or marginally (Cauchy, Peirce, Challis), in forms that many contemporaries may well have thought privately; Thomson rarely put his position into print.

The other main lessons in this second part was the *rarity* of such concerns, and their isolation from each other; and none led to a durable influence on successors. Hence the survey was unavoidably discontinuous. Although many mathematicians for this period were (presumably) believers in some version of Christianity, very rarely did they mention links or uses, even noncognitive ones, in their publications. The point is not that they failed to *declare* their Christian belief, but that it was *immaterial* to their mathematical purposes, even in physical applications, although the newer branches such as mathematical physics are not more detached from Christian interpretation than are the older ones.

A striking English example is Baden Powell (1796–1860), professor of geometry

50. Florensky (1914), esp. ch. 6 (paradoxes) and apps. 14, 18 and 19 (identity), 15 (set theory), 17 (irrational numbers), and 23 (probability theory). Interestingly, he skipped by numerology (pp. 45, 302–204). On his influence on a fellow Christian and set theorist, see Demidoff and Ford (1996).

at Oxford University, who wrote extensively on Cauchy's version of the wave theory of light (¶10) but did not even highlight its etherial aspects never mind its pantheism. His silence is striking because throughout his career he was involved with Christian controversies, defending Anglican orthodoxy against Unitarianism and upholding the distinction between science and religion by allowing Christianity to control the interpretation of scientific theories. He also sought to improve science education at Oxford, but on the grounds of training the mind and emphasizing rigor, especially in the alleged certainty of mathematical theories; however, though a former student of Whately, he did not advocate the study of logic as such.[51]

A similar and more prominent figure is James Clerk Maxwell (1831–1879); his passionate belief guided his stress on the limitations of science, and maybe his advocacy of nonmechanical theories of phenomena and analogies between the physics of things and the mathematics of thought-objects.[52] Nevertheless, he omitted Christianity almost entirely from his scientific writings: even his popular book on mechanics and physics, *Matter and Motion* (1877), is secular in expression, though a posthumous edition prepared by J. J. Larmor was published in 1920 by the Society for Promoting Christian Knowledge!

Conversely, Charles Babbage (1791–1871) had explicitly affirmed his Christian belief and considered possible links between science and religion; but he eschewed mathematical arguments to argue for his stance, and indeed offered a clever nonreligious interpretation of miracles. While sympathetic to deism, he did not associate Christianity with his own mathematical work, although most of it treated new algebras in the 1810s and 1820s, a time and in a country where truth was a serious concern in mathematics (the context later for Boole also, in ¶12).[53] The same lack of discussion is evident among mathematicians in this century who considered Christianity/deism and science.[54]

51. For an excellent study of Powell and his time, see Corsi (1988, esp. chs. 4, 5, 9 and 10).

52. See Maxwell's letters to his wife in Campbell and Garnett (1882, 338–340); interestingly, his biographers rarely even mention his religious position. For discussion of the background, including the place of Scottish reformed theology, see Torrance (1984, ch. 6). Like Cauchy and Boole (notes 26 and 43 respectively), Maxwell wrote poetry though only occasionally with Christian sentiments, to judge by the selection given in Campbell and Garnett (see esp. pp. 594–601, 607–609).

53. See principally Babbage (1838, chs. 9–11) and (1864, ch. 30; kinds MexgC, CtgtM, and CfrmM; and ch. 29 and appendix [miracles]). These books appear respectively as vols. 9 and 11 of his works (Babbage 1989); vol. 1 includes his mathematical papers. On Babbage's views on the mind, see Ashworth (1996).

54. See especially Whitehead (1926), Weyl (1932), and Wiener (1964). The last two books are rather superficial; for example, as with much literature of this kind, even Whittaker's, they are contaminated by pop history. {In addition, two mathematician Gifford lecturers of the 1920s, E. W. Hobson (1856–1933) and E. W. Barnes (1874–1953; sometime bishop of Birmingham), followed

Keyser was a very rare (though waffly) case of *explicit* affirmation of faith. A much more important example, especially from the 1890s onward, is the Frenchman Pierre Duhem (1851–1916). He wrote about being both scientist and Catholic, and partly under the influence of neo-Thomism, which gained currency in France after the Papal decree of 1879 mentioned in ¶13 in connection with Cantor. A principal venue was the *Revue des questions scientifiques,* similar in role to the English *Hibbert Journal* (and also rather forgotten).[55] Again, he argued his position more on general philosophical grounds rather than from especial concern for mathematics.

15. The Prominence of the British Isles

Mention of Duhem breaks the British dominance of this survey. In general the links between mathematics and Christianity were considered far more in Britain than on the Continent, especially in France (despite Duhem).[56] To recall from ¶3 the similarities between the Bible and Euclid's *Elements,* it is surely no coincidence that concern with Euclidean geometry was much greater in Britain during the nineteenth century than in any other country, even to the extent that from the 1870s a professional split developed among mathematics teachers: orthodox followers of the *Elements* wished to preserve the order of propositions and their proofs, but dissenters wanted to admit other orders and proofs of theorems.[57] Similarly (or so it seems), the concerns with relationships between Christianity and sciences in general, such as natural theology, were greater in Britain than on the Continent (Brooke 1991, esp. ch. 6).

Part of the British preoccupation seems to be due to Whewell, whose position (¶9) was the most penetrating of all mathematicians'. Its concern with emulations

---

Stokes (¶10) in partitioning mathematics away from Christianity in their surveys of the sciences (Hobson 1923; Barnes 1933).}

Some later physicists with mathematical expertise affirmed religious positions; for example, Arthur Eddington (1882–1944) and James Jeans (1877–1946). On successor theorists up to recent times see Ruyer (1974). Modern cases include "theories-of-everything" science and anthropic principles, and often ambiguous remarks about God being a mathematician (of course he was: if he existed, he created mathematics anyway, or at least the capacity of mathematicians to produce it).

55. See Paul (1979, ch. 5), Jaki (1984), and Martin (1991). Duhem's relationship to neo-Thomism, somewhat disputed by his historians, will not be examined here. His affirmation of faith may have contributed to his failure to obtain a position in Paris, although he committed other professional sins such as taking seriously the history and philosophy of science, and specializing in applied rather than in pure mathematics.

56. {D. Cohen (2007) treats the religious element in British mathematics of the nineteenth century, but his treatment is often factually faulty.}

57. See Price (1994, esp. chs. 2–3). Those disputes are not to be confused with two far more important developments of the time: exposing the hidden axioms of Euclidean geometry, and popularizing non-Euclidean geometries.

was taken up by Thomas Birks (1810–1883); soon after becoming second wrangler and second Smith's prizeman in mathematics at Whewell's Trinity College, he wrote in 1833 on the "analogy between mathematical and moral calculi," with morality mediated by Christianity. Kind MemuC is evident when he looked "into the nature of moral certainty, we may not find, in the certainty of geometry, an attendant to guide us"; but analogy was restricted by the apparent fact that the axioms of geometry "no one denies" whereas those who wish to deny moral principles "most earnestly cannot always" do so. He also noted the presence of prejudice in moral discussion (allegedly) absent in geometry, but he could not identify an ultimate authority for mathematics corresponding to the "Divine Will" guiding morals.[58] I know of no such comparison in the Continental literature.

Twenty years later, in his career as a vicar, Birks reprinted the text at the end of a more extended comparison of the "revealed truth" with sciences, starting with mathematics and astronomy. The converse kind, CemuM, and CtgtM, are evident: "Wonderful Numberer, the Son of God" for arithmetic, while diagrams and formulae "are parables to teach us the mysteries of the kingdom of heaven" (1855, 14, 20). Indeed, mathematics in general is "bound to the throne of the almighty"; and "all its treasures of knowledge are hidden in Christ," not God (pp. 12, 23). In 1872 he was appointed to the chair of moral philosophy at Cambridge upon the death of Maurice (¶12), who had held it since 1866—an Evangelist to replace Unitarian Maurice.

As was noted in ¶9, Whewell was anxious to combat the rise of doubts over (Orthodox) Christianity in Britain. A leading doubter was the mathematician and scholar Francis Newman (1805–1897), whose confession in 1850 of the phases of his loss of faith led to a wide discussion. However, his motivations began with his inability as an undergraduate at Oxford to accept the 39 articles as truths; later ones included the unconvincing layers of doctrine imposed over "primitive Christianity," secondhand testimony such as Paul's, failed Christian prophecies, and the incompatibility with contemporary geology and biology. But he invoked no mathematical aspects or arguments, not even probabilistic ones.[59]

The greater British than Continental interest may mark a difference between predominantly Protestant and Catholic countries. In this context it is worth noting that almost all major Irish scientists in our period were Protestants (Nudds

58. Birks (1834), delivered December 16, 1833, and reprinted in 1855, 190–225; quotations from pp. 194 and 209.

59. Newman (1854, 3–9); later eds. to 9th (1874). Despite the considerable reaction, this episode is not analyzed at all in Chadwick (1975), which otherwise provides good background material.

and others 1988): it seems that their parents wanted them to advance to a profession maybe new in the history of the family, whereas Catholic youngsters were still usually counseled to see the seminary or nunnery as the prime aim for life (a negative influence, of kind CprsM). {Of especial note is the Protestant George Salmon (1819–1904), who in 1866 even changed appointment at Trinity College Dublin from mathematics to divinity; he kept religion entirely out of his textbooks and research (Gow 2000).}

Attitudes to Christianity in more "mixed" countries such as the German states could be studied usefully from this point of view. An oft-repeated remark by Leopold Kronecker (1823–1891), "the dear God has made the whole numbers, all else is man's work," seems to be only a clever quip made to a general audience, with its weight lying in the constructivistic philosophy of mathematics stated in its second clause more than the Christianity advocated CfrmM-style in the first one:[60] his published writings did not exhibit the latter sentiments.

Kronecker was a converted Jew. As a noncase of CprsM, during the nineteenth century Jews began to appear in mathematics (and science in general) after centuries of exclusion by Christians; but Judaism seemed not to affect the development or practice of mathematics. A significant figure is J. J. Sylvester (1814–1897) because he *practiced* his religion instead of converting to Christianity (and suffered an irregular career as a consequence; Parshall 1998, 2006); but his theories were not accompanied by any religious overtones, even though one of his specialities was invariant theory.

16. Final Considerations

Doubtless the cases reported in the second part of this chapter do not exhaust the occasions when Christianity was invoked in mathematical publications in the period at hand (although the index volumes for mathematics and mechanics of the *Royal Society Catalogue of Scientific Papers* do not disclose any papers on our theme for the literature of the nineteenth century). But the full list must be *minute* relative to the total production. Maybe the private correspondence of mathematicians reveals more concern with Christianity (or deism in general), though I have hardly ever noticed it in the numerous collections that I have used (for other purposes). After around 1750 mathematics seems to have become almost entirely secular, or at least religiously neutral, with Christian belief rarely in print though not out of

60. Kronecker made this remark in 1886 in a lecture at the annual meeting of the Gesellschaft Deutscher Naturforscher und Ärtze; it became known in the obituary (Weber 1893, 19).

mind—in striking contrast with the great growth of involvement of Christian and deist belief in the natural sciences. This situation obtained in all countries, and in both research and teaching; presumably it developed by consensus, though perhaps encouraged by the great prominence of France between the 1780s and the 1830s, and the influence of the Enlightenment and the stance of major figures such as Lagrange and Laplace, and also Condorcet.

Astronomy and physics seem also to have followed largely secular paths, presumably for analogous reasons; but the strong contrast with the nineteenth-century discussions in the natural sciences is hard to understand. If connections with Christianity are open again, why not revive the older ones to mathematics and the physical sciences? Was the Christian "defeat" over heliocentric astronomy in the sixteenth century decisive for *all* branches of mathematics and astronomy, and if so, why?[61] Most of terrestrial mechanics is unaffected by the issue of whether the Sun or the Earth is at rest; indeed, it is usually developed in a geocentric manner in that, for example, a coordinate system is "fixed" relative to the surface of the Earth. Across mathematics, theories continued to be created somehow and seemed to talk about something or other; yet only the eccentrics recorded in the second part of this chapter posited some theologically commodious source or inspiration to explain either feature.

The special complications involved in using mathematics in scientific theories (¶1) may have helped to discourage invoking links, but the near silence is puzzling. Understandably, then, neither topic addressed in my title has been well considered in the history of mathematics; but each deserves further study and maybe even integration into the vigorous general discussion of the history of links between science and religion.

BIBLIOGRAPHY

Ashworth, W. J. 1996. "Memory, efficiency, and symbolic analysis: Charles Babbage, John Herschel, and the industrial mind," *Isis, 87*, 629–653.

Babbage, C. 1838. *The ninth Bridgewater Treatise: A fragment*, 2nd ed., London: Murray. [Repr. as [1989, vol. 9].]

Babbage, C. 1864. *Passages from the life of a philosopher*, London: Longmans, Green, Roberts, and Green. [Repr. as Babbage 1989, vol. 11.]

Babbage, C. 1989. *The works of Charles Babbage*, 11 vols. (ed. M. Campbell-Kelly), London: Pickering.

61. This is the claim made by Odom (1966); but he (unwittingly?) discusses only Newtonian mechanics, and the bulk of his details refer to before 1750.

Baker, K. M. 1975. *Condorcet: From natural philosophy to social mathematics*, Chicago: University of Chicago Press.

Baldini, U. 1993. "Boscovich e la tradizione gesuitica in filosofia naturale: continuità e cambiamento," in P. Bursill-Hall (ed.), *R. J. Boscovich: His life and scientific work*, Rome: Enciclopedia Italiana, 81–132.

Bandmann, F. 1992. *Die Unendlichkeit des Seins: Cantors transfinite Mengenlehre und ihre metaphysischen Wurzeln*, Frankfurt am Main: Lang.

Barbour, I. G. 1997. *Religion and science: Historical and contemporary issues*, New York: Harper.

Barnes, E. W. 1933. *Scientific theory and religion: The world described by science and its spiritual interpretation*, Cambridge: Cambridge University Press. [The Gifford Lectures, 1927–1929.]

Becher, H. 1980. "William Whewell and Cambridge mathematics," *Historical studies in the physical sciences, 11*, 3–48.

Belhoste, B. 1991. *Augustin-Louis Cauchy: A biography*, New York: Springer.

Birks, T. 1834. *Oration on the analogy of mathematical and moral certainty*, Cambridge: Pitt Press.

Birks, T. 1855. *The treasures of wisdom, or thoughts on the connection between natural science and revealed truth*, London: Seeley, Jackson, and Halliday.

Bolt, R. 1960. *A man for all seasons*, London: Heinemann.

Boole, G. 1854. *An investigation of the laws of thought*, Cambridge: Macmillan; London: Walton and Maberly. [Repr. New York: Dover, 1958.]

Breidert, W. 1985. "Einleitung," in his (ed.), G. Berkeley, *Schriften über die Grundlagen der Mathematik und Physik*, Frankfurt am Main: Suhrkamp, 7–69.

Breidert, W. 1999. "Theologie und Mathematik: ein Beitrag zur Geschichte ihrer Beziehung," in M. Toepell (ed.), *Mathematik im Wandel*, Hildesheim, Germany: Franzbecker, 78–88.

Brooke, J. H. 1991. *Science and religion: Some historical perspectives*, Cambridge: Cambridge University Press.

Burchfield, J. 1975. *Lord Kelvin and the age of the Earth*, New York: Science History Publications.

Burkhardt, H. F. K. L. 1908. "Entwicklungen nach oscillirenden Functionen und Integration der Differentialgleichungen der mathematischen Physik," *Jahresbericht der Deutschen Mathematiker-Vereinigung, 10*, pt. 2.

Campbell, L., and Garnett, W. 1882. *The life of James Clerk Maxwell*, London: Macmillan.

Caneva, K. 1980. "Ampère, the etherians, and the Oersted connexion," *British journal for the history of science, 13*, 121–138.

Cantor, G. 1883. "Über unendliche, lineare Punktmannichfaltigkeiten," pt. 5, in *Gesammelte Abhandlungen mathematischen und philosophischen Inhalts* (ed. E. Zermelo), Berlin: Springer, 1932 [repr. Hildesheim: Olms, 1966], 165–209.

Cantor, G., and Hodge, M. J. S. 1981. (Eds.), *Conceptions of ether. Studies in the history of ether theories 1740–1900*, Cambridge: Cambridge University Press.

Cauchy, A.-L. 1823. *Résumé des leçons données à l'Ecole Polytechnique sur le calcul infinitésimal*, Paris: de Bure.

Cauchy, A.-L. 1868. *Sept leçons de physique générale . . .* , Paris: Gauthier-Villars. [Repr. 1885; also in Cauchy, *Oeuvres complètes*, ser. 2, vol. 15, Paris: Gauthier-Villars, 1974, 412–447.]

Chadwick, O. 1975. *The secularization of the European mind in the nineteenth century*, Cambridge: Cambridge University Press.

Challis, J. 1875. *Cambridge mathematical studies, and their relation to modern physical sciences*, Cambridge: Deighton, Bell.

Chase, G. B. 1996. "How has Christian theology furthered mathematics?" in J. M. van der Meer (ed.), *Facets of faith and science*, vol. 2, Lanham, Md.: University Presses of America, 195–216.

Chase, G. B., and Jongsma, C. 1983. (Eds.), *Bibliography of Christianity and mathematics: 1910–1983*, Sioux Center, Iowa: Dordt College Press. [Available on www.messiah.edu/departments/mathsci/acms/bibliog.htm.]

Cohen, D. J. 2007. *Equations from God: Pure mathematics and Victorian faith*, Baltimore: Johns Hopkins University Press.

Cohen, I. B. 1992. "The review of the first edition of the *Principia* . . . ," in P. M. Harman and A. E. Shapiro (eds.), *The investigation of difficult things*, Cambridge: Cambridge University Press, 323–353.

Cohen, I. B., and Westfall R. S. 1995. (Eds.), *Newton*, New York: Norton.

Condorcet, Marquis de. 1994. *Arithmétique politique. Textes rares ou inédits (1767–1789)* (ed. B. Bru and P. Crépel), Paris: Presses Universitaires de France.

Corsi, P. 1988. *Science and religion: Baden Powell and the Anglican debate, 1800–1860*, Cambridge: Cambridge University Press.

Darrigol, O. 2001. "Gods, waterwheels, and molecules: Saint-Venant's anticipation of energy conservation," *Historical studies in the physical sciences, 31*, 285–353.

Daston, L. 1988. *Classical probability in the Enlightenment*, Princeton, N.J.: Princeton University Press.

Dauben, J. W. 1979. *Georg Cantor*, Cambridge, Mass., and London: Harvard University Press. [Repr. Princeton, N.J.: Princeton University Press, 1990.]

Demidoff, S. S., and Ford, C. E. 1996. "N. N. Luzin and the affair of the 'National Fascist Center'" in J. W. Dauben and others (eds.), *The history of mathematics: States of the art*, Boston: Academic Press, 137–148.

Drosnin, M. 1997. *The Bible code*, London: Weidenfeld and Nicolson.

[Euler, L.] 1747. *Rettung der göttlichen Offenbarung gegen die Einwürfe der Freygeister*, Berlin: Sperer. [Repr. in *Opera omnia*, ser. 3, vol. 12, Zurich: Orell Füssli, 1960, 267–286.]

Euler, L. 1768. *Lettres à une Princesse d'Allemagne sur divers sujets de physique et de philosophie*, vol. 2, Saint Petersburg: Academy. [Repr. as *Opera omnia*, ser. 3, vol. 11, Zurich: Orell Füssli, 1960. Many other reprs., eds. and trans. listed on pp. lxi–lxx.]

Ferngren, G. B. 2000. (Ed.), *The history of science and religion in the Western tradition: An encyclopedia*, New York: Cambridge University Press.

Field, J. V. 1997. *The invention of infinity: Mathematics and art in the Renaissance*, Oxford: Oxford University Press.

Fisch, M., and Schaffer, S. 1991. (Eds.), *William Whewell*, Oxford: Oxford University Press.

Fleckenstein, J. O. 1957. "Vorwort des Herausgebers," in Euler *Opera omnia*, ser. 2, vol. 5, Zurich: Orell Füssli, 1957, vii–1.

Florensky, P. A. 1914. *Stolb' i utverzhdernie istini*, Moscow. [English trans. by B. Jakim as *The pillar and ground of truth*, Princeton, N.J.: Princeton University Press, 1997.]

Francoeur, L. B. 1842. *Théorie du calendrier et collection des lois des calendriers des années passées et futures*, Paris: Roret.

Fraser, C. 1983. "J.L. Lagrange's early contributions to the principles and methods of mechanics," *Archive for history of exact sciences, 28*, 197–241.

Galton, F. 1872. "Statistical inquiries into the efficacy of prayer," *Fortnightly review*, new ser. 12, 125–135. [Repr. in *Inquiries into human faculty and its development*, 1st ed., London: Macmillan, 1883, 277–294.]

Gillispie, C. C. 1998. *Pierre Simon Laplace: A life in exact science*, Princeton, N.J.: Princeton University Press.

Gödel, K. 1995. *Collected works*, vol. 3, Oxford: Clarendon Press.

Gouk, P. 1988. "The harmonic roots of Newtonian science," in J. Fauvel and others (eds.), *Let Newton be!*, Oxford: Oxford University Press, 99–125.

Gow, R. 2000. "George Salmon (1819–1904)," in K. Huston (ed.), *Creators of mathematics: the Irish connection*, Dublin: University College Dublin Press, 39–45.

Grant, E. 1974. (Ed.), *A source book in medieval science*, Cambridge, Mass.: Harvard University Press.

Grattan-Guinness, I. 1970. "Berkeley's criticism of the calculus as a study in the theory of limits," *Janus, 56*, 215–227 [printing correction in *57* (1971), 80].

Grattan-Guinness, I. 1977. *Dear Russell—Dear Jourdain*, London: Duckworth; New York: Columbia University Press.

Grattan-Guinness, I. 1982. "Psychology in the foundations of logic and mathematics: The cases of Boole, Cantor and Brouwer," *History and philosophy of logic, 3*, 33–53; also in (no ed.), *Psicoanalisi e storia della scienza*, Florence: Olschki, 1983, 93–121.

Grattan-Guinness, I. 1985. "A Paris curiosity, 1814: Delambre's obituary of Lagrange, and its 'Supplement,'" in M. Folkerts and U. Lindgren (eds.), *Mathemata. Festschrift für Helmuth Gericke*, Munich: Steiner, 493–510; also in C. Mangione (ed.), *Scienza e filosofia. Saggi in onore di Ludovico Geymonat*, Milan: Garzanti, 664–677.

Grattan-Guinness, I. 1990a. *Convolutions in French mathematics, 1800–1840: From the calculus and mechanics to mathematical analysis and mathematical physics*, 3 vols., Basel, Switzerland: Birkhäuser; Berlin: Deutscher Verlag der Wissenschaften.

Grattan-Guinness, I. 1990b. "The varieties of mechanics by 1800," *Historia mathematica, 17*, 313–338.

Grattan-Guinness, I. 1994. (Ed.), *Companion encyclopaedia of the history and philosophy of the mathematical sciences*, 2 vols., London: Routledge.

Guicciardini, N. 1989. *The development of the Newtonian calculus in Britain 1700–1800*, Cambridge: Cambridge University Press.

Guicciardini, N. 1999. *Reading the Principia: The debate on Newton's mathematical methods for natural philosophy from 1687 to 1736*, Cambridge: Cambridge University Press.

Hahn, R. 1986. "Laplace and the deterministic universe," in D. C. Lindberg and R. L. Numbers (eds.), *God and Nature*, Berkeley and Los Angeles: University of California Press, 256–276.

Hallpike, C. R. 1979. *The foundations of primitive thought*, Oxford: Clarendon Press.

Hawkins, T. W. 1975. "Cauchy and the spectral theory of matrices," *Historia mathematica, 2*, 1–29.

Heimann, P. M. 1972. "The *Unseen Universe*: Physics and the philosophy of nature in Victorian Britain," *British journal for the history of science, 6*, 73–79.

Henry, Jr., G. C. 1976. *Logos: Mathematics and Christian theology*, Lewisburg: Bucknell University Press; London: Associated University Presses.

Heuser-Kessler, M.-L. 1991. "Georg Cantors transfinite Zahlen und Giordano Brunos Unendlichkeitsidee," *Selbstorganisation, 2*, 221–244.

Hobson, E. W. 1923. *The domain of natural science: The Gifford Lectures delivered in the University of Aberdeen in 1921 and 1922*, Cambridge: Cambridge University Press.

Iwanicki, J. 1933. *Leibniz et les démonstrations mathématiques de l'existence du Dieu*, Strasbourg: Librairie Universitaire d'Alsace.

Iwanicki, J. 1936. *Morin et les démonstrations mathématiques de l'existence du Dieu*, Paris: Vrin.

Jacob, M. 1991. *Living the Enlightenment*, Oxford: Clarendon Press.

Jaki, S. L. 1984. *Uneasy genius: The life and work of Pierre Duhem*, Dordrecht, Netherlands: Martinus Nijhoff.

Kelvin, Lord. 1882. *Mathematical and physical papers*, vol. 1, Cambridge: Cambridge University Press.

Kennedy, R. 1995. (Ed.), *Aristotelianism and Cartesian logic at Harvard*, Boston: Colonial Society of Massachusetts.

Keyser, C. J. 1904. "The axiom of infinity: A new supposition of thought," *Hibbert journal, 2* (1903–04), 532–553.

Keyser, C. J. 1905. "The axiom of infinity," *Hibbert journal, 3* (1904–05), 380–383.

Keyser, C. J. 1909. "The message of modern mathematics to theology," *Hibbert journal, 7* (1908–09), 370–390, 623–638.

King, D. A. 1994. "Mathematics applied to aspects of religious ritual in Islam," in Grattan-Guinness 1994, vol. 1, 80–84.

Knobloch, E. 1994. "Harmonie und Kosmos: Mathematik im Dienst eines teleologischen Weltverständnisses," *Sudhoffs Archiv, 78*, 19–40.

Kollerstrom, N. 2000. *Newton's Lunar theory, his contribution to the quest for longitude*, Santa Fe, N.Mex.: Green Lion Press.

Kubrin, D. 1967. "Newton and the cyclical cosmos: Providence and the mechanical philosophy," *Journal of the history of ideas, 28*, 325–346. [Repr. in Cohen and Westfall 1995, 281–296.]

Laita, L. M. 1980. "Boolean algebra and its extra-logical sources: The testimony of Mary Everest Boole," *History and philosophy of logic, 1*, 37–60.

Locke, J. 1690. *Essay concerning human understanding*, London: Basset. [Many later eds.]

MacHale, D. 1985. *George Boole—his life and work*, Dublin: Boole Press.

Martin, R. N. D. 1991. *Pierre Duhem: Philosophy and history in the work of a believing scientist*, La Salle, Ill.: Open Court.

McGuire, J. E., and Rattansi, P. M. 1966. "Newton and the Pipes of Pan," *Notes and records of the Royal Society of London, 21*, 108–143. [Part in Cohen and Westfall 1995, 96–108.]

Moigno, F. N. M. 1840. *Leçons de calcul différentiel et de calcul intégral*, vol. 1, Paris: Bachelier.

Moore, J. R. 1979. *The post-Darwinian controversies*, Cambridge: Cambridge University Press.

Nash, R. 1991. *John Craige's mathematical principles of Christian theology*, Carbondale: Southern Illinois University Press.

Neil, S. 1862. "The Right Hon. and Most Rev. Richard Whately," *British controversialist and literary magazine, (3)7*, 1–12, 81–94.

Newman, F. W. 1854. *Phases of faith; or, passages from the history of my creed*, 1st ed., London: Chapman.

Newton, I. 1959. *Correspondence*, vol. 1, Cambridge: Cambridge University Press.

Nudds, J., and others. 1988. (Eds.), *Science and engineering in Ireland 1800–1930: Tradition and reform*, Dublin: Trinity College.

Odom, H. H. 1966. "The estrangement of celestial mechanics and religion," *Journal of the history of ideas, 27*, 533–548.

Parshall, K. H. 1998. *James Joseph Sylvester: Life and work in letters*, Oxford: Clarendon Press.

Parshall, K. H. 2006. *James Joseph Sylvester: Jewish mathematician in a Victorian world*, Baltimore: Johns Hopkins University Press.

Paul, H. 1979. *The edge of contingency: French Catholic reaction to scientific change from Darwin to Duhem*, Gainesville: University Presses of Florida.

Peirce, B. 1855. *Physical and celestial mechanics*, Boston: Little, Brown.

Peirce, B. 1881. *Ideality in the physical sciences* (ed. J. M. Peirce), Boston, 1881.

Peirce, C. S. 1908. "A neglected argument for the reality of God," *Hibbert journal, 7* (1908–09), 90–112. [Repr. in *Collected papers*, vol. 3, Cambridge, Mass.: Harvard University Press, 1935, paras. 6.452–491; and in *The essential Peirce*, vol. 2, Indianapolis: Indiana University Press, 1998, 434–450.]

Peterson, S. R. 1955. "Benjamin Peirce: mathematician and philosopher," *Journal of the history of ideas, 16*, 89–112.

Price, M. H. 1994. *Mathematics for the multitude? A history of the Mathematical Association*, Leicester: Mathematical Association.

Pulte, H. 1990. *Das Prinzip der kleinsten Wirkung und die Kraftkonzeptionen der rationellen Mechanik*, Stuttgart, Germany: Steiner.

Purkert, W., and Ilgauds, H.-J. 1987. *Georg Cantor*, Basel, Switzerland: Birkhäuser.

Pycior, H. 1997. *Symbols, impossible numbers and geometrical entanglements*, Cambridge: Cambridge University Press.

Radelet-De Grave, P., and Benevuto, E. 1995. (Eds.), *Entre mécanique et architecture*, Basel, Switzerland: Birkhäuser.

Richards, J. 1988. *Mathematical visions: the pursuit of geometry in Victorian England*, San Diego, Calif.: Academic Press.

Roscher, W. H. 1904. "Die Sieben- und Neunzahl im Kultus und Mythus," *Abhandlungen der Königlichen Sächsischen Akademie der Wissenschaften zu Leipzig, Philologisch-historische Klasse, 24*, no. 1, 126 pp.

Russell, B. 1904. "The axiom of infinity," *Hibbert journal, 2* (1903–04), 809–812.

Ruyer, R. 1974. *La gnose de Princeton. Des savants à la recherche d'une religion*, Paris: Fayard.

Sakarovitch, J. 1998. *Epures d'architecture. Théorisation d'une pratique, pratique d'une théorie*, Basel, Switzerland: Birkhäuser.

Schweber, H. 1999. "Law and the natural sciences in nineteenth-century American universities," *Science in context, 12*, 101–121.

Smith, C. 1998. *The science of energy*, London: Athlone Press.

Smith, C., and Wise, N. 1989. *Energy and empire: A biography of Lord Kelvin*, Cambridge: Cambridge University Press.

Smith, D. E. 1921. "Religio mathematici," *American mathematical monthly, 28*, 339–349.

Snobelen, C. S. 1999. "Isaac Newton, heretic: the strategies of a Nicodemite," *British journal for the history of science, 32*, 381–419.

Stokes, G. G. 1891–1893. *Natural theology*, 2 vols., London and Edinburgh: A. and C. Black.

Tapp, C. 2005. *Kardinalität und Kardinäle. Wissenschaftshistorische Aufarbeitung der Korrespondenz zwischen Georg Cantor und kathologischen Theologen seiner Zeit.* Stuttgart, Germany: Franz Steiner Verlag.

Torrance, T. F. 1984. *Transformation and convergence in the frame of knowledge*, Belfast: Christian Journals.

Turner, F. M. 1974. *Between science and religion*, New Haven, Conn.: Yale University Press.

Valson, C.-A. 1868. *La vie et les travaux du Baron Cauchy*, vol. 1, Paris: Gauthier-Villars. [Repr. Paris: Blanchard, 1970.]

Van Evra, J. 1984. "Richard Whately and the rise of modern logic," *History and philosophy of logic, 5*, 1–18.

Venn, J. 1880. *Some characteristics of belief scientific and religious*, London and Cambridge: Macmillan.

Venn, J. 1881. *Symbolic logic*, 1st ed., London: Macmillan. [2nd ed. 1894; repr. New York: Chelsea, 1970.]

Venn, J. 1888. *The logic of chance*, 3rd ed., London: Macmillan. [Repr. New York: Chelsea, 1962. 1st ed. 1866, 2nd ed. 1876.]

Warwick, A. 2003. *Masters of theory: Cambridge and the rise of mathematical physics*, Chicago: University of Chicago Press.

Weber, H. 1893. "Leopold Kronecker," *Mathematische Annalen, 43*, 1–25.

Westfall, R. S. 1971. *Force in Newton's physics*, New York: Elsevier.

Weyl, H. 1932. *The open world: Three lectures on the metaphysical implications of science*, New Haven, Conn.: Yale University Press.

Whately, R. 1826. *Elements of logic*, 1st ed., London: Mawman. [Repr. with introduction by P. Dessì, Bologna: CLUEB, 1988.]

Whewell, W. 1833. *Astronomy and general physics, considered with reference to natural theology*, London: Pickering.

Whewell, W. 1835. "On the nature of the truth of laws of motion," *Transactions of the Cambridge Philosophical Society, 5*, 149–172. [Repr. in R. E. Butts (ed.), *William Whewell's theory of scientific method*, Pittsburgh: University of Pittsburgh Press, 1968, 79–100; other relevant texts here.]

Whewell, W. 1837. *The mechanical Euclid*, Cambridge: Deighton.

Whitehead, A. N. 1926. *Religion in the making*, New York: Macmillan.

Whittaker, E. T. 1942. *The beginning and the end of the world*, London: Oxford University Press.

Whittaker, E. T. 1946. *Space and spirit: Theories of the Universe and arguments for the existence of God*, London: Nelson.

Wiener, N. 1964. *God and Golem Inc.: A comment on certain points where cybernetics impinges on religion*, Cambridge, Mass.: MIT Press.

Wills, J. 1860. *An estimate of the antecedent probability of the Christian religion, and of its main doctrines, in six sermons . . . ; being the Donnellan Lectures for 1858*, Dublin.

Wilson, C. A. 1980. "Perturbation and solar tables from Lacaille to Delambre," *Archive for history of exact sciences, 22*, 53–188, 189–304.

Wilson, D. B. 1971. "The thought of late Victorian physics: Oliver Lodge's ethereal body," *Victorian studies, 15* (1971–72), 29–48.

Wilson, D. B. 1984. "A physicist's alternative to materialism: The religious thought of George Gabriel Stokes," *Victorian studies, 28* (1984–85), 69–86.

Worsley, T. 1865. *Christian drift of Cambridge work*, London and Cambridge: Macmillan.

CHAPTER SIXTEEN

Mozart 18, Beethoven 32

Hidden Shadows of Integers in Classical Music

Links between mathematics and music have long been known, and they take several different forms: tuning and temperament, acoustics, influences on notation, mathematical structures in musical forms, and so on. The link discussed here is very ancient, and may well belong to the origins of each science; numerology in music, where mystical properties assigned by a given culture to certain integers, were used to guide the composition and even performance of its music. The cases of Mozart and Beethoven are highlighted.

1. Mathematics in Music

In Western music the case of J. S. Bach is particularly intriguing. Numerology is evident in, for example, his chorale "Das alte Jahr vergangen ist" (BWV 614), which contains 365 notes. But he also practiced gematria, the sister discipline in which integers are assigned in order to the letters of the alphabet of a language and a word or phrase takes the integer given by the sum of those corresponding to its letters. In the 24-letter alphabet of his day (which is explained in ¶4), BACH = 14 and JSBACH = 41, integers that guided works such as the "Crab canon," so called for its palindromic structure. Bach's interest in these practices belongs to a rich tradition of numerology in German cabbalism, which itself was an offshoot of the vogue for Rechenbücher on practical arithmetic.[1]

This paper treats two unknown numerologists and one gematricist. Wolfgang Amadeus Mozart (1756–1791), the subject of ¶2–¶6, practiced both disciplines; the chief example, the opera *Die Zauberflöte* (1791), is followed by the last three

The original version of this essay was first published in J. W. Dauben and others (eds.), *The History of Mathematics: States of the Art,* Boston: Academic Press, 1996, 29–47; a few small additions made. Reprinted by permission of Elsevier Rightslink.

1. Tatlow (1991) rather revises our understanding of Bach's numerology by suggesting that the assignment of numbers may not necessarily have been made by alphabetical order. For other examples of Bach's numerology in the (30) Goldberg variations, see Mellers (1980, 262–289).

symphonies of 1788, and then a general appraisal of his case is offered. Then ¶7–¶9 treat the numerology of Ludwig van Beethoven (1770–1827) in a similar way. Then ¶10 points out some principal differences between the two composers.

The two concluding sections are reflective. Reasons are sought for the deliberate and even anxiously pursued ignorance of these topics among musicologists; then by contrast, aspects of numerological research are described, and further known or candidate composers are mentioned. When general remarks are made, especially in those two sections, only numerology is mentioned, but the comments usually hold for gematria also.

2. Mozart in His Masonic Tradition

In both Mozart and Beethoven, Christianity and Freemasonry provide the basic doctrines from which the numerology sprung. Let us start with the latter, which is not well known in general, since it is supposed to be a secret movement. Yet an astonishing amount has been written about it, and at times even by its practitioners. In the case of Mason Mozart, the literature is very specific, since the Viennese lodge group that he joined in December 1784 had recently started its own *Journal für Freimaurerei*, under the chief stimulus of the mineralogist Ignaz Born (1742–1791). But Freemasonry in general was then running foul of monarchies and governments, especially when the French Revolution of 1789 made them chary of all international movements. The situation in the Habsburg Empire was particularly serious, for Austrian Enlightenment was deeply informed by Freemasonry, and Mozart knew many of its leading Viennese figures (Till 1992).

The crisis peaked in 1790, when Holy Roman Emperor Leopold II planned to ban Freemasonry entirely. In response Mozart and his librettist Emmanual Schikaneder (on his role, see ¶4) prepared *Die Zauberflöte*, an opera that extolled Masonic ideals and virtues. The apparently silly story is in fact a brilliant account of how the virtues of Freemasonry, as apotheosized in the High Priest Sarastro (Saros = the Sun), conquers the dark ignorance of the Königin der Nacht (Moon, the silver reflected light), especially in the training of Tamino (animo = spirit; TAM = Tugend, Arbeit, Menscheitsziele, the principal triad of Masonic virtues) to pass the Masonic trials. Chailley (1967) has explained in detail how this scenario is worked out, and has also given most of the principal integers; Eckelmeyer (1991) is rich on the Masonic and related contexts; and Dalchow, Duda, and Kerner (1966) contains further essential insights. However, none of these texts explores the consequences of the numerology; this is analyzed in Grattan-Guinness (1992, {2003}), from which the examples below are taken.

The line in Freemasonry to which Mozart belonged took Egyptian lore (as then understood) as the main origin of civilized knowledge (hence the opera is set in an Egyptlike land), and Egyptian arithmetic affected some of the choices made of integers. The principal one is 3, with the triads of Masonic virtue (one was just mentioned); it also conveys a masculine connotation, and so does 6. Femininity is encompassed in 5 (hence the 5-note panpipe of birdwatcher Papageno) and 8, while 7 and 18 carry holy ramifications.

Eighteen is especially prominent, due to the property (known to the Egyptians) of the perimeter of a 6 x 3 rectangle equaling its area. For the Masons further attractive features were available: not only

$$6 \times 3 = 6 + 3 + 6 + 3 \text{ but also } 18 \times 2 = 36 = 6^2 = 1 + 2 + 3 + \ldots + 8, \quad (16.1)$$

that is, the triangular integer to 8. For a Mason 18 also governed time, in that the day started in the evening, at 6 o'clock (= 1800 hours).

Roles were assigned to 42 and 55, also significant in Egyptian lore via the so-called pyramidal or triangular integers:

$$42 = 2 \times 21, \text{ where } 21 = 1 + 2 + 3 \ldots + 6; \text{ and } 55 = 1 + 2 + 3 + \ldots 10. \quad (16.2)$$

The other integers are 2 and 10; and 4 carried various important connotations (for example, the number of trials that Tamino had to undertake), above all in the triad 3:4:5. This triad of integers had a special significance for Masons, as the "Masons' square." For the cathedral builders of the Middle Ages, among whom numerology gained great prominence, this right-angled triangle was a principal tool, for it was used to construct not only perpendiculars but also rectangles in the proportions 2:1 (ad quadratum) and 3:1 (ad triangul[at]um); Lund 1921; (§14.5.3). Among their successors the Masons, it was worn as an ornament on their costume, in the form of a suspended L of the shorter sides; a portrait exists of Mozart and his fellow Masons so attired when gathered in their "temple" (Robbins Landon 1991).

3. A Few Examples from *Die Zauberflöte*

Every aspect of *Die Zauberflöte* is controlled by the numerological system just outlined. The music runs on it from beginning to end (literally: from 5, 3, 5 in the Overture before the 36-note subject opens the 3-voiced fugue, to 18, 4, 5, 4, 5 in the final bars). Among other facets are the numberings of both the "Auftritte" (AF) into which music and dialogues are divided and also the musical items (MI); the names of the principal characters—not only Sarastro, Königin, and Tamino, but also Tamino's helper Pamina ("anima" there again); the three-syllable names of these

Figure 16.1. Cadenza for Königin's revenge aria from Mozart's *Die Zauberflöte*

principal characters, as opposed to the four-syllable names of lower orders such as Papageno, Papagena, and Monostasos; the 18 singing parts (counting the chorus as 1); and the scene changes and set designs, which reflect many aspects of Freemasonry (Curl 1991). Even the day for the first performance was carefully chosen: 30 7ber, the end of the 3rd 4ter of the year, but 1 8ber for Masons such as Mozart, since the performance started after 1800 hours. However, to the Masons the year was not 1791 but 5791, dating from the supposed origins of Egyptian knowledge; hence there is no role for $1 + 7 + 9 + 1 = 18$.

I shall give a few musical examples concerning the principal protagonists. The most beautiful case of numerology occurs in the cadenza of the revenge aria of the Königin (AF 8 in Act 2, MI 14). Figure 16.1 shows her line, in which the feminine integers, 5 and 8, are deployed in a wonderful way:

$$5, 8, 5, 8, 5, 8, 8, 8. \tag{16.3}$$

Here are 8 groups of notes comprising 3 5s and 5 8s, with 3 pairs at the beginning and 3 singles at the end; and the sum of the notes is 55. The 5 (not counting the accidentals) is a leitmotif in the work; for example, it appears twice in the 36 notes of the overture theme. Note that for good measure, the cadenza is completed by a $5 + 3$.

The other examples concern the High Priest Sarastro, Saros the Sun God. Born died while the opera was being composed, and maybe Sarastro was conceived in honor of him. In any event, his entire role is controlled by 18 in the most astonishing fashion. He first appears in the Finale of Act 1 at AF 18, on a "triumph-cart" pulled by 6 lions (where the lion symbolized the sun)—that is, as 6×3. In the opera he has 18 entries and sings in 180 bars ($= 6 \times 30$ as well as 10×18, I suspect). But the greatest detail comes in his solos.

Each of Sarastro's two arias has 18-note melodies. Figure 16.2 shows the first, "O Isis und Osiris" (AF1 of Act 2, MI 10), in vocal score, which accurately reflects the orchestral original. Note how the orchestration preserves the 18 scheme by giving only one note to some *but not all tied* quavers, which therefore are to be counted as 1 (for example, "dem" in bar 10 but not "-ris" in bar 7). Then follows 6 9-sequences, then 2 9s, 2 6s, and finally 3 9s on "nehmet sie." The 4-part chorus repeats the last

Figure 16.2. The first verse of Sarastro's "O Isis und Osiris" from Mozart's *Die Zauberflöte*

line of each verse (over 4 bars), giving 10 lines of text in all. With 4 bars for orchestra at the beginning and 3 at the end, the number of bars in the item is 55.

Sarastro's other aria, "In diesen heiligen Hallen" (AF12 of Act 2, MI 15) also has an 18-note melody (if the tied notes on "Hal-" and "die" are counted as one each); this phrase is repeated, and some later passages are on or close to 18 notes. The text comprises 2 verses, each of 6 lines and sung over 24 bars: the words starting each 5th line are sung 3 times, and once again its last phrase is sung 3 times at the end. Counting the opening upbeat in the usual way, the number of bars is 54; *designed* as 2 × 24 plus 6 for the orchestra, but *conceived* as 18 × 3.

In addition to these two arias, Sarastro's priests—18 of them, we are told at the head of Act 2—open scene-change 5. Each priest carrying a pyramid, they sing to the Egyptian gods "O Isis und Osiris" on their own (AF of Act 2, MI 18 [sic]). Their 6 lines of text take 18 + 18 bars, with 4 bars at the end to allow the last words to be sung 3 times. Allowing for upbeats, the orchestra has 6 other bars to itself, giving 42 in all.

4. Mozart as Gematricist

In ¶1 I mentioned that Mozart also deployed gematria in this opera. This is the discovery of Irmen (1991), a most important study on Mozart's Masonic career (for example, Irmen has used the secret Masonic files in Vienna to transcribe the ceremony of Mozart's induction in December 1784).[2] Gematria is "visible" chiefly in the integers corresponding to the names of persons involved in preparing the opera, and in words and phrases in the text. It is also manifest in the numbers of notes assigned to the instruments in parts of the score *in the autograph* (Mozart 1979); a degree of arbitrariness was available to Mozart in the places at which he chose to stop writing out repeated notes and use "&c." or some such sign (thus all printed editions are useless for this analysis).

The most interesting example comes at the start of the first scene, when the story is marvelously encapsulated in the encounter between the frightened Tamino and the Königin's three silver-clad ladies, who (in their veiled ignorance) kill Freemasonry by chopping up into 3 pieces its symbol, the serpent. The 16½ bars of orchestral introduction encode a title page for the opera, in the normal form of words then used following the 24-letter code of the German alphabet:

$$a = 1,\ b = 2, \ldots i = j = 9, \ldots, u = v = 20, \ldots, z = 24: \qquad (16.4)$$

1st violins	129	Zauberfloete
2nd violins + violas	56 + 90	Compositeur
Basses	87	Mozart
All strings	362	Neues Singspiel in zwey Aufzuegen
Oboes	69	[uninterpreted]
Bassoons	71	Poeten
Bassoons and horns	71 + 60	Ludwig Gieseke
Clarinets[3] and typmani	79	Emmanual
All winds	200	Schikaneder der Juengere
All instruments	562	Wolfgang Amadé Mozart, Kapellmeister in wirklichen k.k. Diensten

Note that in this reading both Gieseke and Schikaneder are named as librettists; the question of which *one* was responsible has been a matter of controversy.

2. Irmen's book first appeared in 1988 and is, of course, patronized by musicologists. I am chary of certain claims; in particular, on p. 362 he interprets gematriacally the 143rd item in Mozart's thematic catalog when the numbering is not in Mozart's hand. However, he definitely refutes my original belief that Mozart did not deploy gematria.

3. Irmen misreads Mozart's "Clarinette" for "Trombe" (trumpets). Both parts were deleted on the manuscript at a later stage.

5. Mozart's Last Three Symphonies

One test of a new theory is to try it out in a context from which it was not created and see how it fares. In this case I have been able to come up with a very satisfactory answer to this well-known mystery in Mozart's career: why did he write three symphonies in the summer of 1788 with no commission in hand, when he was short of money? (The fact that they turned out to be his last symphonies is, of course, not involved here.) When one recognizes that his passion for Freemasonry included numerology, then the solution comes readily: he wrote this trio of works to fulfill a major commitment to Masonic ideals. Here are some main features, summarized from Grattan-Guinness (1994c).

1. At the time Mozart was in his 33rd year of life. Although 33 does not feature in *Die Zauberflöte* (unless the 16½ bars just mentioned are 33/2), it is a most important Masonic integer. Christianity was probably the source: in the orthodox interpretation of the Bible it was the supposed length in years of the life of Jesus; in the Apocryphal tradition it served as 3 × 11, where the digit string 1-1 symbolized their belief that Jesus was a rabbi married to Mary Magdelene (Thiering 1992).
2. The 3 keys of the symphonies are Masonic; E♭ (3 flats), the most favored Masonic key, to be used also as the main key of *Die Zauberflöte*; C, the so-called "white" key; and most interestingly of all, G minor, a very special key in Mozart. {As Ruth Tatlow has pointed out to me, these three notes form the basic triad of the key of C minor—that is, of 3 flats.} G was an important symbol in Freemasonry, although (to outsiders like me, anyway) its purpose is unclear. It may have carried three (3?) meanings: "Gott"; "Gnosis," a term in the Apocryphal tradition for knowledge, especially when involving elitist insight; and "gnomon," the triangular device long used to determine the time and latitude, and represented by the Mason's square.
3. The symphony in G minor is a miracle of 18s, for all its main themes are so based. The opening subject is a particularly instructive case. It appears to be a melody of 4 × 10 notes, suitable for this Symphony no. 40, a very important Christian integer; but in fact Mozart never assigned numbers to any of his works (the symphonies were numbered by publisher Breitkopf und Härtel in the early nineteenth century). In the manuscript Mozart followed his practice used elsewhere (such as with Sarastro) of tying pairs of notes to count as one; here, D-B♭ and C-A of the first and third clauses,

with the repeated Cs and Bs of the other two also to be so understood (figure 16.3). Thus we have a quartet not of 10s but of 9s—indeed, the important 36 of (16.1).

The other first subjects also manifest 18s; in the 3 6s of quavers in the andante, before rounding off with an upbeat and a rhythmic 6; as 18 + 18 + 9 in the minuet; and as (7 + 11) + (7 + 11) of the concluding finale. In addition, the second subjects of the second and fourth movements use 18s: respectively, 4 + 4 + 4 + 6 (including a pair of tied quavers as 1) and as (8 + 10).

The other two symphonies show numerology at work, although not as intensively. For example, the E♭ symphony announces its Masonism at once, in the threefold statement of the opening flourish, and the first part of the second subject is again an 18 (= 6 + 6 + 3 + 3), as is the first subject of the second movement (6 + 12).

The 3rd and final symphony, the "Jupiter" (not Mozart's name) in C, announces the importance of 3 very explicitly in the first part of the first subject (figure 16.4a). The number 4 and its multiples feature frequently, especially in the motive C D F E in the first part of the first subject of its last movement (figure 16.4b). This theme goes back to Haydn's Symphony no. 13 (1763), and in Mozart to his Missa Brevis K 192 (1774). Here it will have signified "Cre-do, cre-do"; in addition, the four notes shape in print the cross of Saint Andrew, which has a symbolic meaning in Freemasonry derived from its own Scottish origins (Irmen 1991, 279).

Figure 16.3. Mozart's Symphony in G (1788), the first subject of the first movement

Figure 16.4. Mozart's Symphony in C (1788), the first subjects of the first and fourth movements

6. Motives and Origins in Mozart

Although Mozart joined the Masons at the end of December 1784, numerology begins to appear in works from 1782. Baron van Swieten might have been a crucial acquaintance, made at that time: an aristocratic Masonic intellectual of a kind characteristic of that society; and supporter of Mozart, especially with his rich library, which brought to Mozart access to the music of Händel and Bach. The quartet K452 of that year for piano and 3 wind instruments may be a key initial work in numerology (I leave note-counting as an exercise); not only was the combination of instruments original, but especially because it included wind instruments, which the Masons extolled above all others. This is why there is so such wonderful wind music in Mozart, and an opera by him with the somewhat surprising title *Die Zauberflöte.*

Outside *Die Zauberflöte* the most evident example of numerology is the presence of 18-note melodies. There are many of them, including some of his most famous ones. For example, sing for yourself the opening of "Eine kleine Nachtmusik" (1787), and note the 9 + 9—and also the title (Mozart's own, though not new in music), redolent of the name of the Königin der Nacht still to come and of the secrecy of Freemasonry in general.

When one is oriented to the use of integers in this way, some of the evidence is very obvious. For example, in the manuscript of *Die Zauberflöte* (Mozart 1979) Mozart counted up the numbers of bars of every musical item, even the long finales (which, doubtless following [16.2], are written out on respectively 42 and 55 pages of manuscript paper). He also shows two examples in letters to his wife, Constanze. In 1789 "drucke ich 1095060437082 mal: hier kannst du dich im aus-sprechen über," where the hint and the lack of commas in the string of digits clearly invites the decoding

$$10 + 9 + 50 + 60 + 43 + 70 + 82 = 324 = 18^2, \tag{16.5}$$

the power (square) of holy love. Again, on June 6 (6/6), 1791, while working on *Die Zauberflöte*, "es fliegen 2999 und ein 1/2 bissel von mir," which yields

$$29 + 99 = 128 = 2^7, \tag{16.6}$$

the 7-sacred strength of love between man and woman ("and a 1/2" for the baby whom she was carrying; Mozart *Letters*, vol. 4, 84, 135). The distinctions made here between an integer and a digit string are very clear evidence of Mozart's *conscious* intent—just what one would expect of a numerologist.

7. Beethoven's Numberings

Mozart is a relatively easy case to analyze numerologically, not only for the wealth of examples that he provided but also because of the detailed information available on the Masonic movement to which he belonged. I now take a less straightforward case, another great composer who lived in Vienna in the next generation after Mozart's death and therefore fell under similar social and religious influences.

In Beethoven's numerology some of the same numbers are used as with Mozart, such as 2, 6, and 18; but the quartet 27, 30, 32, and 33 is also prominent. He also used this quartet:

$$3 + 37 = 40, \qquad 3 \times 37 = 111 \text{ (a Trinity number).} \tag{16.7}$$

Again the origins are partly Christian and partly Masonic; but I have not determined the significance to him of 32; maybe it was as 2^5, for the hands and fingers of a man whose own instrument was the piano. There is a possibility from gematria, for in the German alphabet (16.1) LVB gives 32; however, I do not think that Beethoven practiced gematria, so he would not have noticed this property.

Like Mozart though to a lesser extent, Beethoven used these integers to plot out the numbers of notes in a melody, and more often of bars in (a section of) a piece. But the most striking use, quite absent in Mozart, is in his numbering of works of a given kind, and in his placing works at "nice" opus numbers. The word "his" here is worth stressing, for *he chose the numbers himself,* and from the start of his publishing career—quite contrary to the contemporary habit of publishers (if anyone) assigning them. His works as listed by Opus-number order show significant variation from the chronological order of completion, and also a measure of divergence from that of publication.

8. Some Examples from Beethoven's Output

The most intense period of gematriac concern occurred in the early 1820s, when Beethoven was in his early fifties. I shall take two examples from his output; they are summarized from Grattan-Guinness (1994b), where many others are given.

Beethoven wrote 32 numbered piano sonatas. The last 3 were done together, in a consecutive run of Opera 109–111; and the last integer (see [16.7]) does not disappoint us. In 3 flats, it is run on 2, 3, and 32 in a remarkable way. In the first movement, 18 bars of introduction are followed by the first subject stated in a dual manner: its first phrase has 3 + 3 notes, stated twice, to be followed by 5 + 5 notes

Figure 16.5. Beethoven Piano Sonata no. 32, Opus 111, the first subject of the first movement

again stated twice—giving a total of 32 notes. Moreover, the theme is expressed in octaves; that is, with the 2 hands together (figure 16.5). The second and last movement is an arietta on a theme over 4 × 8 bars, as are its first three variations; however, the last one is much freer. Why?

The answer is well known: it is interaction with Beethoven's next major work for piano, already started in 1819 but only finished after (or with) Opus 111. And that work is—the "33 variations" (to quote from its title) on a theme in C by Diabelli, the 33rd major work for piano after the 32 sonatas. I am sure that at this stage of composition Beethoven *decided* that the total of variations was to be 33. The last five were written contemporaneously with the completion of the last movement of Opus 111, which sounds rather like another variation on the theme.[4] Of this quintet no 32 is the most interesting numerologically: in E♭ (that is 3 flats, unusually for the work), lasting 5 × 32 bars, and with a first subject of 32 notes.

This pair of works also shows well his policy of opus numbering. Although they were completed very close together, the 33rd work is nine away from the 32nd, as Opus 120 (= 3 × 40 = 4 × 30, I suspect): not only another nice integer but the *first* such after 111. . . . Again, as with Mozart's symphonies in ¶5, the fact that these piano works were to be the last of their kind is not our concern; however, it is worth noting that he did not add any more during his exceptionally creative final years, presumably having found a good position at which to pause.

During these years Beethoven also worked on his major religious work, the "Missa Solennis" (as I think he wanted it to be known), completing it in 1823. He assigned to it the opus number 123, which as the string 1-2-3 pronounces Christian 3ness in a most obvious way. His deliberate choice of number is confirmed by the

4. The analysis of the manuscripts pertaining to the Diabelli variations made in Kinderman (1987) closely corroborates this conjecture on the chronology of composition of the variations. It also seems that some of the earlier ones were written late. Of particular interest among them is no. 22, which uses a Mozart theme and is structured not on the normal 16 bars but on . . . 18.

fact that when he sent it to his publisher, Schott, he left empty the boring number 122 (Beethoven 1985, 39–40).

Threeness occurs all through the work in both small- and large-scale cases. The latter is evident in the tripartite design of several of its five movements and in major parts of them. Small cases start right at the beginning, for his opening theme started with 3 rising notes stated 3 times, and moreover in bars 6, 8, and 10, twice the Pythagorean ratios of the Mason's square described in ¶3.

I shall describe in detail the "Credo," the 3rd movement and the heart of the work: "I believe," the only first-person declaration in the text. It divides into 3 parts, of which the first and last are 2 of the 3 fugues in B♭—or B (for "Beethoven"?), in the German system of lettering. The first is an allegro of 123 bars, no less, and no more. Divided carefully into 30-bar blocks, it also exhibits a remarkable addition at bar 30: the repetition "et, et" invites one to *add on* the rest of the phrase, which ends—at bar 33. The second statement of "Credo" starts at bar 37 (see [16.7]), and the other changes occur at bars 60 (the start of "Deum de Deo"), 90 (for "Qui propter"), and 120 (the end of the "descendit" part), when 3 orchestral bars complete the movement. The other fugue (a double one) makes up the 3rd part, and it divides into 3 subparts by 33s: the first commences at bar 306 and lasts 66 bars, when there occurs a strange transition bar (372) incorporating a tempo change; then 66 more cover "et vitam," before the concluding "Amen" subpart of 33 bars is launched by the soloists at bar 439 (with a final bar for the orchestra).

Beethoven threw his numerology into this life-work piece, and not only into the music. At the head of the "Kyrie" he wrote a sentiment that is not normally included in editions but is frequently quoted—*and almost always transcribed wrongly.* Figure 16.6 shows his hand (Beethoven 1966):

"Von Herzen—Möge es wieder—zu Herzen gehn!"

The form had been carefully chosen, especially with the correct but somewhat colloquial spelling of the last word: 3 clauses, with dashes stressing the divisions; 2 + 3 + 3 words, 3 + 5 + 4 syllables (the Pythagorean triad, note) and 9 + 12 + 12 letters; and the last sum is 33.

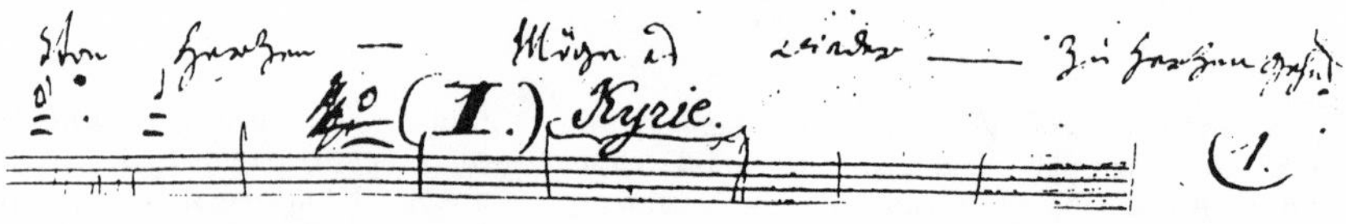

Figure 16.6. Beethoven's sentiment at the head of the manuscript of "Missa solennis"

9. Beethoven as Mason and as Arithmetician

It is not clear whether or not Beethoven was a Mason; but the question is in fact *not* crucial for my analysis, since he clearly went along with its sentiments without having had to have been a member—just like his deism, in fact, which he affirmed without going to church very often.

More important is the question of Beethoven's capacities for effecting arithmetical calculations, for he was notoriously bad at the subject. Musicologists take this fact as sufficient reason to discount the possibility of numerology in him; but mathematical educators (to whom the question properly belongs) recognize that the issue has to be considered in several respects. All that is required of numerologist Beethoven is the mental capacity to do addition: multiplication does not arise, nor scribal competence (which indeed is sometimes wanting in his conversation-book sums). A revealing anecdote tells that on his death bed he attempted to do multiplication (Thayer 1964, 942). The implication that he failed is hardly surprising; far more suggestive is the fact that he *tried* to multiply in the first place. It is well known that people sometimes return to childhood practices when dying; presumably he was reverting to early and lifelong concerns with arithmetic.

10. Comments and Comparisons

Clearly the mathematics involved in numerology is quite elementary: thus no mathematical ability is presumed in any composer who practiced it. Hence Beethoven's difficulties with arithmetic do not refute the interpretation, as we saw. By contrast, Mozart was fond of numbers—in 1770 (his 15th year) he even signed a letter to his sister Nannerl "Freund des Zahlenhausens" (Mozart *Letters*, vol. 1, 311)—but we do not need hypotheses asserting mathematical gifts to understand his later numerological career.

The fact of its occurrence in both composers is indubitable and indeed not surprising when one recalls the history of numerology in (pre)classical music and its place in the Freemasonry of the Austrian Enlightenment. The most extraordinary feature is that nearly two centuries have passed before anybody properly realized its presence in either composer (not at all in Beethoven's case). Some methodological problems arise here, which are discussed in the next section.

11. Numerology in Music: Its Actual Limits

Firstly, numerological doctrines were (and are) almost always held *silently*, whether they occur in music, religion, architecture, painting, or whatever: only in rare cases does the numerologist leave written testimony about the details, or even occurrence of its practice. Neither Mozart nor Beethoven is known to have done so, and to expect to find such sources shows ignorance of the activity. (The structure of *Die Zauberflöte* is so rich that Mozart and Schikaneder/Gieseke may have kept an *aide-mémoire* at the time of its preparation; but, even if so, it would have been destroyed soon after completion.) Thus the scholar has to resort to contextual understanding in order to be aware of the possibility of its occurrence, and of observation in the kinds of place where it is likely to occur—numbers of notes in a melody, opus numbers, or whatever. But such research techniques run quite counter to normal criteria for historical evidence.

Second, there is the tyranny of professionalization within disciplines, especially strong in modern academic life. On the principle of leaving war to the generals, only a musicologist can produce pukka musicology; hence if a mere historian of mathematics comes up with such a suggestion, then it cannot be taken seriously. Given the well-known presence of numerology in the music of earlier periods (noted in the next section), I expected historians of classical music to be aware of at least its basic principles, and of its importance to its adherents; but instead I have been confronted with unawareness even down to the level of not knowing the difference between an integer and a digit string (123 versus 1-2-3)—ignorance that Mozart and Beethoven did not exhibit.

I find also my analysis dismissed on the grounds of some vague (and certainly uninformed) appeal to statistical coincidence. But if you want to think that, for example, Sarastro sings 18-note melodies and all the rest because of some very strong law of strong numbers, then that really is your credulous (and ahistorical) hostility. Further, note that some relatively large integers are involved: one does *not* constantly confront 18, 27, 30, 32, 33, 123, . . .

12. Numerology in Music: Its Possible Scope

Learning about numerology (and gematria) is not too difficult (for a summary history, see Grattan-Guinness 1994a, and §12.2); but the problems involved have to be understood. These doctrines use only the simplest mathematics, but their history is very long, going back to ancient times: the oldest seems to be Jewish gematria, way back in B.C. territory (Scholem 1971). Music was possibly not the prime motivation; it seems to have informed more deeply religious principles and practices of a culture, and determined features of its architecture and art.

Contrary to the view of the instant nonexperts, the associations between integer and property or symbol are *not* arbitrary; but they may be surprising or unexpected, with connotations local to the originating culture—and very likely not often described in print anyway. The real difficulty is quite other: namely, that the *plurality* of doctrines that have evolved over the centuries entail that an integer can have different interpretations in different traditions (Hopper 1938, a masterly survey). For example, 9 in Hinduism has quite distinct connotations from 9 in Dante; and this example is pertinent to Mozart (and his librettist da Ponte), in that Dante's use of 9 and 10 surely guided the Commendatore's music in the finale of Act 2 of *Don Giovanni* (Grattan-Guinness 1992, 231).

In addition, whatever numerological doctrine was being adopted, it was *important* to its followers, not just a "Zahlenspielerei" (to quote a note to me from the editor of the *Mozart Jahrbuch,* apparently the leading organ for Mozart studies). In his case, as a typical example, it affected composition both in internal aspects (such as the numbers of notes in a melody or of bars in a piece) and external ones (such as Masonic or Christian ideals, or dates of first performance).

When alerted to the practices and means of numerological study, a long stream of thought in the history of (not only) Western music is revealed; and a few cases have already been explored to some extent. In the Western tradition, the place of numerology has been well recognized in many figures of the middle ages (for example, Obrecht [van Crevel 1959] and De Vitry [Bent and Howlett, in prep.]), when it penetrated deeply into all aspects of cultural thought. In our century some cases have been recorded, especially Debussy (Howat 1983), Bartok (Lendvai 1971), Schönberg in his serialist phase (Sterne 1983, 1993) and Berg (Carner 1983, 121–142)—usually without Christian or Masonic connections.

But the classical and romantic periods have largely been omitted from this kind of analysis. Yet it is easy to think of further candidates. An obvious one is Haydn, whom Mozart talked into joining the Masons in 1785 (Irmen 1991, 120–125);[5] *Die Schöpfung* (with libretto by van Swieten) is known to carry many numerological and Masonic connotations, and doubtless so do other works.

Among romantic composers, prime "targets" include Schumann and Tchaikovsky, for they are known to have placed codes of other kinds in their music. The chromatic tradition from Chopin to Liszt and Wagner, and maybe on to followers such as Busoni and Reger, is also likely to be a fruitful area. Modern candidates surely include {Shostakovich (Strachan 1999)}, Messaien, Tippett, and Stockhausen, and there is a clear and lovely use of Fibonacci numbers in the structure of the 34 (sic)

5. H. C. Robbins Landon has kindly informed me of the recent discovery of a previously unknown Masonic membership for Haydn in Pressburg.

common chords used in Britten's *Billy Budd* (1951) to represent the fateful conversation between Budd and Captain Vere.[6]

The message of this chapter is that a serious lacuna exists in the understanding of the work of at least two of the greatest composers, and probably in several others. But for reasons discussed in the previous section, this important aspect of musical composition will continue to evade the minds of commentators for a few more centuries. Perhaps the historians of mathematics can take up numerology and gematria in music, as part of their concern with the links between mathematics and music.

BIBLIOGRAPHY

Beethoven, L. van. 1965. *Missa Solemnis Op. 123 Kyrie. Facsimile*, Tutzing, Germany: Schneider.

Beethoven, L. van. 1985. *Der Briefwechsel mit dem Verlag Schott*, Munich: Henle.

Bent, M. and Howlett, D., in prep. [Book in preparation on de Vitry.]

Carner, M. 1983. *Alban Berg . . .*, 2nd ed., London: Duckworth.

Chailley, J. 1972. *The Magic Flute: Masonic opera* (trans. H. Weinstock), London: Gollancz. [French original: "*La flute enchantée" opéra maçonnique*, Paris: Laffont, 1968.]

Curl, J. S. 1991. *The art and architecture of Freemasonry*, London: Batsford.

Dalchow, J., Duda, G., and Kerner, K. 1966. *W.A. Mozart. Die Dokumentation seines Todes*, Pähl, Germany: Bebenburg.

Eckelmeyer, J. 1991. *The cultural context of Mozart's Magic Flute: Social, aesthetic, philosophical*, 2 vols., Lewiston, Queenstown, and Lampeter: Edwin Mullen Press.

Grattan-Guinness, I. 1992. "Counting the notes: Numerology in the works of Mozart, especially *Die Zauberflöte*," *Annals of science, 49*, 201–232.

Grattan-Guinness, I. 1994a. "Numerology and gematria," in I. Grattan-Guinness (ed.), *Companion encyclopaedia of the history and philosophy of the mathematical sciences*, 2 vols., London: Routledge, vol. 2, 1585–1592.

Grattan-Guinness, I. 1994b. "Some numerological features of Beethoven's output," *Annals of science, 51*, 103–135.

Grattan-Guinness, I. 1994c. "Why did Mozart write three symphonies in the summer of 1788?" *Music review, 53*, 1–6.

Grattan-Guinness, I. 2003. "Historical Mozart researches: *Die Zauberflöte* in its historical

6. In the original (1951) version of Britten's opera, these 34 chords complete the 3rd of its 4 acts.

In general, Fibonacci numbers cause some historical complications for the numerologist. They certainly drove Bartok and aspects of Debussy, but otherwise their place seems to me to be exaggerated, especially for the classical era when they were of little interest even to mathematicians. For example, Peter (1983) analyzes intervals, rhythms, keys, and so on of *Die Zauberflöte* in their terms (see esp. pp. 298–300 on Sarastro's aria "O Isis und Osiris"). Similarly, Rutter (1977, 21–28) claims that they run the first movement of Beethoven's 5th Symphony (but on its undoubtedly intentional 5s, see Grattan-Guinness 1994b, 121–124).

context," in R. Franci and others (eds.), *Il sogno di Galois. Scritti di storia della matematica dedicati a Laura Toti Rigatelli per il suo 600 compleano*, Siena, Italy: Dipartimento di Matematica, 125–153.

Hopper, V. F. 1938. *Medieval number symbolism*, New York: Columbia University Press. [Repr. New York: Cooper Square, 1969.]

Howat, R. 1983. *Debussy in proportion: A musical analysis*, Cambridge: Cambridge University Press.

Irmen, H.-J. 1991. *Mozart Mitglied geheimer Gesellschaften*, 2nd ed., Zülpich, Germany: Prisca. [1st ed. 1988.]

Kinderman, W. 1987. *Beethoven's Diabelli variations*, Oxford: Clarendon Press.

Lendvai, E. 1971. *Béla Bartók. An analysis of his music*, London: Kahn and Everill.

Lund, J. L. M. 1921. *Ad quadratum: A study of the geometric bases of classic and medieval religious architecture*, 2 vols., London: Batsford.

Mellers, W. 1980. *Bach and the dance of death*, London: Faber and Faber.

Mozart, W. A. 1962–1975. *Letters. Briefe und Aufzeichnungen*, 7 vols., Kassel, Germany: Bärenreiter.

Mozart, W. A. 1979. *Mozart. Die Zauberflöte, Faksimile der autographen Partitur* (ed. K. H. Köhler); Documenta Musicologica, ser. 2, vol. 7: Kassel, Germany: Bärenreiter. [With an accompanying booklet *Beiheft* that alone carries copyright information.]

Peter, C. 1983. *Die Sprache der Musik in Mozarts Die Zauberflöte*, Stuttgart, Germany: Freies Geistesleben.

Robbins Landon, H. C. 1991. *Mozart and the Masons . . .* , 2nd ed., London: Thames and Hudson.

Rutter, J. 1977. *Elements of music*, Milton Keynes, U.K.: Open University Press. [Course A241, Unit 15.]

Scholem, G. 1971. "Gematria," in *Encyclopaedia judaica*, vol. 4, Jerusalem: Keter, cols. 369–374.

Sterne, C. C. 1983. "Pythagoras and Pierrot: An approach to Schoenberg's use of numerology in the construction of 'Pierrot lunaire,'" *Perspectives on new music, 21* (1982–1983), 506–534.

Sterne, C. C. 1993. *Arnold Schoenberg: The composer as numerologist*, Lewiston, Queenstown, and Lampeter: Edwin Mullen Press.

Strachan. I. 1999. "Shostakovich by numbers," *DSCH journal* (electronic).

Tatlow, R. 1991. *Bach and the riddle of the number alphabet*, Cambridge: Cambridge University Press.

Thayer, A. W. 1964. *Thayer's life of Beethoven* (ed. E. Forbes), Princeton, N.J.: Princeton University Press, 1964.

Thiering, B. 1992. *Jesus the man*, London: Doubleday. [Corgi paperback 1993.]

Till, N. 1992. *Mozart and the Enlightenment*, London: Faber and Faber.

van Crevel, M. 1959. "Secret structure," in J. Obrecht, *Opera omnia*, vol. 1, pt. 6, Amsterdam: Alsbach, xvii–xxv.

CHAPTER SEVENTEEN

Lagrange and Mozart as Critics of Descartes

This chapter traces three lines of positive and negative influence that emanated from Descartes's Discours sur la méthode *(1637), or at least were closely linked to it. Three areas of concern are involved: the relationship between geometry and algebra, celestial mechanics, and physiological psychology. The two critics of the title were contemporaries; reasons for their surprising conjunction will emerge at the end.*

1. Descartes on Geometry and Astronomy

Descartes's *Géométrie*, published in 1637 (his 42nd year) as a supplement to the *Discours*, meant what it said in the title. Although it is best remembered as a major source for the infiltration of algebra into geometry as analytic geometry, the work was basically geometrical in conception, with algebra serving as handmaiden (Bos 1981). For example, means for judging the simplicity of curves were based on traditional geometrical criteria, not on shortness of algebraic expressions. He also offered a protocalculus method to determine the normal to a curve from a double-root condition on a polynomial equation (Grattan-Guinness 1980, ch. 1).

Scraps of Descartes's vorticial theory of planetary motion were presented in the *Discours*, but his main source was *Le monde* (1629–1637). In his *Principia philosophiae* (1644) he outlined some of his principles: that the planets were swept around their paths by a family of concentric rotating vortices; and that God would not permit chaos, so that a major issue was to determine equilibrium, with the planets reaching (somehow) their appropriate vortices. His theory was developed further, including quantitatively, by G. W. Leibniz (1646–1716), whose ideas about the tran-

The original version of this essay was first published in V. Gómez Pin (ed.), *DesCartes. Lo racional y lo real*, Bellaterra: Universidad Autonoma de Barcleona, 1999, 305–310.

sition from one vortex to another one helped to launch energy mechanics via the notion of vis viva (Aiton 1972).

However, Descartes's mechanics were a major target of Isaac Newton (1642/43–1727); even the title of his book *Philosophiae naturalis principia mathematica* (1687) seems to be a retort, perhaps at its largely qualitative nature. He gave strong arguments against the existence of vortices and argued for a quite different mechanical system based on central forces between particles and laws of reaction and of acceleration. He also developed a version of the full calculus, involving the determination of derivative and integral *functions* from a given function, not mere double-root conditions to determine the tangent or normal. I use the names from Leibniz's theory, for they have become standard; he developed it in the 1670s, about a decade after Newton and quite independently, and it involved quite different concepts. Both theories, especially Leibniz's, gave much more place to algebra than Descartes had granted (Grattan-Guinness 1980, ch. 2).

2. Descartes on "the Passions of the Soul"

The *Principia philosophiae* was dedicated to Princess Elizabeth of Bohemia; soon afterward Descartes wrote at her request *Les passions de l'âme*. He also sent drafts of this work to another scholarly royal lady, Queen Christina of Sweden (1626–1689; Bain 1890), with whom he had already corresponded about aspects of the *Principia philosophiae. Les passions* appeared in 1649[1] and was his last publication because he moved that September to Stockholm to serve as court philosopher to Queen Christina: the conditions of employment required him to rise in the early mornings of a Swedish winter to provide tutelage at 5 of the clock, with the result that he rapidly died, in February 1650.

Much of the content of *Les passions* was already in the *Discours*, but the presentation of them was more explicit and systematic. The book was an *exercice* in Descartes's penchant for classification and also a vehicle for his physiological reductionism. For him the soul fulfilled two roles: action or desires, and passions. The latter were transmitted by heat of the blood lying and moving as "spirits" in the crevices of the brain. The conduit was the pineal gland, apparently chosen for its place as a single object in the human frame where so many other parts, such as the brain itself, came in twos.

1. Among the various editions of *Les passions,* the one published in 1968 contains a long though rather overwritten introduction by the editor, J.-M. Monnoyer.

Descartes classified the passions as follows. Six were primitive: love, hatred, desire, joy, and sadness as well as admiration, which had a more rational disposition than the others and so was not accompanied by any changes in the heart or to the blood. From this sextet 34 more "particular" passions were formed, as compounds: for example, gratitude, indignation, and gaiety. The total number of passions was 40: I do not regard as a coincidence that this devout Catholic decided upon this important Christian number for his aggregate (compare §14.4.1).

3. Queen Christina and the Accademia degli Arcadia

Four years after Descartes's death, Queen Christina converted to the Catholic faith; according to her later testimony, he had been an important influence. This action caused her rapid abdication, whereupon she repaired for much of the rest of her life to Rome, the Catholic center. There she founded an academy in her garden (presumed shades of Plato) where literati and intellectuals sought for "better morals," with *Les passions* as a guiding text.

The year after her death in 1689, an Accademia degli Arcadia was founded in Rome to pursue these ideals more formally; they remembered her at an annual celebration. The poet Giovanni Crescimbeni (1663–1728) was the secretary from the founding until his death.[2] Although the members—about 600 men by 1700, and not only in Rome—were drawn from the upper classes, they feigned democracy by wearing the same apparel, that of shepherds which inspired the name of their organization. One of their concerns was a desire to bring more order and probity to the preparation of libretti for operas, then still a rather new art form. Composers early in association included Alessandro Scarlatti (1660–1725), who had conducted Queen Christina's orchestra at her private theater from 1684. Their attention, however, focused on the texts, and their leading protégé came to exercise an extraordinary influence.

4. The Training and Career of a Cartesian Librettist

Another founder of the Accademia was the lawyer Giovanni Gravina (1664–1712), although a split in 1711 caused him to found another academy named after himself. He had as cousin a minor philosopher and poet, Gregorio Caloprese (1650–1714/15), who founded a philosophical school in Calabria in southern Italy, where

2. Crescimbeni (1804) was also a historian of the Accademia. More modern treatments include Acquaro Graziosi (1991), light but with good illustrations.

the instruction was based on *Les passions.* In 1712 Gravina introduced Caloprese to a most promising youth, Pietro Trapassi (1698–1782), who studied under him for a few months. After moving back to Rome, Trapassi fell under the control of Gravina, who Hellenized his name to Metastasio and left him a considerable fortune after dying in 1718. The young man squandered his money, but he absorbed further heavy doses of *Les passions* and began to orient his life toward literary work, especially libretti.

Metastasio's training from Caloprese and Gravina and their Arcadian colleagues made him a Cartesian aesthete, giving him (supposedly) reasoned principles that he applied throughout his career, much of which was spent in Vienna as court poet.[3] His libretti, written in Italian, were normally based on classical or mythological characters. Using stories with happy endings, he showed how reason could triumph over catastrophe, even over personal desires. He expressed an ideal that Crescimbeni had called "mixed beauty": the internal mode that induced in a character love of virtue for general happiness, and the external mode that related to his social situation. His libretti were enormously influential, not only for their number (28) but especially for the fact that several of them were used dozens of times (sometimes in adaptations by others). The greatest composers of the day drew on them: Händel, Gluck (see the next section), Scarlatti, and Pergolesi (whose *l'Olimpiade* [1735] was one of the early classics), J. C. Bach, and Cherubini. On occasion a composer set the same libretto more than once. Curiously and wastefully, ever since libretti have normally been used once only.

A major feature of Metastasian aesthetics was the assumption strongly implicit in *Les passions* and presumably based upon Descartes's reductionism, especially that *the individual (and thus a character) can experience only one passion at a time.* Thus operas written to libretti of this kind (not only by Metastasio) became known as *opera seria*, with passions experienced serially (although at that time the name *dramma per musica* was more common). The structure of the text was rigid; often 3 acts each with 12 scenes. Most arias were solos, although duets between leading characters were permitted. Arias followed a da capo form, although the soloist might be given some freedom to ornament in the reprise. The passion was statically conveyed throughout an aria, which might be written in a specifically associated key (for example, bravery in D, rage in D minor, yearning in G, and love in A). Often the characters exited on

3. The role of Caloprese comes through in Compagnino and others (1973, esp. pp. 73–75, 270–273); also included on pp. 265–314 is a good survey by G. Nicastro of Metastasio, on whom the literature is vast. Birini (1968) is more of a (good) literary than a historical examination of the Arcadia and the poet.

completion of the aria, taking their passion with them; then the next one would be exhibited in the aria following, perhaps with a transition conveyed in intervening recitatives.

5. *Opera Seria* and Its Critics

The main principles of *opera seria* were in place by the 1730s.[4] But one result of the obviously false psychology that Descartes had proposed was that these operas were static and boring and presented unbelievable characters in impossible situations. Efforts to approach reality came from the 1750s, partly influenced by the contemporary tradition of lighter and more realistic *opera buffa*. Two composers stand out.

Jean Philippe Rameau (1683–1764) came late in life to opera but then wrote a century of them, all tragedies. He brought harmonies and even modulations into recitatives, together with changes of meter and syncopation; he also used music of the opera in its overture. His approach was extended and even eclipsed by Christoph Willibald Gluck (1714–1787), who modified not only opera seria but also the French tradition of *tragédie lyrique* in three "reform operas" written in the last decade of his life. He brought back dance, and flight, and machines, which the Accademia degli Arcadia had banned long before; extended the second section of a da capo aria and did not always use that structure; converted recitatives into ariosi; gave a much greater role to the chorus; and made his characters much more humanlike. He did not use Metastasian libretti here; for example, *Iphigénie en Tauride* (1779) was based on Racine. However, the revolution was conservative; several features of opera seria were still in place, such as the use of classical stories and passions in series.

Still more radical changes soon came, under the force of the Enlightenment; and the most noteworthy composer was Wolfgang Amadeus Mozart (1756–1791) with his principal librettist Lorenzo da Ponte (1749–1838). In the three operas on which they collaborated, *La nozze de Figaro* (1786), *Don Giovanni* (1787), and *Così fan tutti* (1790), Cartesian principles were *explicitly* rejected. Characters were real people, even identifiable ones, at least by type; passion*s* were at work simultaneously in *opera parallela* (not a name of that time, but one that could have been coined). The texts contain complicated ensembles with overlapping conversations or switches from

4. For the tradition of opera seria and its musical and historical context, an excellent start can be made from various articles on opera and on the figures mentioned in the text in Sadie (1980), and especially in Sadie (1992). In particular, the article (Neville 1992) on Metastasio in the latter work is an extraordinary source of information in its eleven pages. Händel was perhaps the greatest composer of opera seria, but apparently the Cartesian elements were not noted in England as they were on the Continent; I rely here, but nervously, on the silence over Descartes in Dean (1970).

one to another in rapid tandem; and within some aria or duets, moods or feelings may modify.[5] Correspondingly, the musical language was far more elaborate and contrapuntal.

Opera seria was not dead: Mozart himself produced two fine ones, of which the second, *La clemenza di Tito* (1791), to a Metastasian libretto modified into a more human form by da Ponte's student Caterino Mazzola (1745–1806), went back before Rameau in its musical style. But it was written to order, specifically for the coronation of the new Holy Roman Emperor in Prague. In general the tradition was becoming steadily more passé; Mazzola's adaptation of *Tito* was the 40th use of Metastasio's libretto, but there were only three later ones, up to 1803 (Neville 1992, 355), Indeed, all of his libretti were largely ignored in the new century. Opera became secular, and static seriality was replaced by the dynamic parallelism of (fairly) real life.

6. The Stability Problem in Astronomy

Exactly the same kind of change took place in an important aspect of celestial mechanics (Grant 1852). One belief common to Newton and Descartes was the central place of God, although again there were differences. Catholic Descartes put forward a cosmogony where God perfectly created the world but thereafter let it run on its own (a view that distressed some fellow believers). By contrast, Newton, a heretical Arian, was much concerned with the issue that the planetary system was rendered stable by God's will. His theory of central forces allowed the possibility that the cumulative effect of interplanetary perturbing forces might drive a planet off the convex bounded orbit around the Sun; however, this would never happen, for God would always arrive in time and restore order.

A similar position on stability was maintained by a Protestant, Leonhard Euler (1707–1783), the next great figure in Newtonian celestial mechanics. One of his major innovations, made in the late 1740s, was to bring in a uniform method of handling perturbing forces by using expansions of the distance functions in trigonometric series. Unintentionally, it inspired from J. L. Lagrange (1736–1813) in 1774 a manner of *secularizing* the stability problem: that is, to *prove* stability from the laws of Newtonian mechanics together with the assumption that the planets orbited

5. Da Ponte may well be a key figure in the change of stance over libretti: unfortunately his autobiography (Da Ponte 1929) is almost deliberately uninformative, and of the various biographies, the only one worth reading (Goertz 1985) is not too enlightening on this point. But Mozart's own conscious role in the changes is finely captured in the path-breaking study Till (1992).

around the Sun in the same direction (Gautier 1817; Wilson 1980). (Incidentally, here Newtonian mechanics was *inferior* to the vorticists, who explained coorbitation in terms of the action of the same vortex family.) Transforming the independent variables, he converted the expansions into a sequence of antisymmetric equations, and thus brilliantly reduced the problem to an analysis of the reality of the latent roots of a certain type of matrix (to use the more modern language of the spectral theory of matrices that this very work helped to create). His conditions for stability were necessary but not sufficient; extending them was to be a great problem in both mathematics and celestial mechanics for over a century, and under its influence the theological approach to stability disappeared.

Lagrange presented his theory in a definitive form in his treatise *Méchanique analitique* (1788). The second word of his title, overused in mathematics, referred to his attempt to *algebraize mechanics*, that is, to express all of its principles and theorems (such as stability) by means of general formulae without any recourse to geometric reasoning. Thus Descartes's view of the relationship between these two branches of mathematics was reversed.

However, as with opera seria, negative reactions came in abundance, especially in the new century. Geometrical thinking in mechanics retained much attraction, especially for its unmatchable facility in aiding the creation of new theories; in addition, Lagrange's view that dynamics was reducible to statics was ignored or even replaced by a greater emphasis on dynamics (the naturally wider range of phenomena, in any case), with a rise of energy mechanics, which Leibniz had envisioned (Grattan-Guinness 1990, chs. 8, 16).

The bad news was not over. From 1772 Lagrange had attempted the same kind of reduction of the calculus, taking the Taylor expansion of a mathematical function as a basic principle and producing all required theorems and result by algebraic operations alone. But his approach lacked credibility; indeed, his basic principle was to be shown by counterexample to be false by Augustin-Louis Cauchy (1789–1857). However Descartes was not vindicated; for Cauchy developed a most influential theory from the 1820s, based on neither geometry nor algebra but on the theory of limits that Newton had handled in an unrigorous way (Grattan-Guinness 1980, ch. 3; §13.4).

7. Parallel Processes

Why combine music with mechanics when the two disciplines exercised no influence, positive or negative, on each other? The reason lies in considerations of Zeitgeist, hopefully without the determinism that so often mars its use. Like a new

Renaissance, part of the Enlightenment credo was to reduce the pressure of Christianity on cultural thinking and encourage the idea that mankind can do things: express conflicts among people rather than ponder the ordered sequence of passions from antiquity, and to prove the ordered sequence of orbital paths rather than invoke divine intervention. The individual was seen as an *active agent* in intellectual and artistic life. Thus we see also, among many examples, Romanticism coming into both the arts and the sciences; appraisal of the regularities of life and the world by probability theory rather than yes-no certainties; and above all, the philosophy of Immanuel Kant (1724–1804), based on the principle that we impose our ideas on the world rather than the then-normal vice versa.

Both Lagrange and Mozart lived through the new trends and reacted positively to them.[6] Lagrange was classified as an atheist, although "agnostic" would probably be a more accurate description (interestingly, that word was not created then). Mozart did not allow his Catholicism to accept all the social barriers; after all, Figaro wins over the Count.

BIBLIOGRAPHY

Acquaro Graziosi, M. T. 1991. *L'Arcadia. Trecento anni di storia*, Rome: Polombi.

Aiton, E. J. 1972. *The vortex theory of planetary motions*, London: Macdonald; New York: American Elsevier.

Bain, F. W. 1890. *Christina, Queen of Sweden*, London: W.H. Allen.

Birini, W. 1968. *L'Arcadia e il Metastasio*, Florence: La Nuova Italia.

Borgato, M. T., and Pepe, L. 1981. "Sulle lettere familiari di Giuseppe Luigi Lagrange," *Bollettino di storia delle scienze matematiche, 9*, 193–318.

Bos, H. J. M. 1981. "The representation of curves in Descartes's Géométrie," *Archive for history of exact sciences, 24*, 295–338.

Compagnino, C., and others. 1973. (Eds.), *Il settecento: L'Arcadia a l'età delle reforme*, vol. 6, tome 1, Rome and Bari: Laterza.

Crescimbeni, G. 1804. *Storia dell'Accademia degli Arcadia instituta a Roma d'anno 1690 per la coltivazione della scienza delle lettere e della poesia*, London: Becket.

Da Ponte, L. 1929. *Memoirs* (trans. E Abbott and A. Livingston), New York: Lippincott.

Dean, W. 1970. *Handel and the opera seria*, London: Oxford University Press.

Descartes, R. 1968. *Les passions* (ed. J.-M. Monnoyer), Paris: Gallimard.

Gautier, A. 1817. *Essai historique sur le problème des trois corps*, Paris: Courcier.

Goertz, H. 1985. *Mozarts Dichter Lorenzo de Ponte. Genie und Abenteuer*, Vienna: Österreichischer Bundesverlag.

6. On Lagrange's family background, see the excellent account in Borgato and Pepe (1981).

Grant, R. 1852. *History of physical astronomy, from the earliest ages to the middle of the nineteenth century*, London: Boh. [Repr. New York: Johnson, 1966.]

Grattan-Guinness, I. 1980. (Ed.), *From the calculus to set theory, 1630–1910: An introductory history*, London: Duckworth. [Repr. Princeton, N.J.: Princeton University Press, 2000.]

Grattan-Guinness, I. 1990. "The varieties of mechanics by 1800," *Historia mathematica, 17*, 313–338.

Neville, D. 1992. "Metastasio," in Sadie 1992, vol, 3, 351–361.

Sadie, S. 1980. (Ed.), *The new Grove dictionary of music and musicians*, 20 vols., London: MacMillan.

Sadie, S. 1992. (Ed.), *The new Grove dictionary of opera*, 4 vols., London: MacMillan.

Till, N. 1992. *Mozart and the Enlightenment*, London: Faber and Faber.

Wilson, C. A. 1980. "Perturbation and solar tables from Lacaille to Delambre," *Archive for history of exact sciences, 22*, 53–188, 189–304.

PART 4

LOLLIPOPS

CHAPTER EIGHTEEN

Four Pretty but Little-Known Theorems Involving the Triangle

In §14.5.3 we saw a lovely and simple theorem about the right-angled 3:4:5 triangle, but with an obscure history. Here are four more theorems deserving to be much better known; in two cases their history is unknown.

1. Halving the Right Angle, Halving the Square

Figure 18.1 shows any right-angled triangle *ABC*, with the square *BCDE* set on the hypotenuse *BC*.[1] Then *the line that bisects the right angle also bisects the square on the hypotenuse, and vice versa.*

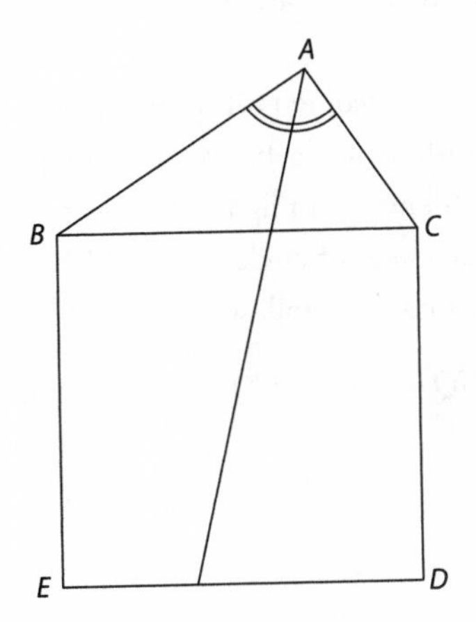

Figure 18.1

All sections are new except ¶18.2, which was first published in the *Mathematical Gazette, 85* (2001), 286–288.

1. The word "hypotenuse" comes from the Greek, where it means something like "extended under." So we can assume that normally they drew their right-angle triangles with this orientation.

This theorem should not be confused with Pythagoras's; neither is required to prove the other. However, a "new demonstration" of Pythagoras incorporating this figure as part of its diagram was given by the Frenchman Olry Terquem (1782–1862) in his "new manual of geometry" (1838, 103).[2] I do not know when the theorem was isolated as such; I learned it from a colleague, one of whose students had found it for himself.

2. A Geometrical Representation of the Classical Means of Two Quantities

A normal component of algebra is the basic properties of the arithmetical, geometrical, and harmonic means of two quantities. The usual teaching strategy is to give the definitions like this: for two quantities p and q, such that $p > q > 0$, the means are

$$\text{Arithmetic:} \quad a = (p + q)/2; \tag{18.1}$$

$$\text{Geometric:} \quad g = \sqrt{(pq)}; \tag{18.2}$$

$$\text{Harmonic:} \quad 2/h = (1/p + 1/q). \tag{18.3}$$

Thereafter basic relationships and inequalities are proved and turn up in various branches of mathematics. They are obviously very useful, but the start is not interesting.

As part of my effort to give geometry its proper place in mathematics education, I give some life to this arithmetic/algebra by presenting a geometrical version that does not seem to be much known. In figure 18.2, p and q together serve as hypotenuse PQ of ΔPQA and diameter of circle PQA, which are our initial concern. The quantities and means are located as follows:

$$PG = p;\ GQ = q;\ PC = CQ = a;\ GA = g;\ \text{and}\ AH = h.$$

2. On the authority of D. T. Whiteside, I attributed this proof in Grattan-Guinness (1997, 567) to the German military officer G. F. von Tempelhoff (1737–1807). However, I find that Pythagoras's theorem is not proved there at all (1769, 194).

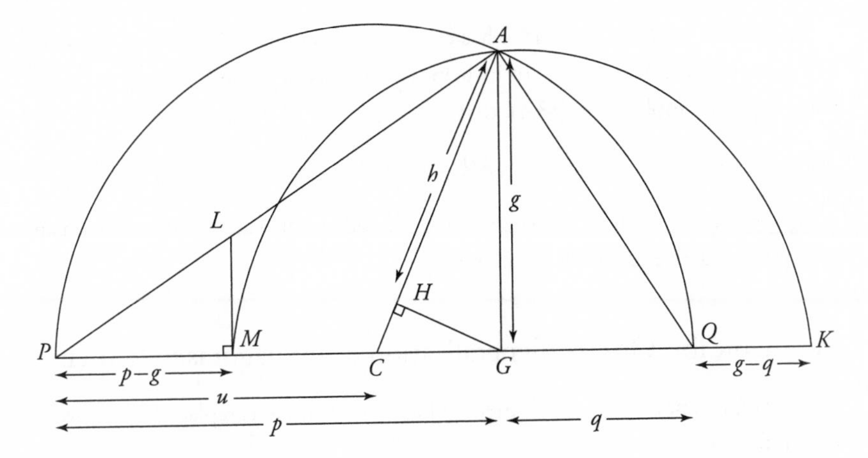

Figure 18.2

All these manifestations are well known *except for the last one,* which constitutes the theorem for this section. It can be proved quite easily from ΔCHG and ΔCGA; for example,

$$GH^2 = (a - q)^2 - (a - h)^2 = g^2 - h^2.$$

$$\text{Hence } 2ah = g^2 + 2aq - q^2$$

$$\text{So from (18.1), } h(p + q) = 2pq,$$

which is a version of (18.3).

These two triangles also provide a ready proof of the fundamental inequality

$$a > g > h; \qquad \text{for} \qquad AC > AG > AH.$$

Further, they deliver the curiously little-recognized result that the geometric mean is also the geometric mean of the other two means (of course, this is simply obtained from [18.1]). By similarity,

$$AG : AC :: AH : AG. \text{ Hence } g^2 = ah.$$

Finally, the usual definitions (18.1) are not the most elegant available. In antiquity a uniform trio was used, and is worth reviving:

$$\begin{aligned} &\text{Arithmetic: } p - a : a - q :: p : p \qquad (\text{so } 2a = p + q) \\ &\text{Geometric: } p - g : g - q :: p : g \qquad (\text{so } g^2 = pq) \\ &\text{Harmonic: } p - h : h - q :: p : q \qquad (\text{so } h(p + q) = 2pq) \end{aligned} \qquad (18.4)$$

The second circle in figure 18.2 has center G and radius $GA = g$. Some of these new quantities can be found with its help (two are marked), and pleasant exercises pursued; for example, from (18.4), that

$$ML = g - q.$$

By these means some worthwhile arithmetic and algebra can be enriched with an easy but instructive use of plane geometry.

3. Thabit ibn Qurra's Generalization of Pythagoras's Theorem

When ΔPQR is not right angled, then Pythagoras's theorem is replaced by the well-known formula

$$QR^2 = PQ^2 + PR^2 - 2PQ \times PR \times \cos \angle QPR \qquad (18.5)$$

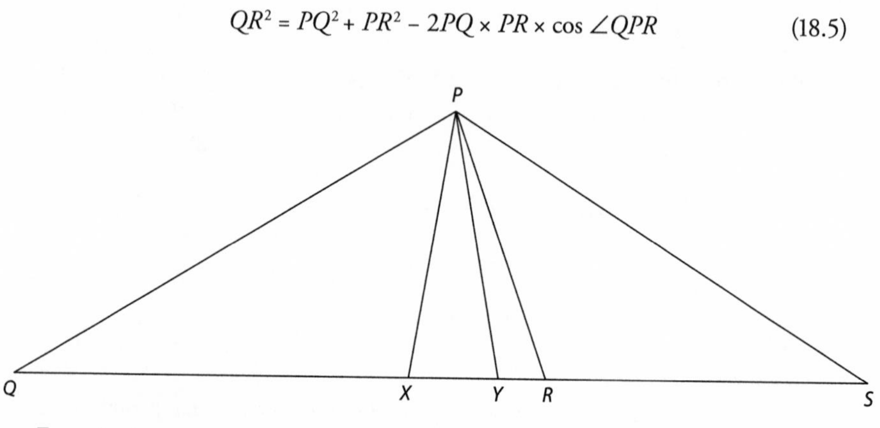

Figure 18.3

(Figure 18.3, where we ignore S for the time being). But this is really an algebraic result; is there not some corresponding geometrical theorem? The Islamic mathematician Thabit ibn Qurra (835?–901) found a beautiful one. Let P be the largest angle in the triangle. His insight was to construct similar triangles within the given one. Take the points X and Y on QR such that ΔPXY is the isosceles triangle for which $\angle PXR = \angle PYQ = \angle QPR$. Then these two consequences follow:

$$\Delta PQR \text{ and } \Delta PQY \text{ are similar; hence } PQ^2 = QR \times QY; \qquad (18.6)$$

$$\Delta PQR \text{ and } \Delta PXR \text{ are similar; hence } PR^2 = QR \times XR. \qquad (18.7)$$

Adding, we have the theorem

$$PQ^2 + PR^2 = QR\,(QY + XR). \qquad (18.8)$$

Figure 18.3 shows the case where P is acute; if it is obtuse, then Y will be to the left of X. When P is a right angle, then of course X and Y coincide, and Pythagoras's theorem returns. Finally, if P is not the largest angle in ΔPQR, then X and/or Y will lie outside QR.

Given the importance of the result as well as its elegance, why has it not long been standard fare in geometry and mathematics education? Yet it is little known either to mathematicians or to historians. For the latter cohort, it gained some circulation from the note by Sayili (1960), and a few general histories have noted it; for example, Cooke (1997, 270). But it still lies largely in the wings.

At least it has some company. For example, consider the isosceles ΔPQS in figure 18.3, and take any point on QS (R, say). Then we have the little-known result

$$PR^2 = PR^2 + QR \times RS. \tag{18.9}$$

It is easily proved in various ways, including by applying (18.6) and (18.7) to this consequence of the isoscelity of ΔPXY:

$$QR = XR + RS. \tag{18.10}$$

Hoehn (2000) does not claim to be its first prover but is rightly astonished that it is so "neglected."

4. Faulhaber's Extension of Pythagoras's Theorem

Pythagoras's theorem has a companion in three dimensions. In figure 18.4 we look down on a tetrahedron, in which each of the three faces that meet at the apex are right-angled triangles. Then we have the result that *the square of the base equals the sum of the squares of the three sides.*

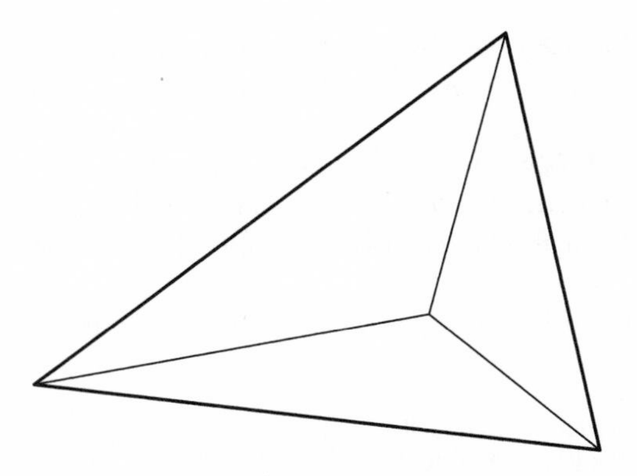

Figure 18.4

For once we have some historical information about origins. It was stated by Johannes Faulhaber (1580–1635) in his *Miracula arithmetica* (1622); he gave no proof, but one can be confected from the Heron-Archimedes formula relating the area of a triangle to its sides. Around the time of its publication, Faulhaber told it to a young visitor, René Descartes (1594–1650), who took it into his own brief.[3] The subsequent neglect of this theorem, however, is unexplained.

5. Jewels under Stones

This last remark applies to all our four cases—and also to the one in §14.5.3, although we also saw reasons there why that case may have been kept secret deliberately. Some are of unknown origins (or so it seems, anyway), and the histories of the neglect of all of them are also fugitive. For example, none of the five results features in the excellent history (Simon 1906) of theorems in elementary geometry presented during the nineteenth century. Similarly, over the years I have presented some or all of the quintet at public lectures before audiences that in total must number some hundreds of mathematicians, historians, educators, and others, asking for information about them; but I have never learned anything, because nobody has ever seen them before, not even Thabit's or Faulhaber's! One recalls the mystery of §12.5–6 concerning the algorithms for multiplication and division that use the place value system in the natural way and yet remained equally unnoticed; I have presented those also, with the same reaction of pleasure but surprise.

Appropriately, then, this book ends on the theme of knowledge and ignorance, which has occurred several times during its course.

BIBLIOGRAPHY

Cooke, R. 1997. *The history of mathematics: A brief course*, 1st ed., New York: Wiley.

Grattan-Guinness, I. 1997. *The Fontana history of the mathematical sciences*, London: Fontana.

Hawlitschek, K. 1995. *Johannes Faulhaber (1580–1635)*, Ulm, Germany: Staatsbibliothek.

Hoehn, L. 2000. "A neglected Pythagorean-like formula," *Mathematical gazette, 84*, 71–73.

Manders, K. 2006. "Algebra in Roth, Descartes and Faulhaber," *Historia mathematica, 33*, 184–209.

Sayili, A. 1960. "Thabit ibn Qurra's generalization of Pythagoras's theorem," *Isis, 51*, 35–37.

3. On this connection between Faulhaber and Descartes, see Hawlitschek (1995, 74, 120–121); and on their respective handlings of algebra, see Manders (2006).

Simon, M. 1906. *Über die Entwicklung der Elementar-Geometrie im XIX. Jahrhundert*, Leipzig, Germany: Teubner.

Terquem, O. 1838. *Nouveau manuel de géométrie*, Paris: Roret.

von Tempelhoff, G. F. 1769. *Anfangs-Gründe der Analysis endlicher Grössen*, Berlin: Arnold Wever.

INDEX

Page numbers in *italics* indicate figures and tables.